PHAST Version 2—A Program for Simulating Groundwater Flow, Solute Transport, and Multicomponent Geochemical Reactions

By David L. Parkhurst, Kenneth L. Kipp, and Scott R. Charlton

Chapter 35 of
Section A, Groundwater, of
Book 6, Modeling Techniques

Techniques and Methods 6–A35

U.S. Department of the Interior
U.S. Geological Survey

U.S. Department of the Interior
KEN SALAZAR, Secretary

U.S. Geological Survey
Marcia K. McNutt, Director

U.S. Geological Survey, Denver, Colorado: 2010

This and other USGS information products are available at http://store.usgs.gov/
U.S. Geological Survey
Box 25286, Denver Federal Center
Denver, CO 80225

To learn about the USGS and its information products visit http://www.usgs.gov/
1-888-ASK-USGS

Suggested citation:
Parkhurst, D.L., Kipp, K.L., and Charlton, S.R., 2010, PHAST Version 2—A program for simulating groundwater flow, solute transport, and multicomponent geochemical reactions: U.S. Geological Survey Techniques and Methods 6–A35, 235 p.

Contents

Figures

Tables

Conversion Factors

SI to Inch/Pound

Multiply	By	To obtain
Length		
centimeter (cm)	0.3937	inch (in.)
millimeter (mm)	0.03937	inch (in.)
meter (m)	3.281	foot (ft)
kilometer (km)	0.6214	mile (mi)
meter (m)	1.094	yard (yd)
Area		
square meter (m^2)	0.0002471	acre
square centimeter (cm^2)	0.001076	square foot (ft^2)
square meter (m^2)	10.76	square foot (ft^2)
Volume		
liter (L)	1.057	quart (qt)
liter (L)	0.2642	gallon (gal)
cubic meter (m^3)	264.2	gallon (gal)
cubic meter (m^3)	0.0002642	million gallons (Mgal)
cubic meter (m^3)	35.31	cubic foot (ft^3)
cubic meter (m^3)	1.308	cubic yard (yd^3)

Flow rate		
cubic meter per second (m³/s)	70.07	acre-foot per day (acre-ft/d)
cubic meter per year (m³/yr)	0.000811	acre-foot per year (acre-ft/yr)
meter per second (m/s)	3.281	foot per second (ft/s)
meter per day (m/d)	3.281	foot per day (ft/d)
meter per year (m/yr)	3.281	foot per year ft/yr)
cubic meter per second (m³/s)	35.31	cubic foot per second (ft³/s)
cubic meter per second per square kilometer [(m³/s)/km²]	91.49	cubic foot per second per square mile [(ft³/s)/mi²]
cubic meter per day (m³/d)	35.31	cubic foot per day (ft³/d)
Mass		
gram (g)	0.03527	ounce, avoirdupois (oz)
kilogram (kg)	2.205	pound avoirdupois (lb)
Pressure		
kilopascal (kPa)	0.009869	atmosphere, standard (atm)
kilopascal (kPa)	0.01	bar
Density		
kilogram per cubic meter (kg/m³)	0.06242	pound per cubic foot (lb/ft³)
Energy		
joule (J)	0.0000002	kilowatthour (kWh)
Hydraulic conductivity		
meter per day (m/d)	3.281	foot per day (ft/d)

Temperature in degrees Celsius (°C) may be converted to degrees Fahrenheit (°F) as follows:
$$°F = (1.8 × °C) + 32$$
Temperature in degrees Fahrenheit (°F) may be converted to degrees Celsius (°C) as follows:
$$°C = (°F - 32)/1.8$$
Vertical coordinate information is referenced to the North American Vertical Datum of 1988 (NAVD 88).

Horizontal coordinate information is referenced to the North American Datum of 1983 (NAD 83).

Altitude, as used in this report, refers to distance above the vertical datum.

Abbreviations

atmosphere	atm
cubic meter	m^3
day	d
equivalent	eq
gram	g
gigabytes	GB
hour	hr
joule	J
kilogram	kg
kilogram of water	kgw
kilometer	km
liter	L
meter	m
mole	mol
milliequivalent	meq
milligram	mg
millimole	mmol
microgram	μg
micromole	μmol
pascals	Pa
parts per million	ppm
second	s
square meter	m^2
year	yr

PHAST Version 2—A Program for Simulating Groundwater Flow, Solute Transport, and Multicomponent Geochemical Reactions

By David L. Parkhurst, Kenneth L. Kipp, and Scott R. Charlton

Abstract

The computer program PHAST (**PHR**EEQC **A**nd H**ST**3D) simulates multicomponent, reactive solute transport in three-dimensional saturated groundwater flow systems. PHAST is a versatile groundwater flow and solute-transport simulator with capabilities to model a wide range of equilibrium and kinetic geochemical reactions. The flow and transport calculations are based on a modified version of HST3D that is restricted to constant fluid density and constant temperature. The geochemical reactions are simulated with the geochemical model PHREEQC, which is embedded in PHAST. Major enhancements in PHAST Version 2 allow spatial data to be defined in a combination of map and grid coordinate systems, independent of a specific model grid (without node-by-node input). At run time, aquifer properties are interpolated from the spatial data to the model grid; regridding requires only redefinition of the grid without modification of the spatial data.

PHAST is applicable to the study of natural and contaminated groundwater systems at a variety of scales ranging from laboratory experiments to local and regional field scales. PHAST can be used in studies of migration of nutrients, inorganic and organic contaminants, and radionuclides; in projects such as aquifer storage and recovery or engineered remediation; and in investigations of the natural rock/water interactions in aquifers. PHAST is not appropriate for unsaturated-zone flow, multiphase flow, or density-dependent flow.

A variety of boundary conditions are available in PHAST to simulate flow and transport, including specified-head, flux (specified-flux), and leaky (head-dependent) conditions, as well as the special cases of rivers, drains, and wells. Chemical reactions in PHAST include (1) homogeneous equilibria using an ion-association or Pitzer specific interaction thermodynamic model; (2) heterogeneous equilibria between the aqueous solution and minerals, ion exchange sites, surface complexation sites, solid solutions, and gases; and (3) kinetic reactions with rates that are a function of solution composition. The aqueous model (elements, chemical reactions, and equilibrium constants), minerals, exchangers, surfaces, gases, kinetic reactants, and rate expressions may be defined or modified by the user.

A number of options are available to save results of simulations to output files. The data may be saved in three formats: a format suitable for viewing with a text editor; a format suitable for exporting to spreadsheets and postprocessing programs; and in Hierarchical Data Format (HDF), which is a compressed binary format. Data in the HDF file can be visualized on Windows computers with the program Model Viewer and extracted with the utility program PHASTHDF; both programs are distributed with PHAST.

Operator splitting of the flow, transport, and geochemical equations is used to separate the three processes into three sequential calculations. No iterations between transport and reaction calculations are implemented. A three-dimensional Cartesian coordinate system and finite-difference techniques are used for the spatial and temporal discretization of the flow and transport equations. The nonlinear chemical equilibrium equations are solved by a Newton-Raphson method, and the kinetic reaction equations are solved by a Runge-Kutta or an implicit method for integrating ordinary differential equations.

The PHAST simulator may require large amounts of memory and long Central Processing Unit (CPU) times. To reduce the long CPU times, a parallel version of PHAST has been developed that runs on a multi-processor computer or on a collection of computers that are networked. The parallel version requires Message Passing Interface, which is freely available. The parallel version is effective in reducing simulation times.

This report documents the use of the PHAST simulator, including running the simulator, preparing the input files, selecting the output files, and visualizing the results. It also presents six examples that verify the numerical method and demonstrate the capabilities of the simulator. PHAST requires three input files. Only the flow and transport file is described in detail in this report. The other two files, the chemistry data file and the database file, are identical to PHREEQC files, and a detailed description of these files is in the PHREEQC documentation.

PHAST Version 2 has a number of enhancements to allow simpler definition of spatial information and to avoid grid-dependent (node-by-node) input definitions. Formerly, all spatial data were defined with rectangular zones. Now wedge-shaped and irregularly shaped volumes may be used to specify the hydrologic and chemical properties of regions within the model domain. Spatial data can be imported from ArcInfo shape and ASCII raster files and from a simple X,Y, Z file format. To accommodate a grid that is not aligned with the coordinate system of the imported files, it is possible to define features in map and grid coordinate systems within the same input file.

The definition of leaky and flux boundary conditions has been modified to select only exterior faces within a specified volume of the model domain and to restrict the boundary conditions to the area of definition, which may include fractions of cell faces. Previously, flux and leaky boundary conditions were applied only to entire cell faces; thus, the area could vary as the model grid was refined and cell sizes changed.

New capabilities have been added to interpolate spatial data to the two- or three-dimensional locations of cells and elements. Two-dimensional interpolation is used to define surfaces for the tops and bottoms of three-dimensional regions within the model domain. Surfaces are created by two-dimensional interpolation of scattered X, Y points with associated elevation data. Within the bounds of the scattered points (the convex hull), natural neighbor interpolation is implemented, which uses an area weighting scheme to assign an elevation to a target point based on elevations at the nearest of the scattered points; outside the convex hull, the elevation of the closest point is assigned to a target point.

Three-dimensional scattered data can be used to define porous-media properties, boundary condition properties, or initial conditions. Three-dimensional interpolation always assigns the value of the closest scattered point to a target point. The interpolation capabilities allow head and chemical data to be saved at the end of one run and used as initial conditions for a subsequent run, even if the grid spacing has been changed.

A new capability has been added to aggregate flows of water and solute into an arbitrarily shaped region and through the boundary-condition cells included in the region. Any number of flow aggregation regions may be defined. These regions need not be mutually exclusive, and regions can be combined to define larger, possibly noncontiguous regions. A facility exists to save heads as a function of time and space for these regions, which then can be used to specify boundary-condition heads in a subsequent run.

A drain boundary condition has been added. Drains function similarly to rivers, except they only discharge water from the aquifer. Drains are one-dimensional features with an associated width or diameter, which will accept water if the head in the aquifer is greater than the elevation of the drain. Drains have no time-dependent parameters.

Other changes implemented in PHAST Version 2 include a method for row scaling of the flow and transport equations to allow the definition of a single convergence tolerance for both flow and transport. Memory requirements for the chemistry processes in parallel processing have been minimized by elimination of most flow and transport storage. A start time other than zero may be defined. PHAST Version 2 includes many bug fixes and some other minor additions to the input files.

Chapter 1. Introduction

The computer program PHAST (**PH**REEQC **A**nd H**ST**3D) simulates multicomponent, reactive solute transport in three-dimensional saturated groundwater flow systems. PHAST is a versatile groundwater flow and transport simulator with capabilities to model a wide range of equilibrium and kinetic geochemical reactions. The flow and transport calculations are based on a modified version of HST3D (Kipp, 1987, 1997) that is restricted to constant fluid density and constant temperature. The geochemical reactions are simulated with the geochemical model PHREEQC (Parkhurst, 1995; Parkhurst and Appelo, 1999), which is embedded in PHAST. Major enhancements in PHAST Version 2 allow spatial data to be defined in a combination of map and grid coordinate systems, independent of a specific model grid (without node-by-node input). At run time, aquifer properties are interpolated from the spatial data to the model grid; regridding requires only redefinition of the grid without modification of the spatial data. The combined flow, transport, and geochemical processes are simulated by three sequential calculations for each time step: first, flow velocities are calculated, then the chemical components are transported, and finally geochemical reactions are calculated. This sequence is repeated for successive time steps until the end of the simulation.

Reactive-transport simulations with PHAST require three input files: a flow and transport data file, a chemistry data file, and a thermodynamic database file. All input data files are built with modular keyword data blocks. Each keyword data block defines a specific kind of information—for example, grid locations, porous-media properties, boundary-condition information, or initial chemical composition in a zone. All spatial data are defined by zones, which can be rectangular boxes, right-triangular wedges, or irregularly shaped volumes based on ArcInfo shapefiles [Environmental Systems Research Institute (ESRI), 1998] or scattered data points. Simulation results can be saved in a variety of file formats, including Hierarchical Data Format (HDF, *http://www.hdfgroup.org/*). Two utility programs are distributed with PHAST: (1) Model Viewer (Windows only), which is used to produce three-dimensional visualizations of problem definition and simulation results, and (2) PHASTHDF, which is used to extract subsets of the data stored in the HDF file.

The PHAST simulator can be run on most computer systems, including computers with Windows, Linux, and Unix operating systems. Coupled reactive-transport simulations are computer intensive, which can result in long run times on a single-processor computer. A parallel (multiprocessor) version of PHAST, implemented with the Message Passing Interface (MPI), is distributed for use on multiprocessor computers and on networked computer clusters (Windows, Unix, or Linux). Using the parallel version of PHAST can greatly reduce run times.

1.1. Enhancements in PHAST Version 2

PHAST Version 2 has a number of enhancements relative to PHAST Version 1 (Parkhurst and others, 2004) to allow simpler definition of spatial information and to avoid grid-dependent (node-by-node) input definitions. Formerly, all spatial data were defined with rectangular zones. Now wedge-shaped and irregularly shaped volumes may be used to specify the hydrologic and chemical properties of regions within the model domain. Spatial data can be imported from ArcInfo shapefiles (ESRI, 1998) and ASCII raster files (ESRI, 2009) and from a simple X,Y,Z file format. The shapefiles and raster files are public-domain formats developed by the Environmental Systems Research Institute. The shapefiles are read by the open-source shapelib software (Maptools.org, 2003). To accommodate a grid that is not aligned with the coordinate system of the imported files, it is possible to define features in map and grid coordinate systems within the same input file.

The definition of leaky and flux boundary conditions has been modified to select only exterior faces within a specified volume of the model domain and to restrict the boundary conditions to the area of definition, which may include fractions of cell faces. Previously, flux and leaky boundary conditions were applied only to entire cell faces and thus the area could vary as the model grid was refined and cell sizes changed.

New capabilities have been added to interpolate spatial data to the two- or three-dimensional locations of cells and elements. Two-dimensional interpolation is used to define surfaces for the tops and bottoms of three-dimensional regions within the model domain. Surfaces are created by two-dimensional interpolation of scattered X, Y points with associated elevation data. Within the bounds of the scattered points (the convex hull), natural neighbor interpolation is implemented, which uses an area weighting scheme to assign an elevation to a target point based on elevations at the nearest of the scattered points; outside the convex hull, the elevation of the closest point is assigned to a target point.

Three-dimensional scattered data can be used to define porous-media properties, boundary condition properties, or initial conditions. Three-dimensional interpolation always assigns the value of the closest scattered point to a target point. The interpolation capabilities allow potentiometric head (groundwater-level elevation) and chemical data to be saved at the end of one run and used as initial conditions for a subsequent run, even if the grid spacing has been changed.

A new capability has been added to aggregate flows of water and solute into an arbitrarily shaped region and through the boundary-condition cells included in the region. Any number of flow aggregation regions may be defined. These regions need not be mutually exclusive, and regions can be combined to define larger, possibly noncontiguous regions. A facility exists to save heads as a function of time and space for these regions, which then can be used to specify boundary-condition heads in a subsequent run.

A drain boundary condition has been added. Drains function similarly to rivers, except they only discharge water from the aquifer. Drains are one-dimensional features with an associated width or diameter, which will accept water if the head in the aquifer is greater than the elevation of the drain. Drains have no time-dependent parameters.

Other changes implemented in PHAST Version 2 include a method for row scaling of the flow and transport equations to allow the definition of a single convergence tolerance for both flow and transport. Memory requirements for the chemistry processes in parallel processing have been minimized by elimination of most flow and transport storage. A start time other than zero may be defined. PHAST Version 2 includes many bug fixes and some other minor additions to the input files.

1.2. Applicability

PHAST is applicable to the study of natural and contaminated groundwater systems at a variety of scales ranging from laboratory experiments to local and regional field scales. Representative examples of PHAST applications include simulation of effluent arsenic concentrations in laboratory column experiments, migration of dissolved organic compounds from a landfill, migration of nutrients in a sewage plume in a sandy aquifer, storage of freshwater in a slightly saline aquifer, and examination of natural mineral and exchange reactions in a regional aquifer.

PHAST is not suitable for some types of reactive-transport modeling. In particular, PHAST is not appropriate for unsaturated-zone flow and does not account for flow and transport of a gas phase or a nonaqueous liquid phase. PHAST is restricted to constant temperature and constant density and does not account for density-dependent flow caused by concentration gradients or temperature variations. Thus, PHAST is not an appropriate simulator for many systems and processes, including transport of volatile organic compounds in a soil zone, systems with two liquid phases, hydrothermal systems, systems with large fluid-density contrasts, or heat storage and recovery in aquifers.

1.3. Simulator Capabilities

The PHAST simulator is a general computer code with various reaction chemistry, equation-discretization, boundary-condition, source-sink, and equation-solver options. Four types of flow and reactive-transport simulations can be performed with PHAST. Listed in order from simple to complex in terms of input-data requirements and computational workload, they are steady-state simulation of groundwater flow, transient simulation of groundwater flow, steady-state simulation of flow followed by reactive transport, and transient simulation of flow with reactive transport. The reactive-transport simulation is always transient, even though it may evolve to steady-state concentration fields after sufficient simulation time. The first two types of simulations are used to model groundwater flow and to calibrate a groundwater flow model. The third type of simulation is used when the flow field can be assumed to be in steady state. The fourth type of simulation is for general reactive-transport simulations with a transient-flow field. Additional details about the four types of simulations are provided in Chapter 2 and Appendix D.

The groundwater flow and transport are defined by initial and boundary conditions for heads and concentrations. The available boundary conditions include (1) specified-head conditions where solution compositions are fixed; (2) specified-head conditions where solution compositions are associated with inflowing water; (3) flux (specified-flux) conditions, such as precipitation recharge, with associated-solution composition for inflowing water; (4) leaky (head-dependent) conditions, such as leakage from a constant-head water table through a confining unit, with associated-solution composition; (5) rivers with associated-solution composition; (6) drains (no recharge), (7) injection of specified-solution compositions or pumpage by wells; and (8) free-surface boundary condition (unconfined flow). A PHAST simulation is composed of a series of simulation periods. Heads and compositions of solutions for boundary conditions are constant over each period, but time series of values may be defined for heads, solution compositions, and pumping rates for a series of time periods to simulate transient boundary conditions.

A wide range of chemical reactions can be coupled to the transport calculations. Heterogeneous equilibrium reactions may be simulated, including equilibrium among the aqueous solution and minerals, ion exchange sites, surface complexation sites, solid solutions, and a gas phase. The most complex simulations include kinetic reactions, which require mathematical expressions that define the rates of reactions as a function of solution composition, time, or other chemical factors. In addition, geochemical simulations, which include all of the above types of reactions plus additional reaction capabilities, may be used to define chemical initial and boundary conditions for the reactive-transport simulations. Any modeling capability available in PHREEQC (including one-dimensional transport simulations) may be used to establish initial and boundary conditions for the reactive-transport simulation of PHAST. The aqueous model, equilibrium conditions, and kinetic rate expressions are defined through a database file that can be modified by the user.

1.4. Simulator Results

PHAST produces a large amount of information internally, including heads, groundwater velocity components, component concentrations, activities and molalities of aqueous species, and saturation indices of minerals for each active node for each time step. The program offers a number of options to select the data to be saved to output files and the frequency at which they are saved. The data may be saved in tabular ASCII files suitable for viewing with a text editor, in ASCII files with tab-delimited columns of data suitable for importing into spreadsheets or plotting programs, and in a binary HDF file.

Data in the HDF file can be visualized on Windows computers with the program Model Viewer (see Appendix A). Model Viewer can be used to display boundary-condition nodes and media properties (hydraulic conductivity, porosity, storage coefficient, and dispersivities) for verification of a problem setup, to view three-dimensional fields of scalar results (head, concentration, moles of minerals, and others), and to view velocity vectors that are located at each node of the grid region. Data can be viewed for selected time steps and data for multiple time steps can be used to produce an animation. Data can be extracted from the HDF file with a Java utility program, PHASTHDF, which runs on any computer with a Java Runtime Environment (see Appendix B).

1.5. Numerical Implementation

PHAST solves a set of partial differential equations for flow and transport and a set of nonlinear algebraic and ordinary differential equations for chemistry. The equations that are solved numerically are (1) the saturated groundwater flow equation for conservation of total fluid mass, (2) a set of solute-transport equations for conservation of mass of each solute component of a chemical-reaction system, and (3) a set of chemical-reaction equations comprising mass-balance equations, mass-action equations, and kinetic-rate equations. The groundwater flow and solute-transport equations are coupled through the dependence of advective transport on the interstitial fluid-velocity field. For constant and uniform density groundwater, this coupling is one way—the fluid-velocity field affects the solute transport, but the solute transport does not affect the fluid-velocity field. The solute-transport equations and the chemical equations are coupled through the chemical concentration terms. The chemical equations are algebraic mass-balance and mass-action equations for equilibrium reactions and ordinary differential equations for kinetic reactions. The chemical equations are fully coupled through the concentration terms and must be solved simultaneously.

By using a sequential solution approach for flow, transport, and reaction calculations, numerical solutions are obtained for each of the dependent variables: potentiometric head, solute-component concentrations, species concentrations, and masses of reactants in each cell. Operator splitting is used to separate the solute-transport calculations from the chemical-reaction calculations (Yeh and Tripathi, 1989). No iterations between transport and reaction calculations within a time step are performed. Finite-difference techniques are used for the spatial and temporal discretization of the flow and transport equations. The linear finite-difference equations for flow and for transport of each solute component are solved either by a direct solver or by an iterative solver.

Operator splitting also is used to separate the chemical equilibrium reactions from the kinetic reactions. The nonlinear chemical equilibrium equations are solved by a Newton-Raphson method (Parkhurst and Appelo, 1999). The kinetic-reaction equations are ordinary differential equations, which are solved by using an explicit Runge-Kutta algorithm or an implicit algorithm for stiff differential equations called CVODE (Cohen and Hindemarsh, 1996). These algorithms evaluate the reaction rates at several (possibly many) intermediate times within the overall time interval equal to the transport time step. An automatic time-step algorithm is used for controlling errors in the integration of the kinetic reaction rates, and the equilibrium equations are solved at each intermediate step of the integration.

A three-dimensional Cartesian coordinate system is used to define the region for reactive-transport simulations. Although flow and transport calculations are always three dimensional, the symmetry of conceptually one- and two-dimensional regions can be used to reduce the computational workload of the geochemical calculations. For these cases, the geochemical calculations are performed on a line or plane of nodes and then copied to the corresponding symmetric parts of the grid region.

1.6. Computer Resources

The computer code for PHAST is written in Fortran-90, C, and C++. The code conforms to American National Standards Institute (ANSI) standards for Fortran-90, C, and C++ with few exceptions and has proven portable to a variety of computer operating systems, including Windows (Windows95 or later), Linux on personal computers, and Unix on Sun, Digital Equipment Corporation (DEC), Hewlett Packard (HP), and International Business Machines (IBM) computers. Mathematical expressions for kinetic reactions and some output values are calculated by using a Basic-language interpreter (with permission, David Gillespie, Synaptics, Inc., San Jose, Calif., written commun., 1997).

Fully dynamic memory allocation is used by PHAST as provided by the Fortran-90, C, and C++ languages. Little effort has been expended to minimize storage requirements. Consequently, the PHAST simulator may require large amounts of memory for execution. The number of nodes defined for the simulation grid is the primary determinant of the amount of memory needed for execution. PHAST can require long CPU (Central Processing Unit) times in addition to large amounts of memory. Processing time is determined mainly by the number of nodes, the number of time steps, and the presence or absence of kinetic reactions in the simulation. For equal numbers of nodes and time steps, a problem with kinetic reactions will run more slowly than a problem without kinetics. The CPU time also increases with the number of components, aqueous species, equilibrium phases, and other reactants that are included in the problem definition.

To reduce the long CPU times of PHAST simulations, a parallel version of PHAST has been developed that runs on a multiprocessor computer or on a collection of computers that are networked. The parallel version requires libraries for MPI, a standard for message passing among processes, be installed on all computers that are used for parallel processing. Multiple processors are used only for the geochemical part of a simulation. In general, the geochemical part of the simulation requires most of the CPU time, so the parallel version is effective in reducing run times. (See Appendix C.)

1.7. Purpose and Scope

The purpose of this documentation is to provide the user with information about the capabilities and usage of the reactive-transport simulator PHAST. Sections on running the simulator, preparing input data files, selecting output data files, and examples are provided. Sections on the output-visualization tool Model Viewer, the interactive output-data extractor PHASTHDF, the parallel-computer version, and the theory and numerical implementation are provided as appendices. The examples include a set of three verification problems for multicomponent reactive transport and three demonstration problems for tutorial purposes. Although the present documentation is intended to provide a complete description of the flow and transport capabilities of PHAST, the documentation for PHREEQC (Parkhurst and Appelo, 1999) is necessary to use the chemical-reaction capabilities of PHAST. Documentation for HST3D Versions 1 and 2 (Kipp, 1987, 1997) provides additional details about the flow and transport calculations.

Chapter 2. Running the Simulator

This chapter contains information needed to run the PHAST simulator as well as some background information on program organization and flow of execution. Specific information on preparation of input data files is contained in Chapter 3 and Chapter 4 and the PHREEQC documentation (Parkhurst and Appelo, 1999).

2.1. Input Files

Three data-input files are needed for simulations that include flow, transport, and reactions: (1) *prefix*.**trans.dat** the flow and transport calculation data, (2) *prefix*.**chem.dat** the chemical-reaction calculation data, and (3) the thermodynamic database (default name *phast.dat*). The identification *prefix* is provided by the user and is used to identify files for a simulation. In this documentation, it is represented by *"prefix"*. (Here, and throughout the documentation, bold type is used for words that must be typed as specified and italic type is used for words that must be specified by the user and for file names.) If only groundwater flow is simulated, only the *prefix*.**trans.dat** file is needed. Chapter 3 and the PHREEQC documentation (Parkhurst and Appelo, 1999) contain the necessary information to create the chemistry data file and to modify the thermodynamic database file. Chapter 4 contains the necessary information to create the flow and transport data file.

2.2. Output Files

Various types of output files can be produced by running the PHAST program. Some of the output files are in a tabular format intended to be displayed or printed. Some files are intended only for postprocessing by visualization or plotting programs.

Table 2.1 lists the file names in alphabetical order with a brief description of the contents of each of the output files. A more detailed description of output files is presented in Chapter 5. Except for the time unit, the units of measure of the output data are always SI (International System) metric, no matter what units were used for the input data. The time unit for output is selected by the user. Chemical concentrations are written in units of molality in all of the output files, except in the *prefix*.**chem.xyz.tsv** and *prefix*.**h5** files, where it is possible for the user to print concentrations in alternative units.

The files with names ending with **.txt** are in tabular format to be viewed in an editor (or to be printed). Many **.txt** files contain values of variables at grid nodes, which are arranged by horizontal or vertical grid slices, as chosen by the user. The files with names ending in **.tsv** are in tab-separated-values format, suitable for importing into spreadsheet and plotting programs. Files with names ending in **.xyz.tsv** in the name are tab-separated-values formatted blocks of data with one row of values for each active node in the grid region; blocks may be written for multiple time steps. All of the output files are written in ASCII format, except for the HDF file and the **.gz** restart file, which are in compressed binary formats. Data written to the binary HDF file (**.h5**) can be visualized by the program Model Viewer (Appendix A) and data can be extracted to an ASCII file with the program PHASTHDF (Appendix B). Data in the **.gz** file are ASCII text that has been compressed with the program *gzip*.

Table 2.1. List of output files that may be generated by a PHAST simulation.

File name	Contents
prefix.**bal.txt**	Fluid and chemical component regional mass balances, total balances, and balances listed by boundary-condition type (text format)
prefix.**bcf.txt**	Boundary-condition fluid and solute flow rates for cells listed by boundary-condition type (text format)
prefix.**chem.txt**	Solution concentrations, distribution of aqueous species, saturation indices, and compositions of equilibrium-phase assemblages, exchangers, kinetic reactants, surfaces, solid solutions, and gas phases from the internal PHREEQC calculation at the beginning of a PHAST run. Data normally are not written to this file during flow, transport, and reaction calculations (text format)
prefix.**chem.xyz.tsv**	Selected initial-condition and transient chemical data (tab-separated-values format by node)
prefix.**comps.txt**	Initial-condition and transient total dissolved concentrations for chemical elements (components) (text format)
prefix.**comps.xyz.tsv**	Initial-condition and transient cell concentrations for each chemical element (component) (tab-separated-values format by node)
prefix **h5**	Grid and boundary-condition information, media properties, potentiometric head field, velocity field, and selected chemical concentration data (binary format)
prefix **head.dat**	Potentiometric heads at the final time step of the simulation in a form that can be read as initial head conditions in subsequent simulations (text format)
prefix **head.txt**	Initial-condition and steady-state or transient potentiometric heads (text format)
prefix **head.xyz.tsv**	Initial-condition and steady-state or transient potentiometric heads (tab-separated-values format by node)
prefix **kd.txt**	Static fluid conductances and transient solute dispersive conductances at cell faces for X-, Y-, and Z-coordinate directions (text format)
prefix **log.txt**	Copies of the *prefix*.**trans.dat** and *prefix*.**chem.dat** input data files, solver statistics, and all error or warning messages (text format)
prefix.**probdef.txt**	Flow and transport problem definition as specified by the input files, including array sizes, grid definition, media properties, initial conditions, static and transient boundary-condition information (text format)
prefix **restart.gz**	Chemical composition of solutions and reactants at all nodes in the active grid region at a point in time, which can be used as initial conditions subsequent simulations (binary format)
prefix.**vel.txt**	Interstitial velocities for X-, Y-, and Z-coordinate directions across cell faces and interstitial velocities in the X-, Y-, and Z-coordinate directions at nodes (text format)
prefix.**vel.xyz.tsv**	Steady-state or transient velocity-vector components interpolated to grid nodes (tab-separated-values format by node)
prefix.**wel.txt**	Well location, well identification number, fluid and solute flow rates, cumulative fluid and solute flow amounts, solute concentrations, and injection and production rates per node for each well (text format)
prefix.**wel.xyz.tsv**	Initial and transient concentration data for wells (tab-separated-values format by well)

Table 2.1. List of output files that may be generated by a PHAST simulation.—Continued

File name	Contents
prefix.**wt.txt**	Steady-state or transient potentiometric heads at the water table (text format)
prefix.**wt.xyz.tsv**	Steady-state or transient potentiometric heads at the water table (tab-separated-values format by X-Y node)
prefix.**zf.tsv**	Fluid and solute flows through zone boundaries and through boundary-condition cells within the zones (tab-separated-values format by zone and component)
prefix.**zf.txt**	Fluid and solute flows through zone boundaries and through boundary-condition cells within the zones (text format)
user defined (**.xyzt**)	Time series of heads for a zone that can be used to define boundary conditions for subsequent simulations (text format)
selected_output	Rarely used. Selected chemical data from the initial call to PHREEQC to process the chemistry data file (Following the initial call to PHREEQC, selected output for initial and transient conditions are written to *prefix*.**chem.xyz.tsv**.)

The types of transient information that may be written include the solution-method information (*prefix*.**log.txt**), the fluid and solute-dispersive conductance distributions (*prefix*.**kd.txt**), the potentiometric head (*prefix*.**head.txt**, *prefix*.**head.xyz.tsv**, *prefix*.**wt.txt**, *prefix*.**wt.xyz.tsv**, and *prefix*.**h5**), component concentration distributions (*prefix*.**comps.txt** and *prefix*.**comps.xyz.tsv**), selected chemical information *(prefix*.**chem.txt**, *prefix*.**chem.xyz.tsv**, and *prefix*.**h5**), the steady-state or final potentiometric head distribution (*prefix*.**head.dat**), the groundwater flow velocity distribution (*prefix*.**vel.txt**, *prefix*.**vel.xyz.tsv**, and *prefix*.**h5**), the regional fluid-flow and solute-flow rates and the regional cumulative-flow results (*prefix*.**bal.txt**), boundary-condition flow rates (*prefix*.**bcf.txt**), well-flow data (*prefix*.**wel.txt**), component concentration data for wells (*prefix*.**wel.xyz.tsv**), and fluid and solute fluxes into and out of specified spatial zones and into and out of the boundary-condition cells within the zones (*prefix*.**zf.txt**, and *prefix*.**zf.tsv**). The selection of the frequency that data are written to these files is explained in Chapter 4.

The potentiometric head field from one simulation may be used to establish the initial condition for a subsequent simulation. Potentiometric head distribution from a steady-state or transient-flow simulation can be written to a file (*prefix*.**head.dat**), which can be read as the initial head conditions for subsequent runs (executions of the PHAST program). The chemical compositions of solutions and reactants at nodes can be written to a file (*prefix*.**restart.gz**), which can be read as the initial conditions for subsequent runs.

2.3. Program Execution

PHAST is run by using a batch file (Windows) or shell script (Unix), each referred to as a "script". The script invokes two programs, (1) PHASTINPUT (file *phastinput* for Unix, *phastinput.exe* for Windows), which converts the flow and transport data file in keyword format into an intermediate input file, *Phast.tmp*, and (2) PHAST (file *phast-ser* for Unix, *phast-ser.exe* for Windows; "ser" for serial or single processor), which performs the flow and reactive-transport simulations. Figure 2.1 illustrates the relation among the input and output files and the execution of PHASTINPUT and PHAST within the script. The script is invoked from a command line with either one or two arguments:

phast *prefix* [*database_name*]

where the first argument is the prefix and the optional second argument is the thermodynamic database file name. The input files for transport and chemistry must be present in the working directory and must be named *prefix*.**trans.dat** and *prefix*.**chem.dat**. If the thermodynamic database file name is not specified as the second argument, the file *phast.dat* in the database subdirectory of the installation directory for PHAST will be used. If the thermodynamic database file name is specified, the file must be present in the working directory or be specified with a complete pathname. PHASTINPUT converts time-series data read in the *prefix*.**trans.dat** to a series of simulation periods over which boundary conditions and print frequencies are constant. Static data and data for each of these simulation periods are written to the file *Phast.tmp*.

When the PHAST simulator is run, the program executes the following steps: (1) The first part of the intermediate file *Phast.tmp* is read and static flow and transport data are initialized. (2) An initial call to PHREEQC is executed, which reads the thermodynamic database file, and then reads and performs the calculations specified in the chemistry data file. (3) Additional static flow and transport data are read from *Phast.tmp*, and, if requested, a steady-state flow simulation is performed. (4) Transient data for the first simulation period are read from *Phast.tmp* and the flow (if not steady state), transport, and reaction calculations are performed for the first simulation period. (5) Step 4 is repeated for each simulation period that is defined. The file *selected_output* is written (if requested) only during step 2, the initial call to PHREEQC; all other output files contain information from steps 3 and 4. The logical flow of the calculation is described further in the section D.7.1. Operator Splitting and Sequential Solution.

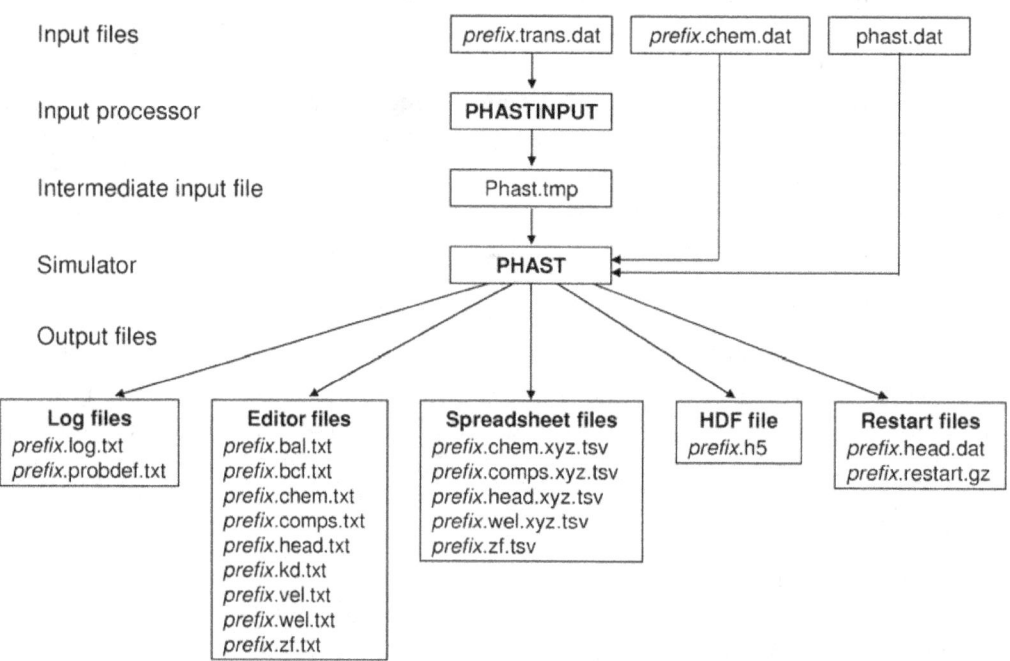

Figure 2.1. The relation between input and output files and the execution of the two programs PHASTINPUT and PHAST.

Chapter 3. Thermodynamic Database and Chemistry Data Files

When the program PHAST is invoked, the thermodynamic database file and the chemistry data file are used to define the chemical model, the initial and boundary-condition solutions, and the equilibrium-phase assemblages, exchangers, surface assemblages, kinetic reactants, solid solutions, and gas phases that are present in each cell of the active grid region. Complete descriptions of the keyword data blocks for these files are found in the PHREEQC manual (Parkhurst and Appelo, 1999) and are not presented in this manual. Both of the chemistry data and thermodynamic database files for PHAST have identical formats to PHREEQC input and database files, and they can be generated or modified by using an interface to PHREEQC. Two interfaces are available: PHREEQC For Windows (*http://www.geo.vu.nl/users/posv/phreeqc/index.html*) and PHREEQCI (Charlton and Parkhurst, 2002; *http://wwwbrr.cr.usgs.gov/projects/GWC_coupled/phreeqc/index.html*).

The thermodynamic database file is used to specify static data for the chemical model, which includes a list of chemical elements; chemical-reaction equations that define aqueous, exchange, surface, and solid species; and the equilibrium constants for the chemical reactions. The chemistry data file is used to specify solution compositions and compositions of reactants (equilibrium phases, exchangers, surfaces, kinetic reactants, solid solutions, and gas phases), either directly, or indirectly through specification of geochemical reactions. Solutions, equilibrium-phase assemblages, ion exchangers, surfaces, kinetic reactants, solid solutions, and gas phases that are specified in the chemistry data file are used to define the initial and boundary conditions and the sets of reactants that are present in each cell of the active grid region. Another important function of the chemistry data file is to select chemical data that are written to output files, such as pH, molalities of aqueous species, and amounts of minerals and kinetic reactants. The chemistry data file can be used to modify the chemical model that is defined in the thermodynamic database file.

3.1. Thermodynamic Database File

The thermodynamic database file is used to specify the reactions and thermodynamic properties of the aqueous and solid chemical species that compose the chemical model of PHAST. Although any keyword data block can be included in the thermodynamic database file, commonly, the file contains the keyword data blocks listed in table 3.1, which are used to define basic thermodynamic properties. These keyword data blocks define master species, reaction stoichiometry, and equilibrium constants for all of the aqueous-phase species, exchange species, surface species, and equilibrium phases that can be modeled with PHAST. Six thermodynamic database files are provided with the program: the default thermodynamic database file, *phast.dat*, which is derived from PHREEQE (Parkhurst and others, 1980) and WATEQ4F (Ball and Nordstrom, 1991), and is the same as *phreeqc.dat* in the PHREEQC distribution; *wateq4f.dat*, which is derived from WATEQ4F (Ball and Nordstrom, 1991) and is consistent with *phreeqc.dat* but contains more trace elements; *minteq.dat* and *minteq.v4.dat*, which are derived from MINTEQA2 (Allison and others, 1990); *llnl.dat*, which is the most comprehensive of the thermodynamic database files and is derived from thermodynamic data (designation thermo.com.v8.r6.230) assembled by Lawrence Livermore National Laboratory (Wolery, 1992a, 1992b; Bethke, 1996; and Johnson and others, 1991); and *pitzer.dat*, which is derived from PHRQPITZ (Plummer and others, 1988). The elements and element valence states that are included in *phast.dat* are listed in table 3.2 along with the PHAST notation and the default formula for gram-formula weight used to convert mass to moles.

Table 3.1. Keyword data blocks commonly used in the thermodynamic database file.

Keyword data block	Function
EXCHANGE_MASTER_SPECIES	Defines names of exchange sites
EXCHANGE_SPECIES	Defines reactions for exchange sites
SOLUTION_MASTER_SPECIES	Defines names of elements
SOLUTION_SPECIES	Defines aqueous species
SURFACE_MASTER_SPECIES	Defines names of surface sites
SURFACE_SPECIES	Defines reactions for surface sites
PHASES	Defines mineral and gas dissociation reactions

3.2. Chemistry Data File

The chemistry data file defines the solutions, equilibrium-phase assemblages, exchangers, surfaces, kinetic reactants, solid solutions, and gas phases that are used in initial and boundary conditions of the reactive-transport simulation. Also defined in the chemistry data file is the set of chemical data to be written to the *prefix*.**h5** and *prefix*.**chem.xyz.tsv** output data files. The keyword data blocks most commonly used in the chemistry data file to define initial and boundary conditions are listed in table 3.3. For details on the formats and options for these and all other data blocks that can be included in the chemistry data file (and thermodynamic database file), it is necessary to refer to the PHREEQC manual (Parkhurst and Appelo, 1999).

3.2.1. Chemical Initial and Boundary Conditions for Reactive Transport

Initial conditions for the reactive-transport simulation are defined with **SOLUTION**, **EQUILIBRIUM_PHASES**, **EXCHANGE**, **SURFACE**, **KINETICS**, **SOLID_SOLUTIONS**, and **GAS_PHASE** data blocks. Solution compositions and speciation calculations are defined with the **SOLUTION** data block. The identity and amount of each phase in an equilibrium-phase assemblage is defined with the **EQUILIBRIUM_PHASES** data block; the composition of an exchange assemblage is defined with the **EXCHANGE** data block; the composition of a surface assemblage is defined with the **SURFACE** data block; sets of kinetic reactions are defined with the **KINETICS** data block, and rate expressions for the kinetic reactions are defined with the **RATES** data block. The composition of a fixed-total-pressure or fixed-volume multicomponent gas phase is defined with the **GAS_PHASE** data block. (The gas phase feature is rarely used; gases are usually defined as fixed partial pressure phases in **EQUILIBRIUM_PHASES** data blocks.) Index numbers are defined for each solution, equilibrium-phase assemblage, exchange assemblage, surface assemblage, kinetic-reactant set, solid-solution assemblage, and gas phase. These index numbers are used in the flow and transport data file to specify initial and boundary conditions by zones. For example, solutions defined by different index numbers can be applied to a series of zones to define the spatially varying initial composition of water within an aquifer.

The units for input in the **SOLUTION** data block are concentration units, which are converted to molality for all chemical calculations. The units for input in the data blocks **EQUILIBRIUM_PHASES**, **EXCHANGE**, **SURFACE**, **KINETICS**, **SOLID_SOLUTIONS** are extensive (moles). For each cell in PHAST, the representative porous-medium volume for chemistry contains one kilogram of water, when saturated. Thus, when defining amounts of solid-phase reactants, the appropriate number of moles is numeri-

Table 3.2. Elements and element valence states included in default thermodynamic database *phast.dat*, including PHAST notation and default formulas for gram formula weight.

[For alkalinity, formula for gram equivalent weight is given]

Element or element valence state	PHAST notation	Formula used for gram formula weight
Alkalinity	Alkalinity	$Ca_{0.5}(CO_3)_{0.5}$
Aluminum	Al	Al
Barium	Ba	Ba
Boron	B	B
Bromide	Br	Br
Cadmium	Cd	Cd
Calcium	Ca	Ca
Carbon	C	HCO_3
Carbon(IV)	C(4)	HCO_3
Carbon(-IV), methane	C(-4)	CH_4
Chloride	Cl	Cl
Copper	Cu	Cu
Copper(II)	Cu(2)	Cu
Copper(I)	Cu(1)	Cu
Fluoride	F	F
Hydrogen(0), dissolved hydrogen	H(0)	H
Iron	Fe	Fe
Iron(II)	Fe(2)	Fe
Iron(III)	Fe(3)	Fe
Lead	Pb	Pb
Lithium	Li	Li
Magnesium	Mg	Mg
Manganese	Mn	Mn
Manganese(II)	Mn(2)	Mn
Manganese(III)	Mn(3)	Mn
Nitrogen	N	N
Nitrogen(V), nitrate	N(5)	N
Nitrogen(III), nitrite	N(3)	N
Nitrogen(0), dissolved nitrogen	N(0)	N
Nitrogen(-III), ammonia	N(-3)	N
Oxygen(0), dissolved oxygen	O(0)	O
Phosphorus	P	P
Potassium	K	K
Silica	Si	SiO_2
Sodium	Na	Na
Strontium	Sr	Sr
Sulfur	S	SO_4
Sulfur(VI), sulfate	S(6)	SO_4
Sulfur(-II), sulfide	S(-2)	S
Zinc	Zn	Zn

cally equal to the concentration of the reactant (moles per liter of water), assuming a saturated porous medium. In terms of moles per liter of water, the concentrations of the solid reactants vary spatially as porosity varies, which makes it difficult to define the appropriate solid reactant concentrations. To avoid this difficulty, PHAST Version 2 has options in the UNITS data block (flow and transport data file) to specify that the number of moles of solid reactants be interpreted as moles per liter of rock. As initial conditions are distributed to the finite-difference cells, the moles of solid reactants are scaled by the factor $(1-\phi)/\phi$, where ϕ is the porosity for the cell. This scaling takes into account the varying porosity and produces units of moles per liter of water. Molality (moles per kilogram of water; mol/kgw) is assumed to equal molarity (mol/L) for all transport calculations.

For unconfined flow (free-surface) calculations, the volume of water in a cell may vary. Thus, it is important to define the rate expressions for kinetic reactions to account for the changing volume of solution. Rate expressions for kinetic reactions introduce moles of reactants; thus, if a known rate expression has units of moles per kilogram of water, it is necessary to multiply by the mass of water present [TOT("water") is the Basic-language function for mass of water] to calculate the proper number of moles of reaction to add to solution (see **KINETICS** data block in Parkhurst and Appelo, 1999).

Reaction calculations specified in the chemistry data file may be used to define initial or boundary conditions. These reaction calculations are processed during the initial call to the PHREEQC module. Reactions are defined by allowing a solution or mixture of solutions to come to equilibrium with one or more of the following reactants: an equilibrium-phase assemblage, an exchange assemblage, a surface assemblage, a solid-solution assemblage, or a multicomponent gas phase. In addition, irreversible reactions, kinetic reactions, and reaction temperature can be specified for reaction calculations. (Although temperature does not affect flow, it does affect chemical reactions. See section D.6. Initial Conditions.) Previously defined reactants can be used in reaction calculations by **USE** data blocks. The composition of the solution, equilibrium-phase assemblage, exchange assemblage, surface assemblage, solid-solution assemblage, or gas phase can be saved after a reaction calculation with the **SAVE** keyword for subsequent use in a reaction calculation or for initial and boundary conditions. The **SAVE** keyword assigns an index number to the specified composition. These index numbers are used in the flow and transport file to define initial and boundary conditions for specified zones within the active grid region.

Some keyword data blocks from PHREEQC that are less commonly used have been omitted from table 3.1 and table 3.3, but all of the functionality of PHREEQC is available in the initial-reaction calculations by the PHREEQC module, including all reaction and one-dimensional transport and advection capabilities. Thus, it is possible to use a reaction calculation to define a water composition that is produced by equilibrating pure water with calcite and dolomite, which can then be used as the initial water composition for an aquifer. It also is possible, for example, to simulate the percolation of a contaminant through a one-dimensional column that represents a confining layer and save the water compositions of the column effluent to be used as time-varying input through a flux boundary condition.

Table 3.3. Keyword data blocks commonly used in the chemistry data file.

[*index*, integer number corresponding to a definition in the chemistry data file; —, index number is not used in flow and transport data file; *keyword*, one of the keywords allowed in a **USE** data block in the chemistry data file]

Keyword data block	Function of data block in initial call to PHREEQC module	Keyword data block using index number in the flow and transport data file
END	Marks the end of definitions for initial and reaction calculations	—
EQUILIBRIUM_PHASES *index*	Defines a mineral assemblage	CHEMISTRY_IC
EXCHANGE *index*	Defines an assemblage of ion exchangers	CHEMISTRY_IC
GAS_PHASE *index*	Defines a gas phase. This option is rarely used because gas transport is not simulated. More commonly, gases are treated with fixed partial pressures in **EQUILIBRIUM_PHASES**.	CHEMISTRY_IC
KINETICS *index*	Defines a set of kinetic reactions	CHEMISTRY_IC
RATES	Defines rate expressions for kinetic reactions	—
REACTION *index*	Adds specified amounts of reactants	—
REACTION_TEMPERATURE *index*	Sets the temperature for reaction calculations	—
SAVE solution *index*	Saves solution composition that is the result of a reaction simulation	CHEMISTRY_IC, FLUX_BC, LEAKY_BC, SPECIFIED_HEAD_BC, RIVER, WELL
SAVE equilibrium_phases *index* **SAVE exchange** *index* **SAVE surface** *index* **SAVE kinetics** *index* **SAVE solid_solution** *index* **SAVE gas_phase** *index*	Saves reaction simulation results to define new compositions of equilibrium phases, exchangers, surfaces, kinetic reactants, solid solutions, and gas phase	CHEMISTRY_IC
SELECTED_OUTPUT	Defines data to be written to the *prefix*.**h5** and *prefix*.**chem.xyz.tsv** files	—
SOLID_SOLUTIONS *index*	Defines an assemblage of solid solutions	CHEMISTRY_IC
SOLUTION *index*	Defines a solution composition	CHEMISTRY_IC, FLUX_BC, LEAKY_BC, SPECIFIED_HEAD_BC, RIVER, WELL
SURFACE *index*	Defines sets of surface complexation reactions	CHEMISTRY_IC
TITLE	Defines a character string to annotate the calculations	—
USE *keyword index*	Selects solution, equilibrium phases, exchanger assemblage, surface assemblage, kinetic reactants, reaction, solid-solution assemblage, or gas phase to be used in a reaction calculation	—
USER_PUNCH	Defines data to be written to the *prefix*.**h5** and *prefix*.**chem.xyz.tsv** files	—

3.2.2. Output of Chemical Data

PHAST simulations produce a large amount of chemical data—including, but not limited to, the molality and activity of every aqueous species, exchange species, and surface species; the moles of every reactive mineral; the saturation index for each of a large number of minerals; and the mole transfer of every kinetic reactant—for every cell for every time step. The **SELECTED_OUTPUT** and **USER_PUNCH** data blocks are used to select a subset of the chemical results to be written to the *prefix*.**h5** and *prefix*.**chem.xyz.tsv** output data files. The **SELECTED_OUTPUT** data block allows selection of certain types of chemical data to be written to these two output files, including pH; pe; total molalities of specified elements; molalities and activities of specified aqueous, exchange, and surface species; saturation indices and partial pressures of specified minerals and gases; amounts and mole-transfers of equilibrium phases and kinetic reactants; and moles of gas components in the gas phase.

The **USER_PUNCH** data block provides another more versatile method to select data for the two output files. The **USER_PUNCH** data block uses an embedded Basic-language interpreter to allow calculation of chemical quantities that can then be written to the output files. The Basic-language interpreter has a full range of programming capabilities, including variables, mathematical functions, FOR loops, and subroutines, which allows calculation of almost any chemical quantity that is needed in the output files. During the reaction calculation for a cell at any given time step in the reactive-transport simulation, virtually all of the information related to the chemical calculation for the cell is available through Basic-language functions—molalities, activities, ionic strength, equilibrium constants, moles of minerals, among others. The calculations using these functions can be written to the output file with a Basic-language command (PUNCH). Thus, for example, it is possible to calculate the number of grams of calcite in a cell from the number of moles of calcite, and write the mass to the two output files *(prefix*.**h5** and *prefix*.**chem.xyz.tsv**). Another example is to calculate the sodium sorption ratio (SAR) from the concentrations of the major cations, and write the result to the output files.

The *selected_output* file also may contain data as defined by **SELECTED_OUTPUT** and **USER_PUNCH**. However, the file only will contain results related to the first call to PHREEQC, which precedes all steady-flow, active-grid initialization and reactive-transport calculations. Frequently, a spreadsheet file of this preliminary calculation is not useful, and the **SELECTED_OUTPUT** data block is located at the end of the *prefix*.**chem.dat file**, following an **END** keyword, which results in no data being written to the *selected_output* file.

If necessary, the chemistry data file can be used to modify or augment the thermodynamic database file. If new elements, aqueous species, exchange species, surface species, or phases need to be included in addition to those defined in the thermodynamic database file, or if the stoichiometry, equilibrium constant, or activity coefficient information from the thermodynamic database file needs to be modified for a simulation, then the keyword data blocks listed in table 3.1 can be included in the chemistry data file. The data read for these data blocks in the chemistry data file will augment or supersede the data read from the thermodynamic database file. In most cases, the aqueous model and thermodynamic data for phases defined in the thermodynamic database file will not be modified, and the keywords listed in table 3.1 will not be used in the chemistry data file.

Chapter 4. Flow and Transport Data File

The flow and transport data file provides information necessary to simulate groundwater flow and solute transport. The file contains the definition of the grid, porous-media properties, initial conditions, boundary conditions, and time step. Selected properties, for example well pumping rate, boundary-condition heads, and boundary-condition solution compositions, may vary over the course of a simulation by definition of time series for these data. The flow and transport data file also is used to specify the units for input data, print frequencies for output files, and information related to the formulation and solution of the flow and transport finite-difference equations. All chemical compositions in the flow and transport data file, which are related to initial conditions, boundary conditions, and chemical reactions, are specified by index numbers that correspond to complete definitions in the chemistry data file (section 3.2. Chemistry Data File).

It requires only a single change—**SOLUTE_TRANSPORT true** to **false**—in the flow and transport data file to switch from a flow, solute transport, and reaction simulation to a flow-only simulation. For flow-only simulations, the flow and transport data file is the only input file necessary for PHAST (the thermodynamic database file and chemistry data file are not used) and all data related to transport and chemistry in the flow and transport data file are ignored.

In this documentation, the term "property" refers to any numerical value used in the definition of a porous-media property, initial condition, or boundary condition. In the flow and transport file, spatially distributed properties are defined by "zone", which may be a rectangular box, a triangular wedge, or an arbitrary volume with a perimeter, top, and bottom referred to as a "prism". Multiple overlapping zones may be used to assemble spatially varying distributions of properties. Time-series data in boundary-condition definitions allow for changes in flow and chemical boundary conditions over the course of a simulation.

All of the data in the flow and transport data file are defined through the keyword data blocks listed in table 4.1. This chapter contains complete descriptions of data input for these keyword data blocks.

4.1. Organization of the Flow and Transport Data File

PHAST uses keyword data blocks in the flow and transport data file to define static properties, initial conditions, and time-varying properties. Some of the data blocks define static data that apply unchanged for the entire simulation, whereas some of the data blocks are used to define both static data and transient data. As indicated in table 4.1, the definition of a flow-only simulation requires, at a minimum, the **GRID**, **HEAD_IC**, **MEDIA**, **SOLUTE_TRANSPORT**, **TIME_CONTROL**, **and UNITS** data blocks. A reactive-transport simulation requires additionally the **CHEMISTRY_IC** data block.

Data blocks that define the input units, grid, porous-media properties, initial conditions, numerical methods, steady-state flow, the selection of cells for writing output data, the free-surface boundary condition, drains, and zones for calculation of flow rates are static and apply to the entire simulation. These data blocks and the data blocks for title and printing initial conditions contain no time-varying properties. The data blocks that define only static information are **CHEMISTRY_IC, DRAIN, FREE_SURFACE_BC, GRID, HEAD_IC, MEDIA, PRINT_INITIAL, PRINT_LOCATIONS, SOLUTE_TRANSPORT, SOLUTION_METHOD, STEADY_FLOW, TITLE, UNITS,** and **ZONE_FLOW**.

Data blocks for other boundary conditions include both static and transient data. The boundary-condition data blocks assign static data—for example, the type and location of all boundary conditions, the thickness of a leaky boundary, the diameter of a well, or the hydraulic conductivity of a riverbed—and transient data, which include time-series property definitions for head, solution composition, flux, and pumping or injection rate. The transient data are properties that are allowed to vary over the course of a simulation; however, the static data cannot be changed. The boundary-condition data blocks that include both static and transient data

Table 4.1. Keyword data blocks for the flow and transport data file.

[Required data block: All, data block is required for all simulations—flow-only and reactive-transport simulations; Transport, the data block is required for all reactive-transport simulations; No, although not required for the minimum simulation, data block may be needed to define the correct flow system or to produce model output]

Keyword data block	Required data block	Function
CHEMISTRY_IC	Transport	Initial conditions for chemistry, and types of reactions for the active grid region
DRAIN	No	Drain boundary condition
END	No	End of definitions for the flow and transport data file
FLUX_BC	No	Flux boundary conditions
FREE_SURFACE_BC	No	Presence or absence of a free-surface boundary condition
GRID	All	Node locations
HEAD_IC	All	Initial head conditions
LEAKY_BC	No	Leaky boundary conditions
MEDIA	All	Porous-media properties—porosity, hydraulic conductivity, specific storage, and dispersivity
PRINT_FREQUENCY	No	Time intervals at which data are written to output files
PRINT_INITIAL	No	Set of initial conditions to be written to output files
PRINT_LOCATIONS	No	Zones for which data will be written to the *prefix*.**chem.xyz.tsv** and *prefix*.**chem.txt** files
RIVER	No	River boundary condition
SOLUTE_TRANSPORT	All	Flow-only or reactive-transport simulation
SOLUTION_METHOD	No	Numerical method for solving the finite-difference equations
SPECIFIED_HEAD_BC	No	Specified-head boundary conditions
STEADY_FLOW	No	Steady state or transient flow conditions
TIME_CONTROL	All	Time step, start time, and times for the ends of simulation periods
TITLE	No	Character string describing the simulation
UNITS	All	Units of input data
WELL	No	Location, open intervals, and pumping (or injection) rates for a well
ZONE_FLOW	No	Flow rates of water and solutes in and out of a specified zone; time series of heads for zone

are **FLUX_BC, LEAKY_BC, RIVER, SPECIFIED_HEAD_BC,** and **WELL.** The **TIME_CONTROL** data block must be included in every simulation to define the start time, the time step for integration (possibly time varying), specified times to end simulation periods, and the end time for the simulation. The data block **PRINT_FREQUENCY** may include time-varying print controls.

Simulation periods are intervals of time over which all time-varying properties and print frequencies are constant. Every time any time-varying property or print frequency changes value, the old simulation period ends and a new simulation period is begun. By default, the end of a simulation period is a time when data are written to output files, but this behavior can be changed by the options in the keyword data block **PRINT_FREQUENCY.**

With one major exception, the order of keyword data blocks is arbitrary within a set of data blocks that defines a simulation. The exception is that the order of the keyword data blocks and the order of definitions within data blocks are important for the definition of spatial data for media properties, boundary conditions, and initial conditions. For these data items, the data are assigned to cells and elements in sequential order, with subsequent definitions overwriting preceding definitions. For a given location in the grid region, the last definition in the data file that applies to that location is the one used, as described below (also see section D.5.8. Boundary-Condition Compatibility).

4.2. Spatial Data

The coordinate system for PHAST is three dimensional; X and Y represent horizontal axes and Z represents the vertical axis (allowing no tilt with respect to the gravity vector). The relation among X, Y, and Z is right-handed, in the sense that if your right thumb is pointed in the positive Z direction, then your fingers curl from the positive X axis to the positive Y axis. To establish a grid, a set of coordinate locations for nodes is defined along each axis—X, Y, and Z. In this report, the subscripts i, j, and k are used to refer to the X, Y, and Z coordinates, respectively. The grid is the set of nodes defined by every combination from the sets of X, Y, and Z coordinate locations. Thus, nodes are distributed along a series of lines and planes parallel to the coordinate axes. The volume that encloses the grid nodes is referred to as the grid region.

The nodes define two sets of spatial volumes, called elements and cells. An element is a rectangular prism defined by exactly eight nodes that are located at the element corners (fig. 4.1). The set of elements fills the entire grid region. To conform to nonrectangular model domains, it is possible to eliminate parts of the grid region by designating selected elements to be inactive. The volume defined by the set of active ele-

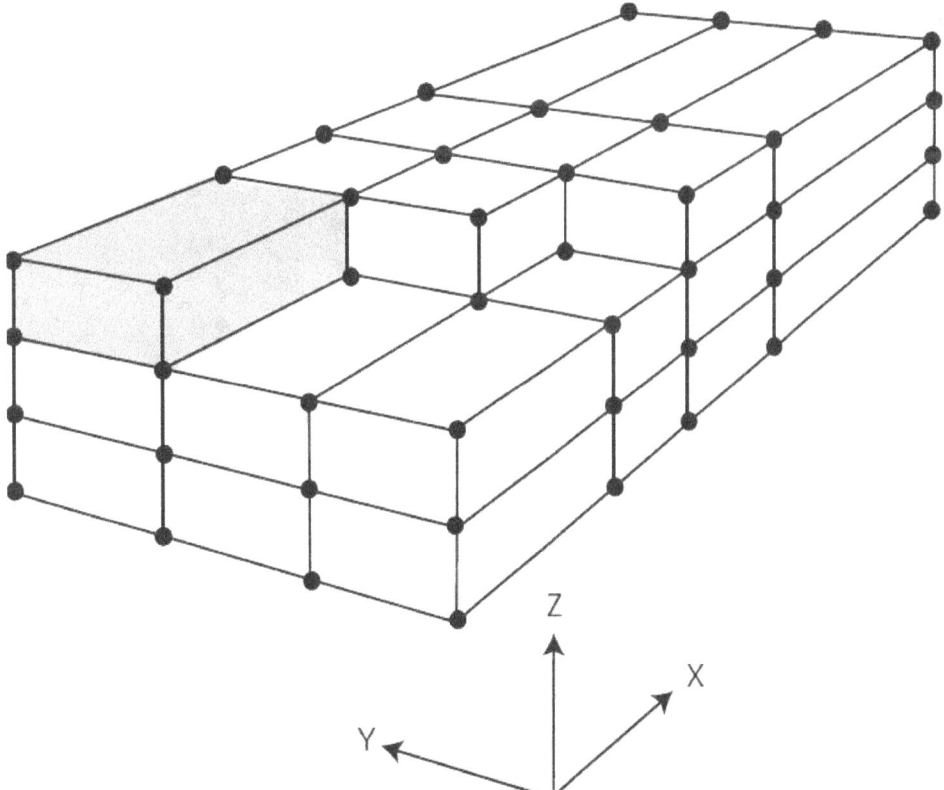

Figure 4.1. The grid region, boundary nodes (dots), and one shaded element.

ments is referred to as the active grid region. All nodes associated with active elements are referred to as active nodes. The volume defined by the set of inactive elements is referred to as the inactive grid region. Nodes interior to the inactive grid region are referred to as inactive nodes. Simulation results are calculated for all active nodes, including nodes internal to and on the boundary of the active grid region. Results are not calculated for any inactive nodes.

Every node in the active grid region is contained within a single cell. Cell faces either bisect the distance between adjacent nodes or lie along a plane of nodes that bounds the active grid region (fig. 4.2). Cell faces are identified by the axis to which they are perpendicular; for example an X face is perpendicular to the X axis. The volume of a cell is composed of parts of each of the elements that meet at the cell node. Only active elements contribute volumes to a cell, so each cell contains parts of one to eight active elements. Cells are full cells (or interior cells) when the cell contains parts of eight active elements. Cells are 1/2, 1/4, or 1/8 cells if one, two, or three faces, respectively, lie along planes of nodes that bound the grid region. These and other fractional cell configurations occur at the boundaries of irregularly shaped active grid regions. No cells are defined for nodes within the inactive grid region.

The "natural" order of elements and cells begins in the left, front, bottom corner of the grid region. Element and cell numbers increase by cycling through X, then Y, and then Z coordinates. This natural order is required whenever node-by-node input (which is not advisable) is used to define properties for a simulation.

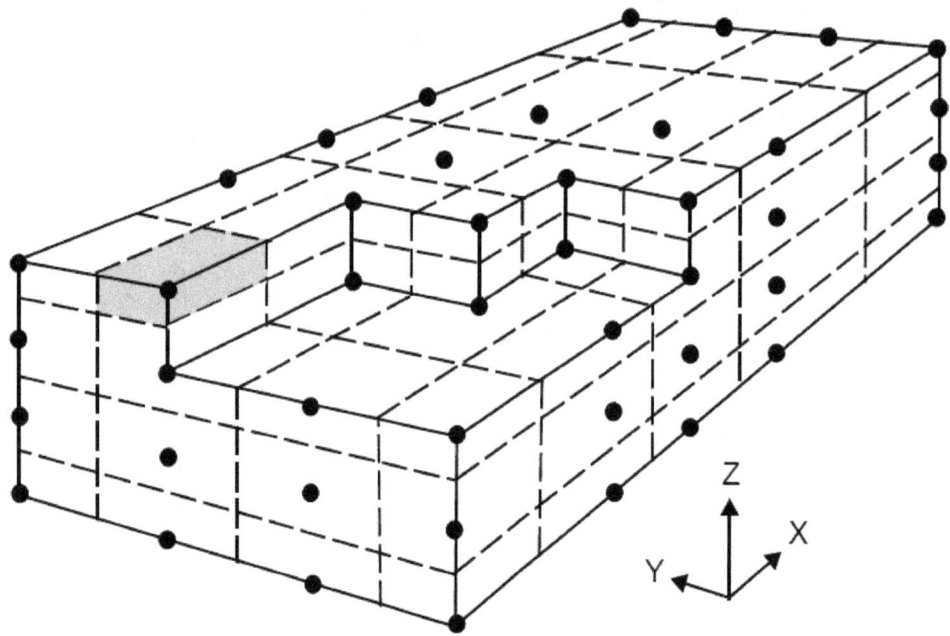

Figure 4.2. The grid region, boundary of the active grid region (solid lines), boundary nodes (dots), cell boundaries (dashed lines), and one shaded cell.

4.2.1. Zones

Spatial properties are distributed by zones, which may be any of four types: (1) the entire model domain, (2) a box—a rectangular parallelepiped, (3) a wedge—one half of a diagonally bisected rectangular parallel-epiped, or (4) a prism—a polygonal perimeter (concave or convex, but nonintersecting) possibly truncated by a top surface and(or) a bottom surface. The term zone is used to refer to any of these four spatial data definitions. (In PHAST Version 1, a zone referred only to a box.) As sets of spatial data are defined, it is permissible to use overlapping zones. For example, a given node could occur within multiple zones that define the initial head distribution, possibly in multiple **HEAD_IC** data blocks. The initial head that is used for the given node in the flow simulation will be the head from the last zone in the flow and transport data file that contains the node and defines the initial head property. The same logic applies to all properties and boundary conditions; the last property value defined for a node will be the one used in the simulation (that is, all properties for a node previous to the last entry will be ignored). Areas of application for leaky and flux boundary conditions are overlaid so that the last definition for each type of boundary condition for an area is the one that is used. The shapes of each type of zone are defined in the following sections.

4.2.1.1. Domain

If a property is defined for the entire model domain, it is applied to every node in the grid region. A domain definition is equivalent to a box that contains the entire grid or a default prism. However, if the grid region is enlarged, no change in the input for the property is needed; the property automatically applies to the new, larger grid region. Like any other zone, if subsequent definitions overlay parts of the domain, the last definition for a node will be used in the simulation.

4.2.1.2. Boxes

A zone that is a rectangular parallelepiped is called a box. A box is defined by two sets of coordinates, the location of the left, front, lower corner (minimum X, minimum Y, and minimum Z values) and the location of the right, back, upper corner (maximum X, maximum Y, and maximum Z values) of a rectangular parallelepiped.

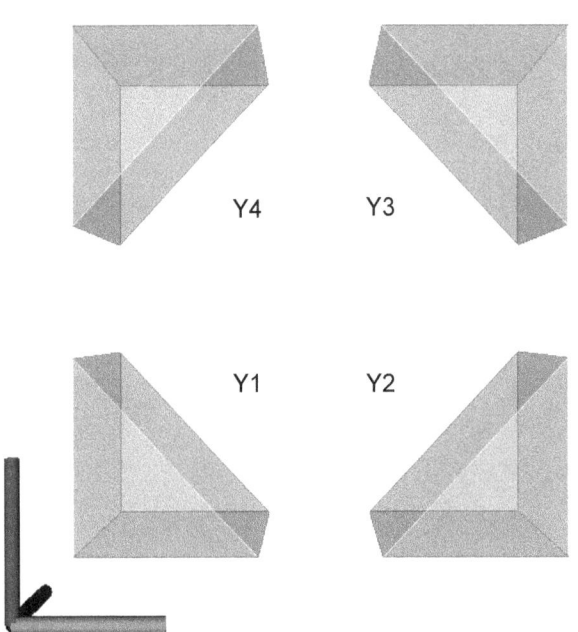

Figure 4.3. Identifying numbers for four wedges with axis of right angle parallel to the Y axis; viewpoint is from the negative Y axis. Positive axes are shown in the lower left: X (red), Y (black), and Z (blue).

4.2.1.3. Wedges

A zone that is a right triangular prism with four of its five sides parallel to the coordinate axes is called a wedge. A wedge is defined by two sets of coordinates plus an orientation identifier. The coordinates are the location of the left, front, lower corner (minimum X, minimum Y, and minimum Z values) and the location of the right, back, upper corner (maximum X, maximum Y, and maximum Z values) of a rectangular box that contains the wedge. The orientation identifier specifies the axis (X, Y, or Z) that is parallel to the axis of the right angle of the triangle (also parallel to the plane containing the hypotenuse face of the wedge) plus a number (1–4) that identifies the orientation of the right angle. Numbers are defined counterclockwise when viewed from the negative axis direction, where 1 corresponds to the right angle appearing at the point with the minimum coordinate values (see fig. 4.3).

4.2.1.4. Prisms

A prism is a zone defined by a perimeter, which is a polygon in the X–Y plane, and optionally, surfaces that define a top and(or) a bottom for the prism. The prisms described here are based on similar volume definitions described in Winston (2006). The polygon may be convex or concave but may not be self intersecting. If a perimeter is not defined, the default perimeter is defined by the X and Y extent of the grid region. If no top or bottom is defined, the vertical extent of the prism is equal to the vertical extent of the grid region. A constant or irregular top or bottom of the prism may be defined by a series of X–Y points that have associated elevations (Z). The top or bottom data may be defined in the input file or read from a file—an ArcInfo shapefile (ESRI, 1998), an ArcInfo ASCII raster file (ESRI, 2009), or a file that contains x, y, z triplets (one per line). Within the convex hull of the scattered data, natural neighbor interpolation (Sakov, 2000) is used to interpolate a target X–Y point. Outside the convex hull of the scattered data, the elevation of the closest scattered data point is used to define the value at the target X–Y point. Natural neighbor interpolation relies on a Delaunay triangulation (Shewchuk, 1996, 2003) to determine area weighting of elevations at near-neighbor points to interpolate an elevation at the target point. Assignment of the closest point elevation finds the closest scattered point by use of a KD tree (Kennel, 2004; Skiena, 2008) and assigns the elevation at that point to the target point. The volume defined by the perimeter, top, and bottom is used to select nodes or element centroids that fall within the prism. A point is within the prism if its X–Y coordinates are inside the perimeter polygon and if its Z coordinate is less than the elevation of the top and greater than the elevation of the bottom. An example of a prism is shown in figure 4.4.

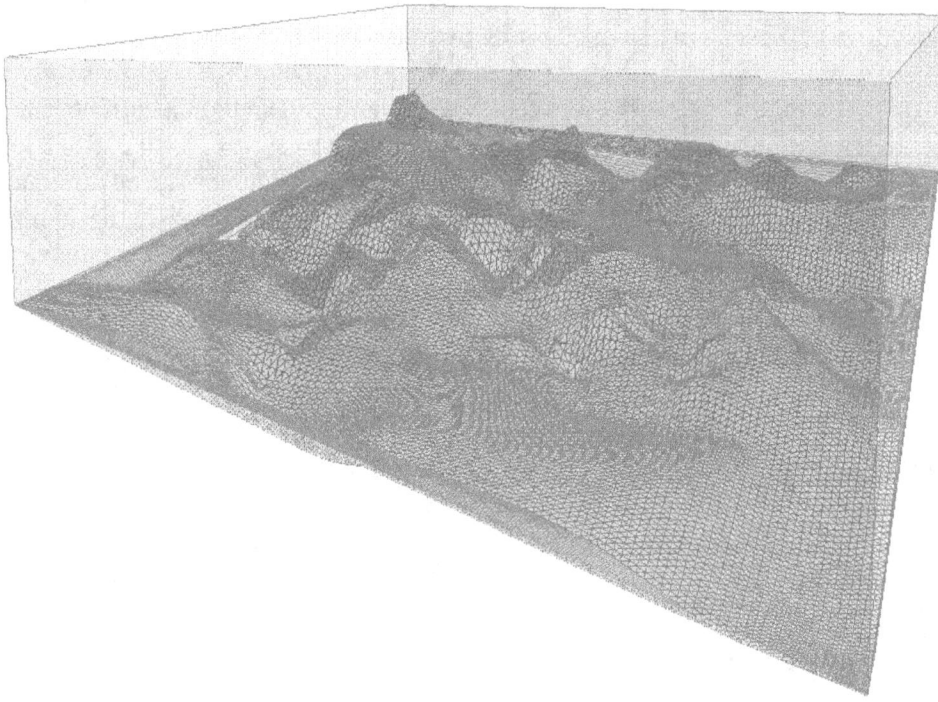

Figure 4.4. Prism defined by rectangular perimeter, default top, and bottom defined by an ArcInfo ASCII raster file (ESRI, 2009) with elevation of impermeable bedrock.

4.2.1.5. Use of Zones for Defining Porous-Media Properties

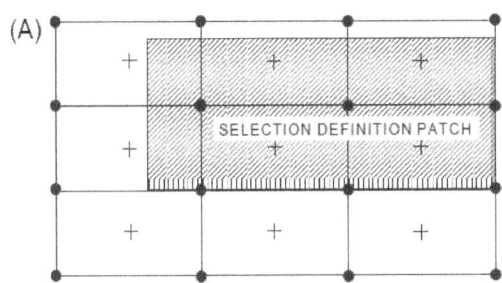

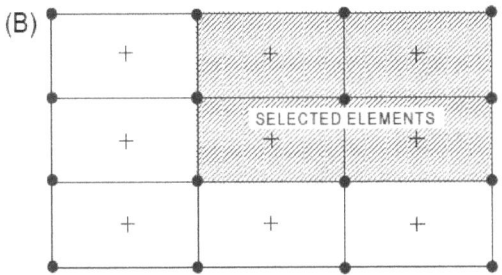

Figure 4.5. (A) Rectangular box used to select elements from a two-dimensional grid, and (B) the elements selected by the box. Lines represent element boundaries. Crosses indicate element centroids.

Porous-media properties, including porosity, hydraulic conductivity, specific storage, dispersivity, and active or inactive status, must be defined for each element. For porous-media properties, the definition of a zone selects each element whose centroid is contained in the zone. An example in two dimensions of the selection of elements by a box is shown in figure 4.5. The properties defined for the zone will be applied to each selected element.

Porous-media properties for a cell are defined by aggregating the properties of the active elements that are contained in the cell (fig. 4.6). A cell contains parts of one to eight active elements; thus, a cell property may be derived from as many as eight different values. Porosity and specific storage are averaged in proportion to the volume of each active element that is contained in the cell to arrive at the porous-media property for the cell. Hydraulic conductivity and dispersivity are calculated for each of the six faces of a cell (for an internal cell) by averaging the properties of the active elements that contain the cell face. The averages are calculated by weighting the element properties in proportion to the area of the cell face that is within each element. A cell face is located within one to four active elements; thus, cell-face conductances are weighted averages derived from the properties of as many as four different elements (fig. 4.6).

Frequently, it is convenient to define a uniform value for a porous-media property by using a zone that includes the entire model domain (either with a domain or box definition). Additional zones may then be defined that overlay different property values in different parts of the grid region to obtain a complete definition of the spatial distribution of the property. Zones are used to define elements that are inactive.

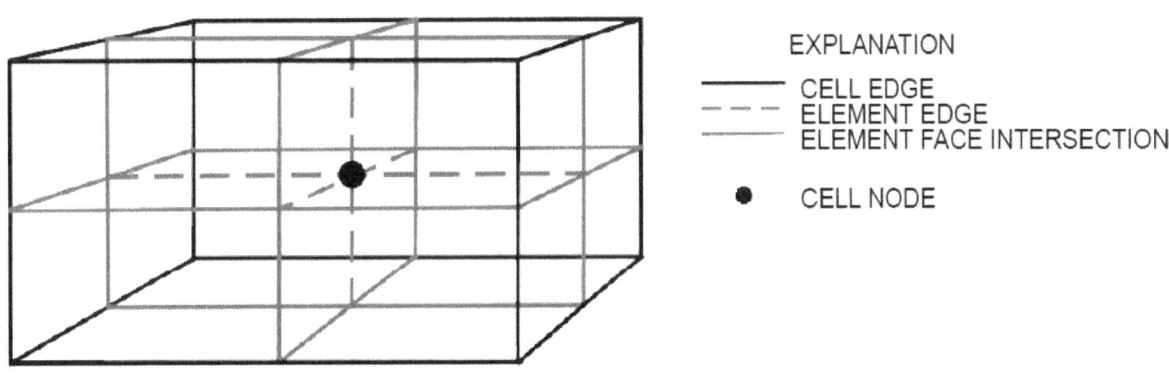

EXPLANATION

— CELL EDGE
-- ELEMENT EDGE
— ELEMENT FACE INTERSECTION

● CELL NODE

Figure 4.6. The parts of eight elements contained within a cell and the parts of four elements located within each cell face.

4.2.1.6. Use of Zones for Defining Initial- and Boundary-Condition Properties

For leaky and flux boundary conditions, zones are used to select the faces of exterior cells that fall, at least partly, inside of the zone. Three-dimensional zones are used to define these boundary conditions, but only the exterior faces with a specified coordinate face are selected. For flux boundaries, the sign of the flux quantity indicates whether the flux is in the positive or negative coordinate direction. Leaky and flux boundary conditions are applied to the part of the exterior cell face that intersects the zone, whether or not the node of the cell is in the zone. In figure 4.7A, a leaky or flux boundary condition would be applied to nine cells. The areas of application for leaky and flux boundary conditions do not change as the grid is coarsened or refined as would happen if these boundary conditions were applied to entire cell faces.

Although leaky and flux boundary conditions are applied to partial cell faces, initial conditions and specified-head boundary conditions are assigned to cells in their entirety. For initial condition and specified-head boundary condition properties, the definition of a zone selects each node contained within the zone. An example in two dimensions of the selection of nodes by a box is shown in figure 4.7. The properties defined for the zone will be applied to each of the cells selected (fig. 4.7B). Zones must be used to define initial conditions for every node in the active grid region. The definitions of well, drain, and river boundary conditions do not use zones.

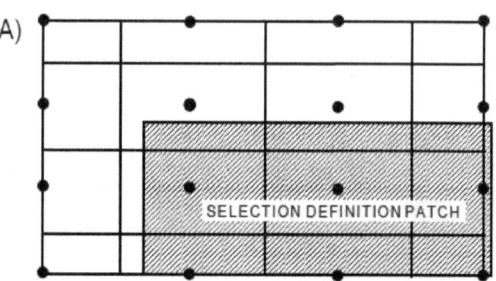

(A)

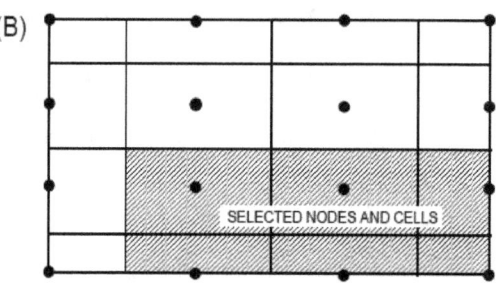

(B)

Figure 4.7. (A) Rectangular zone used for determining areas of application for leaky and flux boundary conditions or to select nodes for initial conditions and specified-head boundary conditions, and (B) the nodes and cells selected by the zone for initial condition and specified-head boundaries. Lines represent cell boundaries.

Multiple zones for each boundary-condition type may be used to specify the complete set of boundary conditions. Different boundary conditions for a single cell and its cell faces may be defined multiple times as part of different zone definitions and different keyword data blocks. The boundary condition that is used for a cell or cell face depends on the last zone definition in the flow and transport data file that defines a boundary condition for that cell or cell face. If the last definition for a cell is specified head, then the cell is a specified-head cell and no other boundary condition applies to that cell. If the last definition is a leaky definition, then the cell will be a leaky boundary cell, and could also be a flux boundary cell (if a flux definition exists for the cell), but the cell cannot be a specified-head cell. Similarly, if the last definition is a flux definition, then the cell will be a flux boundary cell and could also be a leaky boundary cell (if a leaky definition exists for the cell), but the cell cannot be a specified-head cell. If multiple definitions of leaky boundaries apply to a cell face, the area of application of a later definition takes precedence and its area is removed from the areas of application of all earlier leaky boundary definitions for the face. This process is applied to each definition in sequence from last to first, so that no area applies to more than one leaky boundary definition. Flux boundaries are processed in the same way. No checks on areas of application are made between flux and leaky boundaries, so leaky and flux could apply to the same area on the same face, even though that usually does not make physical sense. For more discussion of the co-occurrence of boundary conditions, see section D.5.8. Boundary-Condition Compatibility.

4.2.1.7. Property Definitions

For each zone, spatially distributed properties can be defined by as many as four methods: (1) a single value can be applied to each node or element that is selected by the zone, (2) values for the property can be linearly interpolated along a specified coordinate axis by using property values at two points along the specified axis, (3) values of the property at scattered data points (*x, y, z, value*) can be defined in the input file, from which data will be interpolated to nodes or element centroids by closest point interpolation, and (4) the data for method 3 can be read from a file. When method 2, 3, or 4 is used to distribute properties to nodes, the node location is used for interpolation; for elements, the element centroid is used for interpolation. For method 2, if the coordinate of the node or centroid falls outside the coordinate range defined by the two points, the value from the closest point is applied to the node or element. When method 2 is used for any chemical property, such as composition of solution, kinetic reaction, or equilibrium-phase assemblage, the compositions specified for the two points are mixed in proportion to the distances of the node to the two points along the axis or, if the node is outside the coordinate range defined by the two points, the composition from the closest point is applied.

Two additional methods are retained from PHAST Version 1, but their use is discouraged because the data input is grid dependent (node-by-node) and must be redefined if the number (or location) of grid nodes is changed. A list of values in natural order (section 4.2. Spatial Data) can be defined that includes exactly one value for each node or element that is selected by the volume. This method is practical only for box definitions. Alternatively, a list of values in natural order can be read from a file that includes exactly one value for each node or element that is selected by the volume. For these methods, a value is supplied for each node or element within the volume, regardless of whether the node or element is active or inactive. The property value only has effect for the active nodes or active elements; property values supplied for inactive nodes or elements are ignored.

4.2.2. Rivers, Drains, and Wells

A river is defined by a series of X–Y points that locate the centerline of a river across the land surface. For each river point, the river width, riverbed-leakance parameters, and a time series of head values is defined explicitly or implicitly by interpolation from other river points. Areas of application for a river are calculated from the river length and width information, which are intersected with the Z faces of cells to produce a series of nonintersecting area segments that apply to stacks of cells. For a vertical stack of cells, river leakage will be applied to the cell that contains the water table (free surface). River leakage is calculated from the leakance parameters, the segment areas, and the difference between the head of the water table and the head in the river for each segment.

Drains are similar to rivers, except water only flows from the aquifer to the drain if the head in the aquifer is greater than the elevation of the drain. Water does not flow from the drain to the aquifer. Drains are defined by a series of X–Y points and have elevations, width, and leakance parameters associated with the points. Drains have no time-dependent parameters because the elevations of the drain are fixed and no water enters the aquifer, so no solution composition is needed.

Well location is defined by the X–Y coordinates of the well head. Data required for each well include depth or elevation data to define open (or screened) intervals, well diameter, numerical method for allocating flow to grid layers, a time series of pumping or injection rate, and a time series of solution composition for injection wells.

4.2.3. One-, Two-, and Three-Dimensional Interpolation

Interpolation can be a source of difficulty in assigning properties for model representation. Different methods of interpolation and extrapolation may be better suited to particular kinds of spatial data. For simplicity and robustness, PHAST has just three methods, linear interpolation for one dimension, natural neighbor interpolation for two dimensions, and closest point interpolation for three dimensions. None of these methods allow extrapolation; all three restrict interpolated values to values that are within the range of values at the known points.

One-dimensional interpolation is used for estimation of properties in a single coordinate direction and for parameters along the line segments connecting river points and drain points. Values for target points within the interval of the known points are interpolated linearly by distance along the line segment connecting the known points. For target points that lie outside the interval of the known points, the value at the closest known point is used for the target point. For rivers and drains, known points are required at each end of the river or drain definition, and thus all target points are within an interval defined by known points. For linear interpolation of properties in one coordinate direction, if the interval defined by the known points is smaller than the range of the grid, interpolated values for the extent of the grid will be constant in the given coordinate direction from the edge of the grid to the first known point, sloping along the interval defined by the two known points, and constant from the second known point to the other edge of the grid.

Two-dimensional interpolation is used to define the top and bottom surfaces of prisms. Natural neighbor interpolation requires a set of at least three X–Y points for which the elevation of the surface (top or bottom) is known. Elevations for target points that are within the convex hull of the known points are estimated with natural neighbor interpolation. The convex hull is the smallest convex polygon that encloses all of the known points. Natural neighbor interpolation determines a polygon for each known point that represents the area of the convex hull that is closest to that point. To interpolate to a target point, a new set of closest-to-point polygons is determined that includes both the scattered data points and the target point. Interpolation is carried out by assigning weights to the elevations at the known points based on the fraction of the area of the target-point polygon that overlaps the original (without the added target point) closest-to-point polygon for each known point. This interpolation is not linear along a line segment connecting two known points. The interpolation is continuous for all points within the convex hull and differentiable for all points within the convex hull except at the known points. Outside the convex hull, the value at a target point is assigned from the closest known point. Thus, outside the convex hull, the surface is a series of flat regions determined by closest-to-point polygons that are outside the convex hull. Outside the convex hull, the surface is discontinuous and nondifferentiable at the closest-to-point polygon boundaries. Within and without the convex hull, the range of interpolated elevations is bounded by the range of elevations at the known points.

Three-dimensional interpolation is used to assign properties from known values at a set of scattered three-dimensional points. Closest point interpolation is used, which simply assigns the value at the closest known point (in three dimensional space) to the target point. Thus, space is divided into polyhedrons that enclose the volumes that are closest to the known points. Properties are uniform within each polyhedron and discontinuous at polyhedron boundaries. Again, the range of interpolated values is bounded by the range of data at the known points.

In three dimensions, there is a tendency to bias perception vertically relative to horizontally because sections and visualizations are often vertically exaggerated. PHAST has no capability to bias vertically relative to horizontally in its interpolation scheme; unadjusted Euclidian distance is used to determine the locus of points that is closest to a known point. Thus, if only one known point is defined for an area within a zone, the value at that point will likely apply from top to bottom of the zone because the distance vertically to the bottom of the zone is likely to be less than the distance horizontally to another known point within the zone.

However, addition of more known points vertically and horizontally allows refinement of the property distribution to whatever accuracy is desired. Regardless of the number of points, the three-dimensional interpolation scheme is granular and not continuous. The three-dimensional interpolation scheme does not produce an intermediate interpolated value at a target point that is on the line segment between two adjacent known points; the interpolated value is the value at one of the known points.

4.3. Transient Data

PHAST allows time series of data to be defined for time step, and print frequencies, and certain boundary-condition parameters. Each time one of these values changes, a new simulation period is begun. An unlimited number of simulation periods is allowed. During a single simulation period, boundary conditions, time step, and print frequencies are constant. Properties for boundary conditions that may be transient include head, solution composition, flux, and pumping or injection rate. The following boundary-condition properties are defined by a time series of data: the head and solution composition associated with a specified-head or leaky boundary condition, the flux and solution composition associated with a flux boundary condition, the head and solution composition associated with a river point, and the pumping (or injection) rate and solution composition associated with a well. Many properties of boundary conditions are static; that is, they are fixed for the entire simulation. Examples of static properties include the hydraulic conductivity in a leaky boundary condition, the width of a river associated with a river point, and the locations of the screened intervals of a well.

4.4. Documentation Conventions

Each type of data in the flow and transport data file is defined through a keyword data block. A set of identifiers is used to organize the data that are included in each data block. In the following sections, each keyword data block and its identifiers are described. This section describes conventions that are used in the flow and transport data file and in the documentation of the keyword data blocks. Many of the keyword data blocks require definition of spatial zones to which parameter values are applied. Zones are defined in the same way in all data blocks. The various methods for defining zones are described once in the following section (4.5. Description of Input for Zones) and referenced in each data block that requires zone definitions. Similarly, the values of spatially distributed properties, which are applied to zones, are defined in the same way for all property identifiers in all keyword data blocks. The various methods for defining spatially distributed properties are described once in section 4.6. Description of Input for Properties and referenced for each identifier that defines spatially distributed properties.

The flow and transport data file is completely case insensitive. Data for all data blocks are free format; one or more spaces or tabs are used to delimit input fields. Usually the fields for an identifier must be defined on a single line, but data may be extended to multiple lines in some instances.

In this document, words (or characters) in bold must be included literally when creating input data files (although optional spellings may be permitted). Keywords are always shown in bold uppercase (for example, **CHEMISTRY_IC**); identifiers are shown in bold lowercase and preceded with a hyphen (for example, **-box**). Words in *italics* are input fields that are variable and depend on user selection of appropriate values (for example, **-box** $x_1, y_1, z_1, x_2, y_2, z_2$; the user must insert appropriate coordinates to define the box). Items in brackets ([]) are optional input fields (for example, in the identifier **-u[niform]**, everything in the brackets is optional). Mutually exclusive input fields are separated by a vertical bar and enclosed in parentheses, for example, (**grid** | **map**). In general, the optional characters and fields must be entered in the specified order.

For clarity, commas are used to delimit some input fields in the explanations of data input; however, commas are not used as delimiters in the flow and transport data file. All line numbers associated with example input for the flow and transport data file in this documentation are for identification purposes only. (Input data files for PHAST do not contain any commas or line numbering except in Basic language programs within the chemistry data file.) If multiple lines of the same type are present in an example, an additional letter is included to give the line a unique identification.

The following definitions are used in the flow and transport file:

Keywords—Input data blocks are identified with an initial keyword. This word must be spelled exactly. Several of the keywords have synonyms. For example **SPECIFIED_BC** is a synonym for **SPECIFIED_HEAD_BC**.

Identifiers—Identifiers denote the type of data within a keyword data block. In general, identifiers with their associated data can be in any order within a data block; the exception is spatial-property identifiers, where subsequent definitions overlay previous definitions. Identifiers have two forms: (1) they are spelled completely and exactly or (2) they start with a hyphen followed by enough characters to define the identifier uniquely. The form with the hyphen is always acceptable. Usually, the form without the hyphen is acceptable, but in some cases the hyphen is needed to indicate the word is an identifier rather than an identical keyword. In this documentation, the hyphen is used in the flow and transport data file in all explanations and examples. Note that the hyphen in an identifier never indicates negativity.

Units—In this documentation, the units are identified by the abbreviations "L", length and "T", time. In nearly all cases, the actual units of input for the flow and transport data file are defined by specifications in the **UNITS** data block. The PHAST simulator accepts SI and most U.S. customary units for input. However, PHAST uses SI units internally and all output is in SI units, with the exception of time units. The time unit used for output-file results is selected by the user with the **-time** identifier in the **UNITS** data block.

Comments—The number symbol (#) is used to delimit the beginning of a comment in an input file. All characters in the line that follow this character are ignored. The "#" is useful for adding comments explaining the source of various data or describing the problem setup. In addition, it is useful for temporarily removing lines from an input file. If the entire line is a comment, the line is not echoed to the log file (*prefix*.**log.txt**). If the comment follows input data on a line, the entire line, including the comment, is echoed to the log file.

Logical line separator—A semicolon (;) is interpreted as a logical end-of-line character. This allows multiple logical lines to be entered on the same physical line. Thus, the semicolon should not be used in character fields, such as the title or other comment or description fields.

Logical line continuation—A backslash (\) followed by white space and an end-of-line [carriage return and(or) line feed] causes the input processor to ignore the end-of-line. By ignoring the end-of-line, a single logical line can be written to two (or more) physical lines. Note that only white space (tabs or spaces) can follow the backslash; if nonwhite space follows the backslash, the following line will not be treated as a continuation of the current line.

Repeat counter—An asterisk (*) can be used in node-by-node and element-by-element lists for property definitions to indicate repetitions of the same values. Thus, "3*0.14" is equivalent to "0.14 0.14 0.14" when a property is defined by a list of numbers. However, node-by-node input of property values is discouraged because it is grid dependent.

The methods for defining zones and properties are described in the following sections, after which each keyword data block is described in alphabetical order. Each keyword description includes examples of the use of the keyword and all of its identifiers, which are followed by an explanation of the data required for each identifier. Additional information is presented in notes to further explain the use of the identifiers, the interaction among identifiers, or to aid in the selection of property values. Finally, a list of the example problems that use the keyword is included at the end of the description of each keyword data block.

4.5. Description of Input for Zones

A zone is a region in space for which one or more model properties (porosity, hydraulic conductivity, solution composition, flux, and others) may be specified. A zone can be defined as one of four types: (1) the domain (entire grid region); (2) a box, which is a rectangular parallelepiped; (3) a wedge, which is a diagonally bisected box; or (4) a prism, which is an arbitrarily shaped zone that consists of a polygonal perimeter, possibly truncated by a top and(or) bottom surface. Boxes, wedges, and each part of a prism (perimeter, top, and bottom) may be defined in either grid or map coordinates. The transformation between grid and map coordinates is defined in the **GRID** keyword data block. For cells, properties are applied to the cells with nodes or cell faces (leaky and flux boundary conditions) that are inside the zone or are coincident with the surface of the zone. Similarly for elements, properties are applied to elements with centroids that are inside the zone or coincident with the surface of the zone.

Example

Line numbers in the Example are used for identification purposes only and are referenced in the following Explanation. The line numbers do not indicate a required order for the keyword data block. Line numbers are not used in the flow and transport data file.

```
Line 0:     KEYWORD
Line 1:         -domain
Line 2:             -identifier        property
Line 3:         -box         0      0      0     10     10     10     grid
Line 4:             -description       This box is the whole domain
Line 2a:            -identifier        property
Line 5:         -wedge       0      0      0      5      5      5     X1     grid
Line 4a:            -description       Right angle axis is parallel to X
Line 2b:            -identifier        property
Line 6:         -prism
Line 4b:            -description       Prism uses shape and constant data
Line 7:             -perimeter  shape       map   perimeter.shp
Line 8a:            -top         constant    grid   11.1
Line 8b:            -bottom      shape       map   bottom.shp        11
Line 2c:            -identifier        property
Line 6a:        -prism
Line 4c:            -description default perimeter, Arcraster, points
Line 8c:            -top         arcraster   map   top2.txt
Line 8d:            -bottom      xyz         map   bottom2.xyz
Line 2d:            -identifier        property
Line 6b:        -prism
Line 4d:            -description Default perimeter and top
Line 7b:            -perimeter  points      grid
Line 9:                 0      0
Line 9a:                500      0
Line 9b:                500    900
Line 9c:                 0     900
Line 8e:            -bottom      points      grid
Line 10:                 0      0      1
Line 10a:                0     10      2
line 10b:               10     10      1
Line 2e:            -identifier        property
```

Explanation

Line 0: *KEYWORD*

> *KEYWORD* is one of the keyword data blocks that uses zones to define spatial properties, including **CHEMISTRY_IC**, **FLUX_BC**, **HEAD_IC**, **LEAKY_BC**, **MEDIA**, **PRINT_LOCATIONS**, **SPECIFIED_HEAD_BC**, and **ZONE_FLOW**.

Line 1: -domain

> **-domain**—This identifier selects all nodes of the grid, regardless of the grid dimensions.

Line 2: *-identifier property*

> **-identifier**—One of the identifiers for the *KEYWORD*.
>
> *property*—One of the methods for defining a spatial property, see section 4.6. Description of Input for Properties.

Line 3: -box x_1, y_1, z_1, x_2, y_2, z_2, [(**grid** | **map**)]

> **-box**—The box selects cells, elements or, for leaky, flux, and, optionally, specified-head boundary conditions, cells with exterior faces, that fall within the rectangular parallelepiped. Optionally, **box**, **-box**, **zone**, or **-z[one]**.
>
> x_1, y_1, z_1—X, Y, and Z coordinates for the left (X), front (Y), lower (Z) corner of the box. Units, L, are defined with the **-horizontal_grid** or **-map_horizontal** (X and Y coordinates) and **-vertical_grid** or **-map_vertical** (Z coordinate) identifiers in the **UNITS** data block depending on the coordinate system.
>
> x_2, y_2, z_2—X, Y, and Z coordinates for the right (X), back (Y), upper (Z) corner of the box.
>
> (**grid** | **map**)—Option indicates the coordinate system for the X–Y–Z points defining the box, either grid or map coordinates. Default is **grid**.

Line 4: -description [*description*]

> **-description**—This identifier allows the definition of a character string to define the zone. It has no effect on calculations. Optionally, **description**, or **-de[scription]**.
>
> *description*—Optional character field that identifies the zone.

Line 5: -wedge x_1, y_1, z_1, x_2, y_2, z_2, *orientation*, [(**grid** | **map**)]

> **-wedge**—The wedge selects cells, elements or, for leaky and flux boundary conditions, cells with exterior faces that fall within a right triangular wedge. Optionally, **wedge** or **-w[edge]**.
>
> x_1, y_1, z_1—X, Y, and Z coordinates for the left (X), front (Y), lower (Z) corner of the rectangular parallelepiped that contains the wedge. Units, L, are defined with the **-horizontal_grid** or **-map_horizontal** (X and Y coordinates) and **-vertical_grid** or **-map_vertical** (Z coordinate) identifiers in the **UNITS** data block depending on the coordinate system
>
> x_2, y_2, z_2—X, Y, and Z coordinates for the right (X), back (Y), upper (Z) corner of the rectangular parallelepiped that contains the wedge.
>
> *orientation*—The orientation of the wedge is defined by the location of the right angle of the wedge. It consists of two characters: (1) the axis that is parallel to the axis of the right angle (X, Y, or Z) and (2) the position of the right angle, starting with 1, counterclockwise from the minimum coordinates of the box containing the wedge, when viewed from the negative coordinate direction (see fig. 4.3). Legal values for orientation are X1, X2, X3, X4, Y1, Y2, Y3, Y4, Z1, Z2, Z3, and Z4.
>
> (**grid** | **map**)—Option indicates the coordinate system for the X–Y–Z points defining the wedge, either grid or map coordinates. Default is **grid**.

Line 6: -prism

-prism—A zone with perimeter, top, and bottom that is used to select cells, elements or, for leaky and flux boundary conditions, cells with exterior faces. If a perimeter is not defined, the default perimeter is the rectangular X–Y extent of the grid. If a top is not defined, the top is the plane of the maximum Z coordinate of the grid. If a bottom is not defined, the bottom is the plane of the minimum Z coordinate of the grid. Optionally, **prism** or **-pr[ism]**.

Line 7: **-perimeter (shape | xyz | points) (grid | map)** [*file_name*]

-perimeter—Identifier indicates that data will define an X–Y polygon, either from the external file *file_name* (**shape** or **xyz**) or from X–Y points in succeeding lines of the input file (**points**). Optionally, **perimeter** or **-pe[rimeter]**.

(shape | xyz | points)—Options for source of X–Y points for perimeter. For the **shape** option, X–Y data for the perimeter are read from the ArcInfo shapefile (ESRI, 1998) *file_name*. Logically, the features of the shapefile should be polygons, but files with points or polyline features are also permissible. All points from the shapefile are treated as X–Y data defining a polygon, regardless of the shape type. Shape files may define holes in polygons; however, all holes are ignored in the definition of prisms. The **xyz** option causes data for the perimeter to be read from the file *file_name*. In this file, one X–Y point is defined per line; if present, additional data items on a line are ignored. At least three points must be defined. The **points** option indicates that the points defining the perimeter will be found on succeeding lines of the input file. One X–Y point is defined per line and at least three points must be defined.

(grid | map)—Option indicates the coordinate system for the X–Y points defining the perimeter, either grid or map coordinates. One of these options must be entered.

file_name—Name of file with X–Y points defining the perimeter. If **shape** has been defined, *file_name* must be an ArcInfo shapefile (ESRI, 1998). If **xyz** has been defined, the file must contain X–Y points. *File_name* is not defined with the **points** option.

Line 8: **-(top | bottom) (constant | arcraster | shape | xyz | points) (grid | map)** [(*elevation* | *file_name*) [*attribute*]]

-(top | bottom)—Surfaces for top and bottom are defined in the same way. These identifiers indicate that a surface will be defined that truncates the prism either by a plane perpendicular to the Z axis (**constant**) or by scattered data that will be interpolated to produce a surface (**arcraster**, **shape**, **xyz**, or **points**). Examples of various options for input of top and bottom surfaces are given in the descriptions of lines 8a, 8b, 8c, 8d, and 8e. Optionally, **top** or **-to[p]** and **bottom** or **-b[ottom]**.

Line 8a: **-(top | bottom) constant (grid | map)** *elevation*

-(top | bottom)—Same as line 8.

constant—The **constant** option requires definition of an *elevation* to determine points that are included in the prism. If a top is being defined, points in the prism are less than or equal to *elevation*. If a bottom is being defined, points in the prism are greater than or equal to *elevation*.

(grid | map)—Option indicates the coordinate system for the *elevation*, either grid or map coordinates.

elevation—Maximum (**top**) or minimum (**bottom**) elevation for points that are included in the prism.

Line 8b: **-(top | bottom) shape (grid | map)** *file_name attribute*

-(top | bottom)—Same as line 8.

shape—The **shape** option requires an ArcInfo shapefile (ESRI, 1998) named *file_name* that contains data that will be treated as scattered points. All data points in the file are treated as X–Y–Z data that define a surface, regardless of the shape types associated with the file. An attribute number, *attribute*, is required to define which attribute in the *.dbf* file contains the elevation data. The related *.shx* file is also required.

(**grid** | **map**)—Option indicates the coordinate system for the X–Y–Z points defining the surface, either grid or map coordinates.

file_name—Name of ArcInfo shapefile (ESRI, 1998) (*.shp*) containing the scattered X–Y points.

attribute—Integer number, beginning at 0, defining the position of the elevation attribute within the *.dbf* file. The field name of the attribute selected is printed during data processing.

Line 8c: **-(top** | **bottom) arcraster (grid** | **map)** *file_name*

-(top | **bottom)**—Same as line 8.

arcraster—The **arcraster** option requires an ArcInfo ASCII raster file (ESRI, 2009) named *file_name* that contains gridded elevation data.

(**grid** | **map**)—Option indicates the coordinate system for the X–Y–Z points defining the surface, either grid or map coordinates.

file_name—Name of ArcInfo ASCII raster file (ESRI, 2009) containing the gridded elevation data.

Line 8d: **-(top** | **bottom) xyz (grid** | **map)** *file_name*

-(top | **bottom)**—Same as line 8.

xyz—The **xyz** option requires a text file named *file_name* that contains data that will be treated as scattered points.

(**grid** | **map**)—Option indicates the coordinate system for the X–Y–Z points defining the surface, either grid or map coordinates.

file_name—Name of file with X–Y–Z points defining the surface. One X–Y–Z point (separated by white space) is defined per line.

Line 8e: **-(top** | **bottom) points (grid** | **map)**

-(top | **bottom)**—Same as line 8.

points—The **points** option requires scattered data to be entered in the succeeding lines of the input file. One X–Y–Z point (separated by white space) is defined per line.

(**grid** | **map**)—Option indicates the coordinate system for the X–Y–Z points defining the surface, either grid or map coordinates.

Line 9: *x, y*

x, y—Data point used to define a **perimeter** polygon when using the **points** option. Items are separated with white space. Only one point is defined per line.

Line 10: *x, y, z*

x, y, z—Scattered data point used to define a surface for **top** or **bottom** with the **points** option. Items are separated with white space. Only one point is defined per line.

Notes

Zones are used to define all porous-media properties (**MEDIA**), all initial condition properties (**CHEMISTRY_IC** and **HEAD_IC**), the locations of all specified-head (**SPECIFIED_HEAD_BC**), leaky (**LEAKY_BC**), and flux (**FLUX_BC**) boundaries. Zones also are used to define the areas of cell faces to which flux (**FLUX_BC**) and leaky (**LEAKY_BC**) boundaries are applied. Zones are used to specify cells for which results are printed to output files (**PRINT_LOCATIONS**) and to specify volumes of the aquifer over which budget information (fluxes in and out of water and solutes) will be calculated (**ZONE_FLOW**).

If data are imported from ArcInfo coverages, it may be desirable to have a grid coordinate system that differs from a map coordinate system in terms of angle and origin. Thus, two coordinate systems may be used to define zones: a grid coordinate system and a map coordinate system. The map coordinate system has units of distance defined by the **-map_horizontal** and **-map_vertical** identifiers of the UNITS keyword data block. The origin of the map coordinate system is normally defined by a particular state plane or map projection. The grid coordinate system is defined relative to the map coordinate system with the **-grid_origin** and **-grid_angle** identifiers of the GRID keyword data block. The grid origin is given in map units of distance and the grid angle is defined as the number of degrees counterclockwise from the map X axis to the grid X axis. Units of the grid coordinate system are given by the **-horizontal_grid** and **-vertical_grid** identifiers of the UNITS keyword data block. By default, map and grid coordinate axes coincide (**-grid_origin** is 0, 0, 0 and **-grid_angle** is 0.0). Each box, wedge, and for prisms, each perimeter, top, and bottom, may be defined in either of the two coordinate systems.

For top and bottom surfaces in prisms that are defined by scattered data points, interpolation to grid points is necessary. Only natural neighbor interpolation (Sakov, 2000) is available in PHAST. This interpolation method determines the elevation at a target X–Y point within the convex hull of the scattered data (the smallest convex polygon that encloses all the points) by area weighting elevations at data points determined on the basis of a Delaunay triangulation (Shewchuk, 1996 and 2003). If the target point is outside the convex hull of the scattered data, then the elevation at the closest scattered point as determined by a KD tree (Kennel, 2004; Skiena, 2008) is used for interpolation. A point is inside a prism if the Z coordinate of the point is less than or equal to the interpolated elevation of the top surface, greater than or equal to the interpolated elevation of the bottom surface, and inside the perimeter.

An ArcInfo shapefile (ESRI, 1998) can be used to define the perimeter of a prism. The shapefile is interpreted as a set of X–Y points that define a polygon. The polygon should be non-self-intersecting. Holes in ArcInfo polygon coverages are ignored; all areas within any part of a polygon coverage is considered to be inside the polygon, regardless of whether the part is defined in the clockwise or counterclockwise direction. A shapefile also can be used to define the top or bottom surface of a prism. In this case, the X–Y coordinates are taken from the shapefile and the Z coordinate is one of attributes in the attribute file (.**dbf**). The type of shape assigned in the shapefile is ignored. The set of X–Y–Z points is interpolated by natural neighbor interpolation to define elevations of the top or bottom at grid locations.

Example Problems

Zones are used in every problem definition. Box zones are used in Chapter 6 example problems 1, 2, 3, 4, 5, and 6. A wedge zone is used in example 4. Domain and prism zones are used in examples 5 and 6.

4.6. Description of Input for Properties

The term property is used to describe any of the spatially varying parameters of the model, for example porosity, hydraulic conductivity, solution composition, head of a leaky boundary condition, and many others. Properties are defined over zones (see **4.5. Description of Input for Zones**) and may be constant, linearly varying in a coordinate direction, or interpolated from three-dimensional spatial data. Properties are defined for zones in the following keyword data blocks: CHEMISTRY_IC, FLUX_BC, HEAD_IC, LEAKY_BC, MEDIA, PRINT_LOCATIONS, and SPECIFIED_HEAD_BC. Only the ZONE_FLOW data block accepts zone definitions, but does not use property definitions. For identifiers in CHEMISTRY_IC and the solution identifiers in FLUX_BC, LEAKY_BC, and SPECIFIED_HEAD_BC, the values of the property are integers that correspond to chemical compositions defined in the chemistry data file; for all other identifiers, the values of the property are real numbers corresponding to parameter values. Property definitions follow on the same line as an identifier from one of the keyword data blocks. The property definition either immediately follows the identifier or follows an additional option **mixture**. The **mixture** option can only be used in the **CHEMISTRY_IC** data block and the solution definitions for **FLUX_BC**, **LEAKY_BC**, and **SPECIFIED_HEAD_BC** data blocks.

It is possible to define properties node-by-node for each node within a zone, but use of these methods is discouraged because they are grid dependent. Another grid-dependent feature is the use of a mask, which is a series of 0s (skip) and 1s (include) that define the nodes where the property is assigned. The node-by-node definitions are retained for backward compatibility and are described in Example 2 of this section.

Example 1

Line numbers in the Example are used for identification purposes only and are referenced in the following Explanation. The line numbers do not indicate a required order for the keyword data block. Line numbers are not used in the flow and transport data file.

```
Line 0:    ...              constant    15.5
Line 1:    ...              X           3     0.0    4      100.0
Line 2:    ...              points      grid
Line 3:                                 0.0   0.0    0.0    1e-3
Line 3a:                               100.0  50.0   0.0    2e-3
Line 3b:                               100.0  100.0  0.0    3e-3
Line 4:                                 end_points
Line 5:    ...              xyz         map         points.file
Line 6:    -identifier      restart                 my.restart.gz
Line 7:    0.0              xyzt        grid        xyzt.file
```

Explanation 1

Line 0: ... [**constant**] *value*

   ...—The ellipsis indicates the property definition is preceded by either (1) an identifier for a spatial property (possibly with an additional **mixture** option), or (2) a time value in a time series (with optional time units designation).

   constant—Option specifies a uniform value will be applied to the nodes or element centroids in a zone. The word **constant** is optional.

   value—Parameter value, integer representing a chemical composition, or mixing fraction.

Line 1:... (**X** | **Y** | **Z**) *value*$_1$, *distance*$_1$, *value*$_2$, *distance*$_2$

   ...—See line 0 above.

   (**X** | **Y** | **Z**)—Option specifies that the parameter or chemical composition in the zone will be interpolated in the given coordinate direction. If the target point (a node or element centroid) coordinate is outside the range *distance*$_1$ to *distance*$_2$, the parameter or chemical composition at the target point is assigned the value at the closest endpoint of the range; otherwise the parameter or chemical composition is linearly interpolated to the target point coordinate from the values at the endpoints.

   value$_1$—Parameter value, integer representing a chemical composition, or mixing fraction at coordinate distance *distance*$_1$.

   distance$_1$—Coordinate location along the specified axis (grid coordinates).

   value$_2$—Parameter value, integer representing a chemical composition, or mixing fraction at coordinate distance *distance*$_2$.

distance₂—Coordinate location along the specified axis (grid coordinates).

Line 2: ... points (grid | map)

...—See line 0 above.

points—Option specifies that the parameter values, integers representing chemical compositions, or mixing fractions in the zone will be interpolated from a set of scattered points that are read from subsequent lines. Lines with quadruplets (*x, y, z, value*) follow. Data are separated by white space, and one quadruplet is read per line. Values at target points will be interpolated by closest point interpolation—the parameter or chemical composition at the closest point will be assigned to the target point.

(grid | map)—Specifies the coordinate system for the X–Y–Z points, either grid coordinates, or map coordinates.

Line 3: *x, y, z, value*

x—X coordinate of a point.

y—Y coordinate of a point.

z—Z coordinate of a point.

value—Value of the parameter, integer representing chemical composition, or mixing fraction at the point.

Line 4: end_points

end_points—The character string **end_points** must be included on the next line following the last scattered data point definition of the **points** property option.

Line 5: ... xyz (grid | map) *file_name*

...—See line 0 above.

xyz—Option specifies that the parameter values, integers representing chemical compositions, or mixing fractions in the zone will be interpolated from a set of scattered points that are read from a file. The file contains lines with quadruplets (*x, y, z, value*). Values at target points (nodes or element centroids) will be interpolated by closest point interpolation—the parameter or chemical composition at the closest point will be assigned to the target point.

(grid | map)—Specifies the coordinate system for the X–Y–Z points, either grid coordinates, or map coordinates.

file_name—Name of file containing lines with quadruplets (*x, y, z, value*). Data are separated by white space, and one quadruplet is read per line of the file.

Line 6: -*identifier* restart *file_name*

-*identifier*—Identifier of **CHEMISTRY_IC**; the **restart** option can only be used in the **CHEMISTRY_IC** data block.

restart—Option specifies that chemical compositions in the zone will be interpolated from a set of compositions at scattered points that are read from a PHAST-generated restart file. The restart file is saved at the end of a previous run and contains chemical compositions (solutions, equilibrium phases, exchangers, surfaces, kinetic reactants, solid solutions, and gas phases) at the grid coordinates of the previous run. Chemical compositions at target points (nodes) will be assigned the chemical composition from the closest grid point of the previous run. If the closest point does not have a particular type of composition, exchanger for example, then the target point will not have an exchanger present.

file_name—Specifies the name of the restart file.

Line 7: 0.0 xyzt (grid | map) *file_name*

0.0—A value of 0.0 should be used in this position. The actual times at which boundary conditions change for a time-varying boundary-condition property, for example, the head of a specified-head boundary condition, is determined by the time values contained in the XYZT file.

xyzt—Option specifies that the parameter values, integers representing chemical compositions, or mixing fractions in the zone will be interpolated from scattered points at a series of time planes that are read from a file. The file contains lines with quintuplets (x, y, z, time, value). The times must be in ascending order. Boundary-condition values will be changed at each discrete time included in the file. Values at target points (nodes) will be interpolated by closest point interpolation from the set of X–Y–Z points included in the file for a given time.

(grid | map)—Specifies the coordinate system for the X–Y–Z points, either grid coordinates or map coordinates.

file_name—Name of file containing lines with quintuplets (x, y, z, time, value). Data are separated by white space, and one quintuplet is read per line of the file.

Notes 1

For linear interpolation along a coordinate axis and data in the restart file, all distances must be defined in the grid coordinate system. Data for other methods of interpolation can be defined in either the grid or map coordinate system.

Data defined in example 1 are grid independent in the sense that the grid may be refined or coarsened without redefining parameter values. However, any parameter distributions defined by three-dimensional closest point interpolation (xyz, points, and restart) have a fixed granularity; that is, the zone over which a parameter or chemical composition is uniform is defined by the set of scattered points and does not change or become more smooth as the grid is refined.

The value of a property for any boundary condition (for example, the flux in a flux boundary condition, or the head for a specified-head or leaky boundary condition) is determined by interpolation of the property to the node of the cell. Specified-head cells are those cells with nodes that fall within the defining zone, with an option to select only cells with exterior faces. Leaky- and flux-boundary cells are those cells with exterior faces (perpendicular to the specified coordinate direction) that fall at least partially within the defining zone. Areas of application for boxes and wedge definitions are straightforward intersections of triangles or rectangles (zone boundaries) with cell faces (rectangles).

When the zone is a prism, areas of application for leaky and flux boundary conditions are more complicated. For X or Y cell faces, 10 points are distributed along a horizontal edge of the face. For each point, the X–Y coordinates are used to determine whether the point is inside the perimeter of the prism. For each point inside the prism perimeter, points are eliminated if the top of the cell face is below the bottom of the prism or if the bottom of the cell face is above the top of the prism. For the points that remain, slices of the cell face, bounded by the distance between points and the top and bottom of the prism, are summed to determine the area of application for the boundary condition. For Z cell faces, a mesh of 100 rectangles is generated that has 10 rectangles along each edge. The centroid of each rectangle is checked to determine whether it is inside the prism. The sum of the areas of rectangles with centroids that are within the prism is the area of application for the boundary condition.

The restart file written by PHAST is in *gzip*-compressed format, and the name ends with the suffix *.restart.gz*. Writing of the restart file is controlled by the -restart identifier of the PRINT_FREQUENCY data block. The file can be uncompressed with the *gunzip* command to generate an ASCII file. Either the compressed or uncompressed version of the restart file can be read by PHAST; no input option is needed to distinguish the format. The restart file contains the locations, in grid coordinates, of all the chemically active nodes. If the calculation is logically one- or two-dimensional, as defined by the -chemistry_dimensions of the **GRID** data block, the chemically active nodes are restricted to active nodes in an appropriate one- or two-dimensional slice. The presence or absence of each type of reactant (solutions, equilibrium phases, exchangers, surfaces, kinetic reactants, solid solutions, and gas phases) at each node is included in the restart file along with a complete description of each chemical composition at each node. In a run using a restart file, compositions at target points (nodes) within a specified zone are assigned by closest point interpolation from the grid locations of the previous run, which are saved in the restart file. If the closest point in the restart file lacks a particular kind of reactant, an exchanger for example, then no exchanger composition is assigned to the target point for the new run.

An XYZT file can be used to define a time series for a possibly spatially varying property of a boundary condition. The XYZT option can only be used to define a time series of data for a boundary condition. The file contains a set of X-Y-Z points for a series of times; the order of the X-Y-Z points is irrelevant, but the times in the file must be in ascending order. For a given time plane, the values at the X-Y-Z points are used to define a spatially varying distribution of the property value. Values at nodes are interpolated by closest point interpolation from the set of X-Y-Z points. This XYZT file is intended to simplify definition of complicated time-series data for boundary conditions. In particular, the **ZONE_FLOW** data block can be used to write heads for a subset of a model domain to an XYZT file at a series of times. A subdomain can be defined with specified-head boundaries at the edges of the subdomain and the XYZT file can be used to define the heads for the specified-head boundaries. In this way a refined subdomain model can be developed that captures the flow conditions of the larger model domain. See example 5 (6.6. Example 5: Simulation of Groundwater Flow for a Sewage Wastewater Plume at Cape Cod, Massachusetts) for an example of generating an XYZT file and example 6 (6.7. Example 6: Simulation of Ammonium Transport and Reactions for a Sewage Wastewater Plume at Cape Cod, Massachusetts) for its use in boundary-condition definitions.

Only one option may precede the property definition—**mixture** i j—where i and j refer to chemical compositions defined in the chemistry data file. This option can be used only in the solution definitions for **FLUX_BC**, **LEAKY_BC**, and **SPECIFIED_HEAD_BC** data blocks and with identifiers of the **CHEMISTRY_IC** data block. When **mixture** is used, the property defines a mixing fraction f, for composition i; the mixing fraction for composition j is $1 - f$.

Example 2

Line numbers in the Example are used for identification purposes only and are referenced in the following Explanation. The line numbers do not indicate a required order for the keyword data block. Line numbers are not used in the flow and transport data file.

```
Line 0:    KEYWORD
Line 1:    -box    100   0     0     100   100   10    grid
Line 2a:        -identifier  by_node
Line 3a:            < 1.1 1.5 2.0 7*2.5 >
Line 2b:        -identifier  file        node.by.node.file
Line 2c:        -identifier  mixture 2 4
Line 3b:            < 5*1.0 0.8 0.6 0.4 0.2 0.0 >
Line 4a:        -mask        file        mask.file
Line 0a:   MEDIA
Line 1a:   -box    0     0     0     100   100   10    grid
Line 2a:        -identifier  by_element
Line 3c:            < 11*0.2 11*0.21 99*0.22 >
Line 2b:        -identifier  file        elt.by.elt.file
Line 4b:        -mask        by_element
Line 3d:            < 11*1 11*0 11*1 11*0 77*1 >
```

Line 0: *KEYWORD*

    KEYWORD is any of the keywords that define spatial data through zone and property definitions. All spatially distributed properties are applied by cell (node) except those in the **MEDIA** data block, which are applied by element.

Line 1: **-box**

    -box—A box zone definition as defined in section 4.5. Description of Input for Zones. Whereas it may be possible to use zone definitions other than a box for grid-dependent (node-by-node) data, it is difficult to know the set of nodes or element centroids in natural number order that fall within a zone that is not a rectangular parallelepiped.

Line 2: *-identifier* (**by_node** | **by_element** | **file** | **mixture**)

    -identifier—An identifier of *KEYWORD* that requires spatial data.

    (**by_node** | **by_element** | **file** | **mixture**)—Option specifies one of the three ways that grid-dependent spatial data for the zone may be defined: (1) a value or integer representing a chemical composition can be read from the input file for each node or element centroid in the zone (**by_node** or **by_element**); (2) a value or integer representing a chemical composition for each node or element centroid can be read from a specified file (**file**); or (3) a mixture of two chemical compositions can be defined for each node in the zone (**mixture**). Values must be defined for each node or element centroid whether it is active or not. Additional information is required for each option as described for lines 2a–2c.

Line 2a: *-identifier* (**by_node** | **by_element**)

    -identifier—An identifier of *KEYWORD* that requires spatial data.

    (**by_node** | **by_element**)—Option specifies that property values or index numbers representing a chemical composition will follow in the input file. A value must be specified for each node (active or inactive) or element centroid in the zone in natural order. Values are separated by white space; multiple values may be included on each line. The list of values must be enclosed in angle brackets "< >". An error message is returned if there are more or fewer values than the number of nodes or centroids within the zone.

Line 2b: *-identifier* **file** *file_name*

    -identifier—An identifier of *KEYWORD* that requires spatial data.

    file—Option specifies that property values or index numbers representing a chemical composition will be read from a file.

    file_name—Specifies the name of the file containing parameter values or index numbers representing solution compositions. A value must be specified for each node (active or inactive) or element centroid in the zone in natural order. Values are separated by white space; multiple values may be included on each line. An error message is returned if there are more or fewer values read from the file than the number of nodes or centroids within the zone.

Line 2c: *-identifier* **mixture** *index$_1$ index$_2$*

    -identifier—An identifier of *KEYWORD* that requires a spatial distribution of chemical composition.

    mixture—Option specifies that the chemical compositions in the zone will be defined by mixing the composition referenced by *index$_1$* with the composition referenced by *index$_2$* (2 and 4 in Line 2c). A mixing fraction, *f*, for the composition referenced by *index$_1$* is specified for each node in the zone. The fractions must be in the range of 0 to 1 (inclusive), and the fraction $1 - f$ is the mixing fraction for the composition referenced by $index_2$. A mixing fraction must be specified for each node (active or inactive) in the zone in natural order. Line 2c defines compositions at nodes in the box as mixtures between composition 2 and 4, where the mixing fraction of composition 2 at each node is read in natural numbering order from the Line 3s that immediately follow Line 2c.

Line 3: < *list of values* >

< *list of values* >—A list of values is included in the input file. The list must be preceded by an opening angle bracket "<" and ended with a closing angle bracket ">". Multiple values may be placed on a line and must be separated by white space. Multiple lines may be used to complete the list. A repeat counter may be used to define consecutive values that are equal; "5*1.0" indicates five values of 1.0 and is the same as "1.0 1.0 1.0 1.0 1.0". An error message is returned if there are more or fewer values than the number of nodes within the zone.

Line 4a: **-mask file** *file_name*

-mask—An array is used to include and exclude nodes or element centroids from the zone definition.

file—A file containing a list of 1s (or any positive number) and 0s (or any nonpositive number); one number for each node or element centroid (active or inactive) in the zone is read from the file in natural order.

file_name—Specifies the name of the file containing a list of 1s and 0s. A value must be specified for each node or element centroid (active or inactive) in the zone in natural order. Values are separated by white space; multiple values may be included on each line. It is an error if there are more or fewer values read from the file than the number of nodes or centroids within the zone.

Line 4b: **-mask (by_node | by_element)**

-mask—An array is used to include and exclude nodes or element centroids from the zone definition.

(by_node | by_element) —A list of 1s (or any positive number) and 0s (or any nonpositive number) follows, one number for each node or element centroid (active or inactive) in the zone in natural order. The list items are separated by white space and may extend over multiple lines. The list must be enclosed in angle brackets "<>". See Line 3d of the example for a list definition of a mask.

Notes 2

Use of the features described in Example 2 is discouraged because the features are grid dependent (node-by-node). When using these features, if the grid is refined and the spatially distributed property definitions are not changed, the simulator will fail. These features described in Example 2 are retained to provide backward compatibility but are no longer supported or maintained. For complicated simulations, use of the grid-dependent features requires a computer program to select and format data in natural-number order. It is advisable that such a program simply list the X–Y–Z coordinates and values in a format described in Example 1 above. Such a listing of data could then be used in a grid-independent fashion.

Example Problems

Properties are defined in every example of Chapter 6: **1**, **2**, **3**, **4**, **5**, and **6**.

4.7. Description of Keyword Data Blocks

Keyword data blocks for the flow and transport are described in this section. The keywords are presented in alphabetical order. Each keyword description begins on a new page and headers on each page identify the keyword that is being described.

CHEMISTRY_IC

This keyword data block is used to define the initial chemical conditions in the grid region, including the initial compositions of solutions, and, optionally, the initial compositions of equilibrium-phase assemblages, exchangers, surfaces, kinetic reactants, solid solutions, and, rarely, gas phases (usually gases are defined as fixed partial pressures in **EQUILIBRIUM_PHASES**). The compositions are defined by index numbers that correspond to definitions in the chemistry data file with **SOLUTION**, **EQUILIBRIUM_PHASES**, **EXCHANGE**, **SURFACE**, **KINETICS**, **SOLID_SOLUTIONS**, and **GAS_PHASE** data blocks or through initial geochemical calculations that are subsequently saved with a **SAVE** data block. Chemistry initial conditions also can be defined as specified mixtures of chemical compositions and boundary-condition solutions can be defined as mixtures. The **CHEMISTRY_IC** data block is mandatory for all reactive-transport simulations.

Example

Line numbers in the Example are used for identification purposes only and are referenced in the following Explanation. The line numbers do not indicate a required order for the keyword data block. Line numbers are not used in the flow and transport data file.

```
Line 0:    CHEMISTRY_IC
Line 1:         -box    0.0    0.0    0.0    1000.0       100.0        10.0
Line 2:                 -solution                 constant    1
Line 3:                 -equilibrium_phases       X           1    0.0    5     10.0
Line 4:                 -exchange                 points      grid
Line 5:                     0  0 0 1
Line 5a:                    5  0 0 10
Line 5b:                   10 0 0 7
Line 6:                     end_points
Line 7:                 -kinetics                 xyz         map    kin.file
Line 8:                 -solid_solutions          restart     project.restart.gz
Line 9:                 -surface    mixture 2 4 xyz     grid surface.file
Line 10:                -gas_phase  mixture 3 5 points      grid
Line 11:                    0  0 0 1.0
Line 11a:                   5  0 0 0.5
Line 11b:                  10 0 0 0.0
Line 6a:                    end_points
```

Explanation

Line 0: **CHEMISTRY_IC**

  CHEMISTRY_IC is the keyword for the data block; no other data are included on this line.

Line 1: *zone*

  zone—A zone definition as defined in section 4.5. Description of Input for Zones. Line 1 defines a box zone.

Line 2: **-solution** [**mixture** *i j*] *property*

  -solution—This identifier is used to specify the solution compositions initially present in the cells of *zone*. For solute transport modeling, all cells in the active grid region must have a solution composition defined. Optionally, **solution** or **-s[olution]**.

mixture *i j*—Optionally, solution composition can be defined as a mixture of two solutions numbered *i* and *j* in the chemistry data file. For a **mixture**, the *property* defines mixing fractions, *f*, of solution *i*; mixing fraction for solution *j* is *1 − f*.

property—Property input as defined in section 4.6. Description of Input for Properties. Property values are index numbers or mixing fractions (when **mixture** *i j* is used) that define the solution composition in the *zone*. Line 2 illustrates definition of a single composition that applies to the entire zone.

Line 3: **-equilibrium_phases** [**mixture** *i j*] *property*

-equilibrium_phases—This identifier is used to specify the equilibrium-phase assemblages initially present in the cells of *zone*. Optionally, **equilibrium_phases**, **-eq[uilibrium_phases]**, **pure_phases**, **-p[ure_phases]**, **phases**, or **-p[hases]**.

mixture *i j*—Optionally, equilibrium-phase assemblages can be defined as a mixture of assemblages numbered *i* and *j* in the chemistry data file. For a **mixture**, the *property* defines mixing fraction, *f*, of assemblage *i*; mixing fraction for assemblage *j* is *1 − f*.

property—Property input as defined in section 4.6. Description of Input for Properties. Property values are index numbers or mixing fractions (when **mixture** *i j* is used) that define the equilibrium-phase assemblages in the *zone*. A negative index number for a property value indicates that no equilibrium phases are present. Line 3 demonstrates an equilibrium-phase assemblage that varies linearly between composition 1 and composition 5 in the X coordinate direction.

Line 4: **-exchange** [**mixture** *i j*] *property*

-exchange—This identifier is used to specify the exchange compositions initially present in the cells of *zone*. Optionally, **exchange** or **-ex[change]**.

mixture *i j*—Optionally, exchange compositions can be defined as a mixture of compositions numbered *i* and *j* in the chemistry data file. For a **mixture**, the *property* defines mixing fraction, *f*, of exchange composition *i*; mixing fraction for exchange composition *j* is *1 − f*.

property—Property input as defined in section 4.6. Description of Input for Properties. Property values are index numbers or mixing fractions (when **mixture** *i j* is used) that define the exchange compositions in the *zone*. A negative index number for a property value indicates that no exchanger is present. Line 4 defines exchange compositions at nodes in *zone* by closest point interpolation from a scattered set of points (*x, y, z, index number*) that are read from succeeding lines in the input file.

Line 5: *x, y, z, index*

x—X coordinate of a point in space.

y—Y coordinate of a point in space.

z—Z coordinate of a point in space.

index—Index number for a chemical composition.

Line 6: **end_points**

end_points—The character string **end_points** must be included on the next line following the last scattered data point definition of the **points** property option.

Line 7: **-kinetics** [**mixture** *i j*] *property*

-kinetics—This identifier is used to specify the kinetic reactions initially present in the cells of *zone*. Optionally, **kinetics** or **-k[inetics]**.

mixture *i j*—Optionally, kinetic reactions can be defined as a mixture of kinetic reactions numbered *i* and *j* in the chemistry data file. For a **mixture**, the *property* defines mixing fraction, *f*, of kinetic reactions *i*; mixing fraction for kinetic reactions *j* is *1 − f*.

property—Property input as defined in section 4.6. Description of Input for Properties. Property values are index numbers or mixing fractions (when **mixture** *i j* is used) that define the kinetic reactions in the *zone*. A negative index number for a property value indicates that no kinetic reactions are present. Line 7 defines kinetic reactions at nodes in *zone* by closest point interpolation from a scattered set of points (*x, y, z, index number*) that are read from a file with lines formatted as in Line 5.

Line 8: **-solid_solutions** [**mixture** *i j*] *property*

> **-solid_solutions**—This identifier is used to specify the solid-solution assemblages initially present in the cells of *zone*. Optionally, **solid_solution, solid_solutions** or **-soli[d_solutions]**.
>
> **mixture** *i j*—Optionally, solid-solution assemblages can be defined as a mixture of solid-solution assemblages numbered *i* and *j* in the chemistry data file. For a **mixture**, the *property* defines mixing fraction, *f*, of solid-solution assemblage *i*; mixing fraction for solid-solution assemblage *j* is *1 − f*.
>
> *property*—Property input as defined in section 4.6. Description of Input for Properties. Property values are index numbers or mixing fractions (when **mixture** *i j* is used) that define the solid-solution assemblages in the *zone*. A negative index number for a property value indicates that no solid solutions are present. Line 8 defines solid-solution assemblages at nodes in *zone* by closest point interpolation from compositions read from a restart file.

Line 9: **-surface** [**mixture** *i j*] *property*

> **-surface**—This identifier is used to specify the surface compositions initially present in the cells of *zone*. Optionally, **surface** or **-su[rface]**.
>
> **mixture** *i j*—Optionally, surface compositions can be defined as a mixture of surface compositions *i* and *j* in the chemistry data file. For a **mixture**, the *property* defines mixing fraction, *f*, of surface composition numbered *i*; mixing fraction for surface composition *j* is *1 − f*.
>
> *property*—Property input as defined in section 4.6. Description of Input for Properties. Property values are index numbers or mixing fractions (when **mixture** *i j* is used) that define the surface compositions in the *zone*. A negative index number for a property value indicates no surface is present. Line 9 defines surface compositions at nodes in *zone* as mixtures between composition 2 and 4, where the mixing fraction of composition 2 at a node is interpolated by closest point interpolation from a scattered set of mixing fractions read from a file with lines formatted as in Line 5.

Line 10: **-gas_phase** [**mixture** *i j*] *property*

> **-gas_phase**—This identifier is used to specify the gas-phase compositions initially present in the cells of *zone*. Optionally, **gas_phase** or **-g[as_phase]**.
>
> **mixture** *i j*—Optionally, gas-phase compositions can be defined as a mixture of gas-phase compositions *i* and *j* in the chemistry data file. For a **mixture**, the *property* defines mixing fraction, *f*, of gas-phase composition *i*; mixing fraction for gas-phase composition *j* is *1 − f*.
>
> *property*—Property input as defined in section 4.6. Description of Input for Properties. Property values are index numbers or mixing fractions (when **mixture** *i j* is used) that define the gas-phase compositions in the *zone*. A negative index number for a property value indicates no gas phase is present. Line 10 defines gas-phase compositions at nodes in *zone* as mixtures between composition 3 and 5, where the mixing fraction of composition 3 at a node is interpolated by closest point interpolation from a scattered set of mixing fractions read from subsequent lines in the input file formatted as in Line 11.

Line 11: *x, y, z, fraction*

x—X coordinate of a point in space.

y—Y coordinate of a point in space.

z—Z coordinate of a point in space.

fraction—Mixing fraction for the composition represented by *i*, the first index number following **mixture**; mixing fraction for composition *j* will be 1 − *fraction*.

Notes

The **CHEMISTRY_IC** data block defines initial conditions and is required for a solute-transport simulation. The compositions of reactants (solutions, equilibrium-phase assemblages, exchangers, surfaces, kinetic reactants, solid solutions, and gas phases) in the chemistry data file or mixtures of two reactant compositions are used for initial conditions. The units for input in the **SOLUTION** data block are concentration units; internally, all chemical calculations use molality for the concentration unit of solutions. The appropriate unit for input in the data blocks **EQUILIBRIUM_PHASES, EXCHANGE, SURFACE, KINETICS, SOLID_SOLUTIONS** in the chemistry data file is moles of reactant per liter of pore space. For each cell in PHAST, the representative porous-medium volume for chemistry contains one kilogram of water, when saturated. Thus, when defining amounts of solid-phase reactants, the appropriate number of moles is numerically equal to the concentration of the reactant (moles per liter of water), assuming a saturated porous medium. In terms of moles per liter of water, the concentrations of the solid reactants vary spatially as porosity varies, which makes it difficult to define the appropriate solid reactant concentrations. To avoid this difficulty, PHAST Version 2 has options in the UNITS data block (flow and transport data file) to specify that the number of moles of solid reactants be interpreted as moles per liter of rock. As initial conditions are distributed to the finite-difference cells, the moles of solid reactants are scaled by the factor $(1-\phi)/\phi$, where ϕ is the porosity for the cell. This scaling takes into account the varying porosity and produces units of moles per liter of water. Molality (mol/kgw) is assumed to equal molarity (mol/L) for all transport calculations.

All nonnegative index numbers must correspond to index numbers for entities (solutions, equilibrium-phase assemblages, exchanges, surfaces, kinetic reactants, solid solutions, and gas phases) that are defined in the chemistry data file. Negative index numbers indicate that the given chemical entity is not present in the zone or at the specified location. For solute transport modeling, initial solution definitions are mandatory for all nodes in the active grid region. Equilibrium phases, exchangers, surfaces, kinetic reactants, solid solutions, and gas phases are optional. By default, all entities are absent from all cells. Multiple overlapping zones can be used within the **CHEMISTRY_IC** data block to define the initial conditions for the grid region. The index number for a property for a single node may be defined multiple times as part of different zone definitions. The index number used in the reactive-transport simulation for that entity for that node is the last index number defined for it in the flow and transport data file.

If spatial interpolation along a coordinate axis is used to define compositions for cells, the composition of each cell is determined by linear interpolation of the end-member compositions represented by two index numbers. The fractions applied to the end members are determined by the location of the node relative to the locations specified for the end members along the coordinate axis. For each type of initial condition the following are linearly interpolated: for solutions, moles of each element and temperature; for equilibrium-phase assemblages, moles of each mineral phase; for exchange assemblages, moles of exchange sites and moles of exchanged ions for each exchanger; for surface assemblages, moles of each type of surface site, surface area for each surface, and moles of sorbed elements for each surface; for kinetic reactants, the moles of each reactant; for solid solutions, moles of each component of each solid solution; and for gas phases, moles of each gas component.

When using kinetic reactions (**KINETICS** in the chemistry data file), Basic-language programs must also be defined in **RATES** data blocks in the chemistry data file or the thermodynamic data file. Rate equations must be written to use seconds as the unit of the time step (the TIME variable). PHAST converts time steps from the user input units for flow and transport calculations, so that seconds are used for all for kinetic calculations in the geochemical module.

Example Problems

The **CHEMISTRY_IC** data block is used in Chapter 6 example problems 1, 2, 3, 4, and 6.

DRAIN

This keyword data block is used to describe a single drain boundary condition. The data block is optional and only is used if a drain boundary condition is needed for the simulation. A drain boundary condition is similar to a river boundary condition, except that water only flows to the drain when the head in the aquifer is greater than the elevation of the drain; water does not flow from the drain to the aquifer. A drain is defined by a series of X–Y points, which define line segments that locate the drain within the model domain. For each drain point, the drain width, elevation, and leakance parameters must be defined, either by explicit definitions or by interpolation from other drain points. Leakage out of the aquifer through the drain is calculated from these parameters. Multiple **DRAIN** data blocks are used to define multiple drains in the grid region.

Example

Line numbers in the Example are used for identification purposes only and are referenced in the following Explanation. The line numbers do not indicate a required order for the keyword data block. Line numbers are not used in the flow and transport data file.

```
Line 0:      DRAIN 1 Agricultural drain
Line 1:          -xy_coordinate_system                    grid
Line 2:          -z_coordinate_system                     grid
Line 3:          -point  155.0    3633. # X, Y location of first drain point
Line 4:              -z                                   18.7
Line 5:              -width                               10.
Line 6:              -hydraulic_conductivity              1.5e-2
Line 7:              -thickness                           1.
Line 3a:         -point  165.0    3663.  # X, Y location of 2nd drain point
Line 3b:         -point  175.0    3603.  # X, Y location of last drain point
Line 4a:             -z                                   17.9
Line 5a:             -width                               10.
Line 6a:             -hydraulic_conductivity              1.5e-2
Line 7a:             -thickness                           1.
```

Explanation

Line 0: **DRAIN** *number* [*description*]

> **DRAIN** is the keyword for the data block.
>
> *number*—Positive number to designate this drain.
>
> *description*—Optional character field that identifies the drain.

Line 1: **-xy_coordinate_system (grid | map)**

> **-xy_coordinate_system**—Coordinate system for X–Y points that define the drain. If **-xy_coordinate_system** is not defined for a drain, the default is **grid**. Optionally, **xy_coordinate_system** or **-xy[_coordinate_system]**
>
> **(grid | map)**—**Grid** indicates that the X–Y points are defined in the grid coordinate system with units defined by **-horizontal_grid** in the **UNITS** data block. **Map** indicates that the X–Y points are defined in the map coordinate system with units defined by **-map_horizontal** in the **UNITS** data block. The transformation from grid to map coordinates is defined in the **GRID** data block.

Line 2: **-z_coordinate_system (grid | map)**

-z_coordinate_system—Coordinate system for elevation of drain points. If **-z_coordinate_system** is not defined for a drain, the default is **grid**. Optionally, **z_coordinate_system** or **-z_[coordinate_system]**

(grid | map)—**Grid** indicates that the elevations of drain points are defined in the grid coordinate system with units defined by **-vertical_grid** in the UNITS data block. **Map** indicates that the elevations of drain points are defined in the map coordinate system with units defined by **-map_vertical** in the UNITS data block. The transformation from grid to map coordinates is defined in the GRID data block.

Line 3: **-point** *x, y*

-point—This identifier is used to specify the X–Y coordinate location of a drain point. Line 3 may be repeated as many times as needed to define the entire length of the drain. At least two points must be defined. Units, L, are defined by the **-horizontal_grid or -map_horizontal** identifier in the UNITS data block, depending on the value of **-xy_coordinate_system**. Optionally, **point**, **-p[oint]**, **node**, or **-n[ode]**.

x—X coordinate location of a drain point.

y—Y coordinate location of a drain point.

Line 4: **-z** *elevation*

-z—This identifier is used to specify the elevation of the drain at the X–Y point.

elevation—Vertical elevation of the drain. Units, L, are defined by the **-vertical_grid or -map_vertical** identifier in the UNITS data block, depending on the value of **-z_coordinate_system**.

Line 5: **-width** *width*

-width—This identifier is used to specify the width of the drain at the point. Units, L, are defined by the **-drain_width** identifier in the UNITS data block. Optionally, **width**, or **-w[idth]**.

width—Width of the drain at the drain point.

Line 6: **-hydraulic_conductivity** *hydraulic_conductivity*

-hydraulic_conductivity—This identifier is used to specify the hydraulic conductivity for a layer adjacent to the drain at the point. Optionally, **hydraulic_conductivity**, **k**, **-h[ydraulic_conductivity]**, or **-k**.

hydraulic_conductivity—Hydraulic conductivity of the of the layer adjacent to the drain. Units, L/T, are defined by the **-drain_hydraulic_conductivity** identifier in the UNITS data block.

Line 7: **-thickness** *thickness*

-thickness—This identifier is used to specify the thickness of the layer adjacent to the drain at the point. The thickness is used to define the gradient in head from the aquifer to the drain. Optionally, **thickness**, or **-t[hickness]**.

thickness—Thickness of the layer adjacent the drain. Units, L, are defined by the **-drain_thickness** identifier in the UNITS data block.

Notes

Drains are boundary conditions that are applied to unconfined aquifer systems. These boundary conditions only accept water from the aquifer and never contribute water to the aquifer. Drains accept water when the head in the aquifer is greater than the elevation of the drain. Multiple drains are defined by using multiple **DRAIN** data blocks, where each drain is uniquely identified by the integer following the **DRAIN** keyword.

The coordinate system for the X–Y location of a drain point may be either of the two available coordinate systems, grid or map, as defined by the **-xy_coordinate_system** identifier. The coordinate system for the elevation may be grid or map as defined by the **-z_coordinate_system** identifier. All data are converted to the grid coordinate system and then converted to SI units for the simulation calculations.

A series of drain points can be defined in upstream or downstream order. Parameters may be defined explicitly for a drain point, or they may be interpolated from other drain points. The elevation, width, hydraulic conductivity, and thickness must be defined for the first and last points of the drain. For a point between the first and last points, a parameter can be defined explicitly with its identifier (**-z**, for example), or it will be interpolated from the nearest drain points upstream and downstream where the parameter has been explicitly defined. Linear interpolation is performed by the ratio of the distance of the interpolation point from one of the explicitly defined points and the total distance along the line segments that lie between the two explicitly defined points. No time-dependent parameters are associated with a drain.

The elevation, width, hydraulic conductivity, and thickness at each point are used to quantify leakage to the drain from the aquifer as described in section D.5.6. Drain Boundaries. The specified widths and calculated distances between drain points are used to determine areas of application for the drain boundary. These areas are subdivided into subareas through which water can flow to a single cell. An X–Y–Z coordinate location is associated with each subarea, which allows the subarea to be associated with the appropriate cell in the grid.

Example Problems

The **DRAIN** data block is used in Chapter 6 example problems **5** and **6**.

END

This keyword has no associated data. It is used optionally to designate the end the data input for a simulation. No input lines are processed after encountering the **END** keyword in the flow and transport data file.

Example Problems

The **END** keyword is used in Chapter 6 example problems 1, 2, 3, 4, 5, and 6.

FLUX_BC

This keyword data block is used to define flux boundary conditions. For flow-only simulations, only the fluid flux for each cell with a flux boundary condition is required. For reactive-transport simulations, an associated-solution composition, which may be a mixture of two solutions, is required. The flux and associated solution may vary with time independently over the course of the simulation. This keyword data block is optional and only is needed if flux boundaries are to be included in the simulation.

Example

Line numbers in the Example are used for identification purposes only and are referenced in the following Explanation. The line numbers do not indicate a required order for the keyword data block. Line numbers are not used in the flow and transport data file.

```
Line 0:    FLUX_BC
Line 1:            -box   0.0    0.0    0.0    1000.0      100.0       10.0
Line 2:                   -face   Z
Line 3:                   -flux
Line 4:                          0      points grid
Line 5:                                 0.0    0.0   0.0   -3.5e-3
Line 5a:                                10.0   10.0  0.0   -5.2e-3
Line 6:                                 end_points
Line 4a:                         100    -2.0e-3
Line 7:                   -associated_solution
Line 8:                          0      X     2     0.0    6    10.0
```

Explanation

Line 0: **FLUX_BC**

> **FLUX_BC** is the keyword for the data block; no other data are included on this line.

Line 1: *zone*

> *zone*—A zone definition as defined in section 4.5. Description of Input for Zones. Line 1 of the above Example defines a box zone.

Line 2: **-face (X | Y | Z)**

> **-face**—This identifier is used to specify the coordinate direction of the flux. Optionally, **face** or **-fa[ce]**.
>
> **(X | Y | Z)**—The coordinate direction of the flux.

Line 3: **-flux**

> **-flux**—This identifier is used to specify a time series of volumetric fluxes for those external cell faces that are completely or partly within the *zone* and are perpendicular to the coordinate direction defined by **-face**. Volumetric fluxes are applied only to cells associated with active nodes and only to cell faces on the active grid boundary. A flux is applied to that part of a cell face that is within the *zone*. The magnitude and direction of the flux is defined by a time series of properties (one or more line 4s). The first *time* in the series must be zero. Optionally, **flux** or **-fl[ux]**.

Line 4: *time [units] property*

> *time*—Simulation time (T) at which the flux property definition (*property*) will take effect. Units may be defined explicitly with *units*; default units are defined by **-time** identifier in **UNITS** data block.

units—Units for *time* can be "seconds", "minutes", "hours", "days", or "years" or an abbreviation of one of these units.

property—Property input as defined in section 4.6. Description of Input for Properties. Property values are the volumetric flux for the cell-face areas selected by the *zone*. The flux is a signed quantity that indicates whether the flux is in the positive or negative coordinate direction (-**face**). Units, L/T, are defined by -**flux** identifier in the **UNITS** data block. Lines 4 through 6 use the **points** property option to define flux at two points in space. The flux value for a an exterior cell face is interpolated by closest point interpolation to the cell node and applied to the area of the cell face that falls within the zone as discussed in the notes of section 4.6. Description of Input for Properties.

Line 5: *x, y, z, value*

x—X coordinate of a point.

y—Y coordinate of a point.

z—Z coordinate of a point.

value—Flux at the specified point.

Line 6: **end_points**

end_points—The character string **end_points** must be included on the next line following the last scattered data point definition of the **points** property option.

Line 4a: *time* [*units*] *property*

time—Same as line 4.

units—Same as line 4.

property—Same as line 4, except line 4a defines a uniform flux for all exterior cell faces (-face Z) in the zone.

Line 7: -**associated_solution**

-**associated_solution**—This identifier is used to specify a time series of solution composition for the flux of water into the active grid region through the flux boundary. The water composition is defined by the property definition of line 8. A time series of solution properties may be defined by using multiple line 8s. The first *time* in the series must be zero. Optionally, **associated_solution**, -**a**[**ssociated_solution**], **solution**, or -**s**[**olution**].

Line 8: *time* [*units*] [**mixture** *i j*] *property*

time—Same as line 4.

units—Same as line 4.

mixture *i j*—Optionally, solution compositions can be defined as a mixture of solution compositions *i* and *j* in the chemistry data file. For a **mixture**, the *property* defines mixing fraction, *f*, of solution composition *i*; mixing fraction for solution composition *j* is *1 − f*.

property—Property input as defined in section 4.6. Description of Input for Properties. Property values are index numbers or mixing fractions (when **mixture** *i j* is used) that define the solution composition associated with water flowing into the active grid region through the flux boundary. Line 8 illustrates a linearly varying solution composition.

Notes

Flux boundary conditions are described in detail in section D.5.3. Flux Boundary. Flux boundary conditions are applied only to cell faces that bound the active grid region. The -**face** identifier is used to specify the exterior cell faces within the zone that receive the flux. Fluxes are applied to exterior cell faces perpendicular to the coordinate direction defined with -**face**. A flux is applied only to that part of a cell face that is within the zone as described in the notes to section 4.6. Description of Input for Properties. For unconfined flow simulations, if the flux boundary applies to faces perpendicular to the X or Y axis, flux is multiplied by the fraction of the cell that is saturated. The sign of the flux quantity indicates whether the flux is in the positive or negative coordinate direction. Note that the hyphen in **-flux** does not indicate a negative quantity; it only indicates that the word is an identifier. If the flux is out of a flux-boundary-condition cell, the composition of the water is equal to the composition of the water in the cell; if the flux is into a flux-boundary-condition cell, the composition of the water is equal to the composition defined by the **-solution** identifier.

Multiple zones may be used within any **FLUX_BC** data block to define flux boundary conditions for selected cell faces within the grid region. Different boundary conditions for a single cell and its cell faces may be defined multiple times as part of different zone definitions and different keyword data blocks. The last boundary condition defined for a cell determines the type of boundary conditions that apply to the cell. If the last boundary condition defined is a specified-head boundary, then the cell is exclusively a specified-head boundary. If the last definition is flux or leaky, then the cell may have a combination of flux and leaky boundary conditions. If multiple flux boundary conditions are defined for a cell, the areas of application are revised so that no areas overlap; later-defined conditions take precedence over earlier definitions. Similarly, areas of application for leaky boundaries are revised to avoid overlaps. However, no attempt is made to avoid overlaps between leaky and flux boundaries. See section D.5.8. Boundary-Condition Compatibility for more details.

Example Problems

The **FLUX_BC** data block is used in Chapter 6 example problems **4**, **5**, and **6**.

FREE_SURFACE_BC

This keyword data block is used to define the presence of a free surface and unconfined flow conditions. If the **FREE_SURFACE_BC** data block is not included, confined flow is simulated.

Example

Line numbers in the Example are used for identification purposes only and are referenced in the following Explanation. Line numbers are not used in the flow and transport data file.

```
Line 0:    FREE_SURFACE_BC          true
```

Explanation

Line 0: **FREE_SURFACE_BC** [(**True** | **False**)]

> **FREE_SURFACE_BC**—This keyword data block is used to include or exclude a free-surface boundary condition in the simulation. By default, confined flow is simulated. Optionally, **FREE_SURFACE**.
>
> (**True** | **False**)—A value of **true** indicates that unconfined flow is simulated. A value of **false** indicates that confined flow is simulated. If neither **true** nor **false** is entered on the line, **true** is assumed. Optionally, **t[rue]** or **f[alse]**.

Notes

This option is used to simulate confined (**FREE_SURFACE_BC false**) or unconfined (**FREE_SURFACE_BC true**) flow. If unconfined flow is simulated, the specific storage in each cell is automatically set to zero, which is equivalent to setting the fluid and matrix compressibilities to zero. A free-surface boundary condition is required for simulation of river and drain boundary conditions. See section D.5.8. Boundary-Condition Compatibility for more details. The free-surface boundary condition is described in detail in section D.5.7. Free-Surface Boundary.

Example Problems

The **FREE_SURFACE_BC** data block is used in Chapter 6 example problems **1**, **2**, **3**, **4**, **5**, and **6**.

GRID

This keyword data block is used to define the finite-difference node locations for the simulation grid. It is also used to define the transformation between a map coordinate system and the grid coordinate system. This data block is mandatory for all simulations. The **GRID** data block contains only static data.

Example

Line numbers in the Example are used for identification purposes only and are referenced in the following Explanation. The line numbers do not indicate a required order for the keyword data block. Line numbers are not used in the flow and transport data file.

```
Line 0:     GRID
Line 1:         -uniform         X      0.    1000. 6
Line 2:         -nonuniform      Y      0.    100.
Line 3:                                 400.  800.  1000.
Line 1a:        -uniform    Z    0.    10.   2
Line 4:         -overlay_uniform Z 0.0 1.0 11
Line 5:         -overlay_nonuniform Z 8.2 8.4 8.5 8.6 8.8
Line 6:         -snap Z   0.05
Line 7:         -area_tolerance           1e-7
Line 8:         -chemistry_dimensions     XZ
Line 9:         -print_orientation        XZ
Line 10:        -grid_origin              275000      810000      0
Line 11:        -grid_angle               90
```

Explanation

Line 0: **GRID**

> **GRID** is the keyword for the data block; no other data are included on this line.

Line 1: **-uniform (X | Y | Z)** *minimum, maximum, number of nodes*

> **-uniform**—This identifier is used to specify uniformly spaced nodes for the coordinate direction. Line 1 defines 6 equally spaced nodes in the X coordinate direction.
>
> > **-Nonuniform** and **-uniform** are mutually exclusive for a coordinate direction. Optionally, **uniform** or **-u[niform]**.

> **(X | Y | Z)**—Coordinate direction with uniformly spaced nodes.

> *minimum*—Minimum node coordinate in the specified coordinate direction. Units, L, are defined by **-horizontal_grid** (X and Y coordinates) and **-vertical_grid** (Z coordinate) identifiers in the **UNITS** data block.

> *maximum*—Maximum node coordinate in the specified coordinate direction. Units, L, are defined by **-horizontal_grid** (X and Y coordinates) and **-vertical_grid** (Z coordinate) identifiers in the **UNITS** data block.

> *number of nodes*—Number of nodes in the specified coordinate direction. The number of cells in this direction is equal to the number of nodes. The number of elements in this direction is one less than the number of nodes. The minimum number of nodes is two.

Line 2: **-nonuniform (X | Y | Z)** *list of node coordinates*

> **-nonuniform**—This identifier is used to specify nonuniform node spacing for the coordinate direction. **-Nonuniform** and **-uniform** are mutually exclusive for a coordinate direction. Optionally, **nonuniform** or **-n[onuniform]**.

(**X** | **Y** | **Z**)—Coordinate direction with nonuniform node spacing.

list of node coordinates—List of node coordinates in the specified coordinate direction. List of node coordinates must be in ascending order. Although only one coordinate is required for **-nonuniform** input, altogether, at least two nodes must be defined for a coordinate direction by **-nonuniform**, **-overlay_uniform**, and **-overlay_nonuniform** identifiers. Units, L, are defined by **-horizontal_grid** (X and Y coordinates) and **-vertical_grid** (Z coordinate) identifiers in the UNITS data block. List of node coordinates for **-nonuniform** may continue on successive lines. Lines 2 and 3 define 5 unequally spaced nodes in the Y coordinate direction; Line 3 is a continuation of the list of nodes begun on Line 2.

Line 3: *list of node coordinates*

list of node coordinates—Continuation of list of node coordinates. This line must be preceded by the **-nonuniform** identifier or another continuation line for the list.

Line 4: **-overlay_uniform** (**X** | **Y** | **Z**) *minimum, maximum, number of nodes*

-overlay_uniform—This identifier is used to specify additional uniformly spaced nodes for the specified coordinate direction. Either **-nonuniform** or **-uniform** is required for each coordinate direction; optional **-overlay_uniform** and **-overlay_nonuniform** identifiers may be defined for each coordinate direction. Multiple overlays may define the same node or nodes that are close together; nodes closer than the *snap_distance* for the coordinate (Line 6) will be merged into a single node. Optionally, **overlay_uniform** or **-o[verlay_uniform]**.

(**X** | **Y** | **Z**)—Coordinate direction with additional uniformly spaced nodes.

minimum—Minimum node coordinate for additional uniformly spaced nodes in the specified coordinate direction. Units, L, are defined by **-horizontal_grid** (X and Y coordinates) and **-vertical_grid** (Z coordinate) identifiers in the UNITS data block.

maximum—Maximum node coordinate for the additional uniformly spaced nodes in the specified coordinate direction. Units, L, are defined by **-horizontal_grid** (X and Y coordinates) and **-vertical_grid** (Z coordinate) identifiers in the UNITS data block.

number of nodes—Number of additional nodes for the specified coordinate direction.

Line 4 indicates that beginning with a node at 0.0 and ending with a node at 1.0, 11 equally spaced nodes will be added to the grid in the Z direction.

Line 5: **-overlay_nonuniform** (**X** | **Y** | **Z**) *list of node coordinates*

-overlay_nonuniform—This identifier is used to specify additional nonuniformly spaced nodes in the specified coordinate direction. Either **-nonuniform** or **-uniform** is required for each coordinate direction; optional **-overlay_uniform** and **-overlay_nonuniform** identifiers may be defined for each coordinate direction. Multiple overlays may define the same node or nodes that are close together; nodes closer than the *snap_distance* for the coordinate (Line 6) will be merged into a single node. Optionally, **overlay_nonuniform** or **-overlay_n[onuniform]**.

(**X** | **Y** | **Z**)—Coordinate direction with additional nonuniformly spaced nodes.

list of node coordinates—List of one or more node coordinates in the specified coordinate direction. List of node coordinates must be in ascending order. Units, L, are defined by **-horizontal_grid** (X and Y coordinates) and **-vertical_grid** (Z coordinate) identifiers in the UNITS data block. List of node coordinates may continue on successive lines.

Line 5 indicates that additional nodes are placed at 8.2, 8.4, 8.5, 8.6, and 8.8 in the Z direction.

Line 6: **-snap** (**X** | **Y** | **Z**) *snap_distance*

-**snap**—This identifier is used to specify the minimum distance between nodes in the coordinate direction. Optionally, **snap** or -**s**[**nap**].

(**X** | **Y** | **Z**)—Coordinate direction for which *snap_distance* is defined.

snap_distance—Minimum distance between nodes in the specified coordinate direction. After coordinate lists are combined and sorted for a coordinate direction, the nodes are processed in sort order. Nodes closer than *snap_distance* to a previous node will be eliminated. Default is 0.001 of specified units. Units, L, are defined by -**horizontal_grid** (X and Y coordinates) and -**vertical_grid** (Z coordinate) identifiers in the **UNITS** data block.

Line 7: -**area_tolerance** *fractional_area*

-**area_tolerance**—This identifier is used to specify the minimum fractional area of a cell face needed for a flux or leaky boundary condition to apply to a cell. Optionally, **area_tolerance** or -**a**[**rea_tolerance**].

fractional_area—If the area of application for a leaky or flux boundary condition is less than this fraction of the cell face area, the boundary condition will not apply to the cell. Default is 1e-7 (unitless).

Line 8: -**chemistry_dimensions** [**X**] [**Y**] [**Z**]

-**chemistry_dimensions**—This identifier is used to specify coordinate directions for which chemical calculations are performed. If, conceptually, the transport calculation is one or two dimensional, the geochemical calculations can be performed for a single line or plane of nodes, and the resulting chemical compositions can be copied to the remaining symmetric lines or plane of nodes. Omitting chemical calculations in one or two coordinate directions saves substantial amounts of computation time. By default, geochemical calculations are performed on all active nodes of the three-dimensional grid region. Optionally, **chemistry_dimensions**, or -**c**[**hemistry_dimensions**].

[**X**] [**Y**] [**Z**]—The coordinate directions for which chemical calculations are performed. For example, "**Z**" indicates that chemical calculations are performed for one line of nodes in the Z direction, "**XZ**" indicates chemical calculation are performed on the set of active nodes in one X–Z plane, and "**XYZ**" indicates chemical calculations are performed on all active nodes. Warning: initial and boundary conditions must be consistent with the symmetry assumed for one- or two-dimensional calculations or erroneous results will be produced.

Line 9: -**print_orientation** (**XY** | **XZ**)

-**print_orientation**—This identifier is used to specify the orientation of the planes of data to be written to files with names that end with **.txt** and have node-by-node data. These files contain spatial data written as a series of planes. By default, writing is by X–Y planes. Optionally, **print_orientation** or -**p**[**rint_orientation**].

(**XY** | **XZ**)—Only two orientation options are allowed, either **XY**, indicating X–Y planes are printed or **XZ,** indicating X–Z planes are printed.

Line 10: -**grid_origin** *x_offset y_offset z_offset*

-**grid_origin**—This identifier is used to specify the origin of the grid in map coordinates. Data in map coordinates are typically read from ArcInfo shapefiles (ESRI, 1998) as part of the definition of a prism zone. The **grid_origin** option gives the origin of the grid (grid location 0, 0, 0) in terms of the map coordinate system. By default, the grid origin coincides with the map origin. Optionally, **grid_origin** or -**grid_o**[**rigin**].

x_offset—X location of the grid origin in map coordinates.

y_offset—Y location of the grid origin in map coordinates.

z_offset—Z location of the grid origin in map coordinates.
Line 11: **-grid_angle** *degrees_counterclockwise*

-grid_angle—This identifier is used to specify the angle of the grid relative to the map coordinate system. Data in map coordinates are typically read from ArcInfo shapefiles (ESRI, 1998) as part of the definition of a prism zone. The **grid_angle** option gives the angle of the X axis in degrees counterclockwise relative to the X axis of the map coordinate system (typically the east-west axis). By default, the grid angle is zero so that the X axis of the grid is in the same direction as the X axis of the map coordinate system. Optionally, **grid_angle** or **-grid_a[ngle]**.

degrees_counterclockwise—Angle of the grid X axis in degrees counterclockwise from the map X axis.

Notes

Node coordinates must be defined for all three coordinate directions for all simulations. Each coordinate direction must be defined with **-uniform** or **-nonuniform**, which are mutually exclusive identifiers for each coordinate direction. The grid may be refined by adding additional nodes with the identifiers **-overlay_uniform** and **-overlay_nonuniform**. All of the nodes defined are merged into a single list and sorted; nodes within the distance defined by **-snap** of a previous node in the list are eliminated.

Flow and transport calculations always involve the entire active grid region and are thus three dimensional. The minimum number of cells (nodes) in any coordinate direction is two. The identifier **-chemistry_dimensions** is used to save computation time for one- and two-dimensional problems. If the calculation represents a one- or two-dimensional flow system and initial and boundary conditions are appropriate for one- or two-dimensional transport, the symmetry of the transport simulation can be used to reduce the number of geochemical calculations. The geochemical calculations can be performed on a single line or plane of nodes and the results copied to the other lines or plane of nodes. If the **-chemistry_dimensions** identifier is not included, geochemical calculations are performed for all active nodes. For one- and two-dimensional geochemical calculations, one or two coordinate directions are listed for the **-chemistry_dimensions** identifier; the number of nodes for the coordinate directions not listed must be exactly two.

The use of **-chemistry_dimensions** can lead to erroneous results if the initial and boundary conditions do not allow a truly one- or two-dimensional transport simulation. For example, an X–Y simulation would not have zero flow in the Z direction if a free-surface boundary condition were used, which violates the two-dimensional flow assumption. Also, one- or two-dimensional flow is not sufficient to ensure the same symmetry for transport. For example, for steady one-dimensional flow in the X direction, there are four cells in each Y–Z plane. If a contaminant is introduced into only one cell in a Y–Z plane, concentrations are not equal in each cell of the Y–Z plane and the transport system is not one dimensional. Thus, boundary and initial conditions must be chosen carefully to ensure one- or two-dimensional transport. The program does not check for conceptual errors in the boundary conditions. Operationally, a one- or two-dimensional simulation can be tested by removing the **-chemistry_dimensions** identifier and checking that the results are the same as the one- or two-dimensional simulation.

PHAST allows two coordinate systems to be used when entering spatial data—a map coordinate system and a grid coordinate system. Data imported from a GIS (geographic information system) typically have coordinates, such as UTM (Universal Transverse Mercator), that are derived from a particular map projection. PHAST uses a simple translation and rotation transformation defined by **-grid_origin** and **-grid_angle** to convert data in map coordinates to data in grid coordinates. The **-grid_origin** identifier gives the location of the grid origin (0, 0, 0) in terms of the map coordinate system. (Note that the active grid region may or may not include the origin.) The **-grid_angle** identifier allows the grid to be at an angle relative to the map coordinate system, with the angle of the grid defined in degrees counterclockwise relative to the map coordinate system. These two identifiers are needed only if some of the spatial data used in the model are not defined with grid coordinates.

No set of discretization rules exists that will guarantee an accurate numerical solution with a minimum number of nodes and time steps, even for the case of uniform coefficients in the differential equations. However, the following guidelines need to be considered. See Appendix D for more details.

1. If using the backward-in-space or backward-in-time differencing (**SOLUTION_METHOD** data block), verify that the grid-spacing and time-step selection do not introduce excessive numerical dispersion.

2. If using centered-in-space and centered-in-time differencing, examine the results for spatial and temporal oscillations that are caused by the time or space discretization being too coarse.

3. With reactive transport, it is important to assess the effects of spatial and temporal discretization error on the reaction chemistry that takes place by refining the spatial and temporal discretization.

4. The global-balance summary table (*prefix*.**bal.txt** file) may indicate that the time step is too long by exhibiting large errors for conservative constituents.

Example Problems

The **GRID** data block is used in Chapter 6 example problems 1, 2, 3, 4, 5, and 6.

HEAD_IC

This keyword data block is used to define the initial head conditions in the grid region. This data block is mandatory for all simulations.

Example 1

Line numbers in the Example are used for identification purposes only and are referenced in the following Explanation. The line numbers do not indicate a required order for the keyword data block. Line numbers are not used in the flow and transport data file.

```
Line 0:    HEAD_IC
Line 1:         -box         0.0   0.0   0.0   1000.0      100.0        10.0
Line 2:              -head    150
Line 1a:        -wedge       0.0   0.0   0.0   1000.0      100.0        10.0   Z1
Line 2a:             -head    xyz  grid wt.dat
```

Explanation 1

Line 0: **HEAD_IC**

   HEAD_IC is the keyword for the data block; no other data are included on this line.

Line 1: *zone*

   zone—A zone definition as defined in section 4.5. Description of Input for Zones. Line 1 defines a box zone and line 1a defines a wedge zone.

Line 2: **-head** *property*

   -head—This identifier is used to specify initial heads for the zone. Optionally, **head** or **-h[ead]**.

   property—Property input as defined in section 4.6. Description of Input for Properties. Property values are initial head values for the simulation. Line 2 illustrates a constant head for the zone. Line 2a illustrates a head field that is read from a file (*wt.dat*). The file has quadruplets of values (*x, y, z, head*), one per line, where *x, y,* and *z* are in grid coordinates; values of head at grid nodes in the zone will be interpolated by closest point interpolation from the scattered data points. Units of head, L, are defined by the **-head** identifier in the **UNITS** data block.

Example 2

Line numbers in the Example are used for identification purposes only and are referenced in the following Explanation. Line numbers are not used in the flow and transport data file.

```
Line 0:    HEAD_IC
Line 1:         -water_table by_node
Line 2:              < 150.0 150.5 151.0 151.5 196*152.0 >
```

Explanation 2

Line 0: **HEAD_IC**

   HEAD_IC is the keyword for the data block; no other data are included on this line.

Line 1: **-water_table** *property*

-water_table—This identifier is used to specify hydrostatic potentiometric heads for the active grid region. Hydrostatic potentiometric heads are specified by an array of heads, one for each node in the X–Y plane. Optionally, **water_table** or **-w[ater_table]**.

property—Heads for the entire X–Y plane can be entered with either of two methods for defining a spatially distributed property: (1) **by_node** followed by a value for each node (active or inactive) in the X–Y plane in natural order (list of values may extend over multiple lines); the list of values must be enclosed in angle brackets "< >", or (2) **file** followed by a file name, a value for each node (active or inactive) in the X–Y plane is read from the file in natural order (list of values may extend over multiple lines). Line 2 illustrates method 1. Units, L, are defined by **-head** identifier in the **UNITS** data block.

Notes

The **HEAD_IC** data block defines the initial head conditions. The initial head condition applies only at the beginning of the simulation, and it is not possible to redefine a head condition later in the simulation. Initial heads are mandatory for all nodes in the active grid region.

When using zones to define the initial head, multiple zones may be used within the **HEAD_IC** data block to define the initial conditions for the entire grid region. Heads for a single node may be defined multiple times as part of different zone definitions. The initial head used in the flow or reactive-transport simulation for that node is the last head defined for the node in the flow and transport data file.

Use of the **-water_table** identifier is discouraged because the input for the identifier is grid dependent (node-by-node). Instead, use **-head** with heads defined on a horizontal plane with an XYZ file or **points** property definition. When using the **-water_table** identifier, it is not permissible to use zones to define initial head conditions. Water-table heads are defined only for the uppermost X–Y plane of nodes. For each vertical stack of nodes in the grid region, the initial water-table head of the uppermost node is assigned to all nodes in the stack.

Example Problems

The **HEAD_IC** data block is used in Chapter 6 example problems **1**, **2**, **3**, **4**, **5**, and **6**.

LEAKY_BC

This keyword data block is used to define leaky boundary conditions. Conceptually, a leaky boundary layer is located outside of the active grid region, and flow through the layer is determined by the thickness of the layer, the hydraulic conductivity of the layer, and the difference in head between the specified head at the exterior side of the layer and the heads in the leaky boundary cells of the active grid region. For flow-only simulations, only the parameters related to flow are required. For reactive-transport simulations, an associated-solution composition, which may be a mixture of two solutions, is required. The head at the exterior side of the layer and the composition of the associated solution may vary independently over the course of the simulation. This keyword data block is optional and is needed only if leaky boundary conditions are included in the simulation.

Example

Line numbers in the Example are used for identification purposes only and are referenced in the following Explanation. The line numbers do not indicate a required order for the keyword data block. Line numbers are not used in the flow and transport data file.

```
Line  0:    LEAKY_BC
Line  1:        -box            5      0      0      10    10    10
Line  2:            -face              Z
Line  3:            -thickness         X      1.0    0     5.0   10
Line  4:            -hydraulic_conductivity       points      grid
Line  5:                0.0    0.0    0.0    1e-5
Line  5a:               10.0   10.0   10.0   1e-4
Line  6:                end_points
Line  7:            -elevation  xyz    grid          bathymetry
Line  8:            -z_coordinate_system      map
Line  9:            -head
Line 10:                0      xyzt   grid  head.xyzt
Line 11:            -associated_solution
Line 12:                0      mixture 2     4     xyz   mixf.dat    grid
Line 12a:               100    points       grid
Line 13:                       0.0    5.0    0.0   1
Line 13a:                      10.0   5.0    0.0   2
Line  6a:               end_points
```

Explanation

Line 0: **LEAKY_BC**

> **LEAKY_BC** is the keyword for the data block; no other data are included on this line.

Line 1: *zone*

> *zone*—A zone definition as defined in section 4.5. Description of Input for Zones. Line 1 defines a box zone.

Line 2: **-face (X | Y | Z)**

> **-face**—This identifier is used to specify the cell face for the leakage. Leakage is through exterior cell faces in the zone that are perpendicular to the coordinate direction. Optionally, **face** or **-fa[ce]**.
>
> **(X | Y | Z)**—The cell face for leakage.

Line 3: **-thickness** *property*

>> **-thickness**—This identifier is used to specify the thickness of the leaky boundary layer. Optionally, **thickness** or **-t[hickness]**.

>> *property*—Property input as defined in section 4.6. Description of Input for Properties. Property values are the thicknesses for the leaky-boundary-condition cells selected by the zone. Line 3 illustrates a linearly varying thickness in the X direction. Units of leaky boundary thickness, L, are defined by the **-leaky_thickness** identifier in the **UNITS** data block.

> Line 4: **-hydraulic_conductivity** *property*

>> **-hydraulic_conductivity**—This identifier is used to specify the hydraulic conductivity of the leaky boundary layer. Optionally, **hydraulic_conductivity**, **-hy[draulic_conductivity]**, **k**, or **-k**.

>> *property*—Property input as defined in section 4.6. Description of Input for Properties. Property values are the hydraulic conductivity for the leaky-boundary-condition cells selected by the zone. Lines 4 through 5a define a set of two points that are used to interpolate the hydraulic conductivity of the leaky boundary layer. Interpolation to nodes is done by closest point interpolation. Units of leaky-boundary-layer hydraulic conductivity, L/T, are defined by the **-leaky_hydraulic_conductivity** identifier in the **UNITS** data block.

> Line 5: *x, y, z, value*

>> *x*—X coordinate of a point.

>> *y*—Y coordinate of a point.

>> *z*—Z coordinate of a point.

>> *value*—Leaky-boundary-layer hydraulic conductivity at the specified point.

> Line 6: **end_points**

>> **end_points**—The character string **end_points** must be included on the next line following the last scattered data point definition of the **points** property option.

> Line 7: **-elevation** *property*

>> **-elevation**—This identifier is used to specify the elevation of the top of the leaky boundary layer. It is only used for leaky boundaries on positive Z faces and only when unconfined flow is simulated. Optionally, **bottom** or **-b[ottom]**.

>> *property*—Property input as defined in section 4.6. Description of Input for Properties. Property values are the elevation of the top of the leaky layer for the leaky-boundary-condition cells selected by the zone. Line 7 illustrates interpolation from an X–Y–Z file named *bathymetry*, which contains lines with quadruplets of X, Y, Z, and elevation. Units of leaky boundary elevation, L, are defined by the **-z_coordinate_system** identifier defined for this zone, either **grid** or **map**.

> Line 8: **-z_coordinate_system** (**grid** | **map**)

>> **-z_coordinate_system**—This identifier is used to specify the coordinate system for the data defined with **-elevation**. The coordinate-system definition is only used for leaky boundaries on positive Z faces and only when unconfined flow is simulated. Leakage is through exterior cell faces in the zone that are perpendicular to the coordinate direction. Optionally, **face** or **-fa[ce]**.

>> (**grid** | **map**)—Specifies the coordinate system for **-elevation is** either grid coordinates or map coordinates.

> Line 9: **-head**

>> **-head**—This identifier is used to specify a time series of head on the exterior side of the leaky boundary layer. The head is defined by the property definition on line 7. A time series of

head properties may be defined by using multiple line 7s. The first *time* in the series must be zero. Optionally, **head** or **-he[ad]**.

Line 10: *time* [*units*] *property*

> *time*—Simulation time (T) at which the head property definition (*property*) will take effect. Units may be defined explicitly with *units*; default units are defined by **-time** identifier in **UNITS** data block.

> *units*—Units for *time* can be "seconds", "minutes", "hours", "days", or "years" or an abbreviation of one of these units.

> *property*—Property input as defined in section 4.6. Description of Input for Properties. Property values are the heads for the leaky-boundary-condition cells selected by the zone. Line 7 defines a time series of head conditions for the leaky boundary condition that are extracted from the file *head.xyzt*. The file contains heads at scattered points at one or more time planes. Units of head, L, are defined by the **-head** identifier in the **UNITS** data block.

Line 11: **-associated_solution**

> **-associated_solution**—This identifier is used to specify a time series of solution composition for the flux of water into the active grid region through the leaky boundary. In this example, the solution composition time series is defined by the property definitions on Lines 12, 12a, 13, and 13a. The first *time* in the series must be zero. Optionally, **associated_solution, -a[ssociated_solution], solution,** or **-s[olution]**.

Line 12: *time* [*units*] [**mixture** *i j*] *property*

> *time*—Same as line 7.

> *units*—Same as line 7.

> **mixture** *i j*—Optionally, solution compositions can be defined as a mixture of solution compositions *i* and *j* in the chemistry data file. For a **mixture**, the *property* defines mixing fraction, *f*, of solution composition *i*; mixing fraction for solution composition *j* is $1 - f$.

> *property*—Property input as defined in section 4.6. Description of Input for Properties. Property values are index numbers or mixing fractions (when **mixture** *i j* is used) that define the solution composition associated with water flowing into the active grid region through the leaky boundary. Line 12 defines mixtures of solutions 2 and 4 that apply beginning at time 0. Mixing fractions at scattered points are read from a file named *mixf.dat*; values of mixing fraction of solution 2 will be interpolated at the nodes of the leaky-boundary-condition cells by closest point interpolation. Line 12a defines a set of solution compositions that apply beginning at time 100. Solution compositions at a set of points follow in the input file (Lines 13 and 13a); Line 6a terminates the set of points. The set of points will be used for closest point interpolation to the boundary-condition cells.

Line 13: *x, y, z, value*

> *x*—X coordinate of a point.

> *y*—Y coordinate of a point.

> *z*—Z coordinate of a point.

> *value*—Associated solution at the specified point.

Notes

Leaky boundaries are applied only to cell faces that bound the active grid region. The **-face** identifier is used to specify the exterior cell faces within the zone that receive the leakage. Leakage occurs at exterior cell faces perpendicular to the coordinate direction defined with **-face**. Leakage is applied to that part of an exterior cell face within the zone, as described in the notes to section 4.6. Description of Input for Properties.

For unconfined flow simulations, if the leaky boundary applies to X and Y faces (faces perpendicular to the X or Y axis), leakage is multiplied by the fraction of the cell that is saturated.

For unconfined flow simulations, leakage through upper Z faces is handled analogously to a river (D.5.5. River Boundary). For a vertical stack of cells, leakage is applied to the cell that contains the water table. The same limitation for flux from a river boundary to the aquifer is applied to the leakage from the leaky boundary to the aquifer. An elevation of the top of the leaky bed is defined (**-elevation**) along with a thickness of the bed (**-thickness**), from which the bottom of the leaky bed is calculated. The maximum flux from the leaky boundary to the aquifer occurs when the head in the aquifer is at the elevation of the bottom of the leaky bed. The leakage does not increase as the head in the aquifer drops below the bottom of the leaky bed. The units of leaky-bed elevation (**-elevation**) are either grid or map units as defined by the **-z_coordinate_system** for the leaky-boundary-condition zone. By default, elevations are assumed to be in grid units. The identifiers **-elevation** and **-z_coordinate_system** are only used when flow is unconfined and leakage is through the upper Z face of a cell; under all other conditions, these identifiers need not be defined.

Leakage at a leaky-boundary-condition cell may be into or out of the active grid region and depends on the relationship between the head defined for the leaky boundary condition and the head in the cell. If the head in the cell is higher than the head specified for the leaky boundary condition, leakage is out of the cell; if the head in the cell is lower, leakage is into the cell. For solute transport simulations, if leakage is out of the cell, the water composition of the outflow is equal to the composition in the cell; if leakage is into the cell, the water composition of the inflow is equal to the composition of the solution defined with the **-solution** identifier.

Multiple zones may be used within any **LEAKY_BC** data block to define leaky boundary conditions for selected exterior cell faces within the grid region. Different boundary conditions for a single cell and its cell faces may be defined multiple times as part of different zone definitions and different keyword data blocks. The last boundary condition defined for a cell determines the type of boundary conditions that apply to the cell. If the last boundary condition defined is a specified-head boundary, then the cell is exclusively a specified-head boundary. If the last definition is flux or leaky, then the cell may have a combination of flux and leaky boundary conditions. If multiple leaky boundary conditions are defined for a cell, the areas of application are revised so that no areas overlap and later-defined conditions take precedence over earlier definitions. Similarly, areas of application for flux boundaries are revised to avoid overlaps. However, no attempt is made to avoid overlaps between leaky and flux boundaries. See section D.5.8. Boundary-Condition Compatibility for more details. More discussion on leaky boundary conditions is found in section D.5.4. Leaky Boundary.

Example Problems

The **LEAKY_BC** data block is used in Chapter 6 example problem **4**.

MEDIA

This keyword data block is used to define media properties, including hydraulic conductivities, porosities, specific storages, and dispersivities. These spatial properties are applied to elements, not cells (see sections 4.2. Spatial Data and D.1.2. Spatial Discretization). The **-shell** identifier is used to define quasi-two-dimensional features by selecting a set of elements that surrounds a zone definition. This feature could be used to define the hydraulic conductivity of a fault or lake bottom. The **MEDIA** data block is mandatory for all simulations and contains only static data.

Example

Line numbers in the Example are used for identification purposes only and are referenced in the following Explanation. The line numbers do not indicate a required order for the keyword data block. Line numbers are not used in the flow and transport data file.

```
Line 0:   MEDIA
Line 1:        -box           0      0      0      10    10    10
Line 2:            -active    0
Line 3:            -Kx        xyz         map   kx.dat
Line 4:            -Ky        xyz         map   ky.dat
Line 5:            -Kz        xyz         map   kz.dat
Line 6:            -porosity  points      grid
Line 7:                5.0  0.0  0.0   0.2
Line 7a:               5.0  10.0 0.0   0.3
Line 8:            -specific_storage      X     1e-5  0     1e-4   10
Line 9:            -longitudinal_dispersivity      2.0
Line 10:           -horizontal_dispersivity        2.0
Line 11:           -vertical_dispersivity          0.2
Line 1a:       -prism
Line 12:           -perimeter  shape       map   pond_edge.shp
Line 13:           -bottom     shape       map   pond_bottom.shp   5
Line 14:           -shell      10     10    1
Line 3a:           -kx         1e-4
Line 4a:           -ky         1e-4
Line 5a:           -kz         1e-4
```

Explanation

Line 0: **MEDIA**

> **MEDIA** is the keyword for the data block; no other data are included on this line.

Line 1: *zone*

> *zone*—A zone definition as defined in section 4.5. Description of Input for Zones. Line 1 defines a box zone; lines 1a, 12, and 13 define a prism zone.

Line 2: **-active** *property*

> **-active**—This identifier is used to specify elements within the zone to be active or inactive. Optionally, **active** or **-a[ctive]**.

> *property*—Property input as defined in section 4.6. Description of Input for Properties. Property values are indicators of whether elements are active or inactive. If the interpolated value for an element centroid is less than or equal to 0, the element is inactive; if the interpolated value is greater than zero, the element is active. Line 2 defines all elements with centroids

inside the zone to be inactive. Note that the other parameters defined for the zone are applied to the elements and will be in effect if any elements are defined later to be active.

Line 3: **-Kx** *property*

-Kx—This identifier is used to specify the hydraulic conductivity in the X direction for the elements in the zone. Optionally, **Kx**, **Kxx**, or **-Kx[x]**.

property—Property input as defined in section 4.6. Description of Input for Properties. Property values are hydraulic conductivities in the X direction. Line 3 specifies that hydraulic conductivities at scattered points are read from a file and interpolated for elements at the element centroids. Units, L/T, are defined by the **-hydraulic_conductivity** identifier in the **UNITS** data block.

Line 4: **-Ky** *property*

-Ky—This identifier is used to specify the hydraulic conductivity in the Y direction for the elements in the zone. Optionally, **Ky**, **Kyy**, or **-Ky[y]**.

property—Property input as defined in section 4.6. Description of Input for Properties. Property values are hydraulic conductivities in the Y direction. Line 4 specifies that hydraulic conductivities at scattered points are read from a file and interpolated for elements at the element centroids. Units, L/T, are defined by the **-hydraulic_conductivity** identifier in the **UNITS** data block.

Line 5: **-Kz** *property*

-Kz—This identifier is used to specify the hydraulic conductivity in the Z direction for the elements in the zone. Optionally, **Kz**, **Kzz**, or **-Kz[z]**.

property—Property input as defined in section 4.6. Description of Input for Properties. Property values are hydraulic conductivities in the Z direction. Line 5 specifies that hydraulic conductivities at scattered points are read from a file and interpolated for elements at the element centroids. Units, L/T, are defined by the **-hydraulic_conductivity** identifier in the **UNITS** data block.

Line 6: **-porosity** *property*

-porosity—This identifier is used to specify the porosity for the elements in the zone. Optionally, **porosity** or **-p[orosity]**.

property—Property input as defined in section 4.6. Description of Input for Properties. Property values are porosities for the zone. Lines 6, 7, and 7a specify that porosities at two scattered points are read; porosities for elements are interpolated at the element centroids by closest point interpolation. Porosity is dimensionless.

Line 7: *x, y, z, value*

x—X coordinate of a point.

y—Y coordinate of a point.

z—Z coordinate of a point.

value—Porosity at the specified point.

Line 8: **-specific_storage** *property*

-specific_storage—This identifier is used to specify the specific storage for the elements in the zone. Optionally, **specific_storage**, **storage**, **-s[pecific_storage]**, or **-s[torage]**.

property—Property input as defined in section 4.6. Description of Input for Properties. Property values are specific storages for the zone. Line 8 specifies a linearly varying distribution of specific storage in the X direction. Units, 1/L, are defined by the **-specific_storage** identifier in the **UNITS** data block.

Line 9: **-longitudinal_dispersivity** *property*

-longitudinal_dispersivity—This identifier is used to specify the dispersivity in the direction of the flow velocity for the elements in the zone. Optionally, **longitudinal_dispersivity**, **dispersivity_longitudinal**, **long_dispersivity**, or **-l[ongitudinal_dispersivity]**, **-d[ispersivity_longitudinal]**, **-l[ong_dispersivity]**.

property—Property input as defined in section 4.6. Description of Input for Properties. Property values are dispersivities in the direction of the velocity for elements in the zone. Line 9 specifies a constant longitudinal dispersivity for the zone. Units, L, are defined by the **-dispersivity** identifier in the **UNITS** data block.

Line 10: **-horizontal_dispersivity** *property*

-horizontal_dispersivity—This identifier is used to specify the horizontal transverse (to the velocity) dispersivity for the elements in the zone. Optionally, **horizontal_dispersivity**, **dispersivity_horizontal**, **-h[orizontal_dispersivity]**, or **-dispersivity_h[orizontal]**.

property—Property input as defined in section 4.6. Description of Input for Properties. Property values are horizontal transverse dispersivities for elements in the zone. Line 10 specifies a constant horizontal transverse dispersivity. Units, L, are defined by the **-dispersivity** identifier in the **UNITS** data block.

Line 11: **-vertical_dispersivity** *property*

-vertical_dispersivity—This identifier is used to specify the vertical transverse (to the velocity) dispersivity for the elements in the zone. Optionally, **vertical_dispersivity**, **dispersivity_vertical**, **-v[ertical_dispersivity]**, or **-dispersivity_v[ertical]**.

property—Property input as defined in section 4.6. Description of Input for Properties. Property values are vertical transverse dispersivities for elements in the zone. Line 11 specifies a constant vertical transverse dispersivity. Units, L, are defined by the **-dispersivity** identifier in the **UNITS** data block.

Line 12: *zone continued*

zone continued—Perimeter of the prism zone (see 4.5. Description of Input for Zones).

Line 13: *zone continued*

zone continued—Bottom surface of the prism zone (see 4.5. Description of Input for Zones).

Line 14: **-shell** *x_width, y_width, z_width*

-shell—This identifier modifies the selection of elements that are assigned the porous-media properties to a set of elements that surrounds the zone. First, all nodes are selected that are outside the zone but are adjacent to a node inside the zone. Then, all elements that touch a selected node are assigned the porous-media properties. Except at edges of the grid, the shell is at least two elements wide. Additional elements with centroids within given distances of one of the selected nodes may be included by defining the *x_width, y_width*, and *z_width* parameters. Optionally, **shell** or **-she[ll]**.

x_width, y_width, z_width—Any element is included in the shell if its centroid is within a box with dimensions *x_width, y_width*, and *z_width* in the X, Y, and Z coordinate directions that is centered on a shell node. Elements touching a shell node are included in the shell regardless of the values of *x_width, y_width*, and *z_width*. Units, L, are defined by the **-horizontal_grid** (X and Y) and the **-vertical_grid** (Z) identifiers in the **UNITS** data block. Default 0, 0, 0.

Notes

Media properties are defined by element, not by cell. The number of elements in each coordinate direction is one fewer than the number of nodes. Multiple zones may be used within the **MEDIA** data block to define media properties within the grid region. Different media properties for a single element may be defined multiple times as part of different, overlapping zone definitions. The individual media property that is used for an element is the last definition that defines that media property for that element centroid.

By definition of active and inactive zones, it is possible to model a configuration other than the entire rectangular grid region that is defined by the **GRID** data block. Elements are specified to be inactive by the use of the **-active** identifier in the **MEDIA** data block. Elements specified to be inactive are logically removed from the grid region. The parts of cells that are contained in inactive elements are not included in the active grid region. If all elements that join at a node are inactive, then that node is inactive and is removed from the simulation. The degree to which the active grid region conforms to a desired three-dimensional shape is limited by the spatial discretization of the grid and the effort necessary to define the shape accurately. Use of prism zones defined with ArcInfo shapefiles (ESRI, 1998) or XYZ files simplifies the process of adapting grids to conform to complicated domain shapes.

Thin zones of media properties may be difficult to define in PHAST, especially when redefining grids. Fixed zones may include element centroids for one grid arrangement, but not for another. Also, the properties at nodes are averaged from the properties of the elements that meet at the node, which means that, to ensure a specific property value at a node not on a boundary, it is necessary to assign that property to all of the ele-

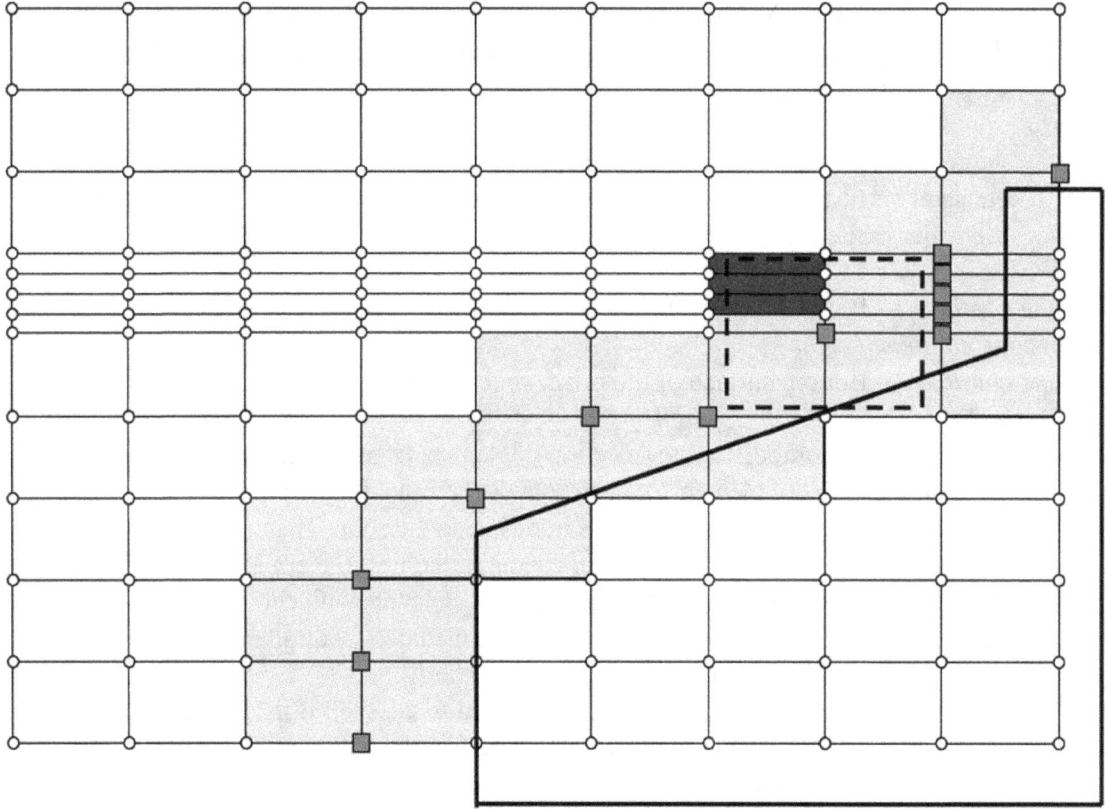

Figure 4.8. Two-dimensional example of the selection of elements using a shell definition. The zone is defined as the polygonal area enclosed by the solid black line. Small circles represent nodes of the grid. Squares are shell nodes that are adjacent to nodes inside the zone definition. All elements containing a square node are selected to be in the shell (light shaded elements). Elements with element centroids within a specified distance of a square node, represented by the dashed rectangle, are also selected to be in the shell (dark shaded elements).

ments that meet at the node. The **-shell** identifier, which applies only in the **MEDIA** data block, is an attempt to simplify definitions of media properties, particularly for thin zones. The **-shell** identifier is used to select elements that are near the boundary of a zone, rather than elements that are internal to the zone. A shell is defined as follows (fig. 4.8): first, all nodes are selected that are outside the zone but are adjacent to nodes inside the zone. All elements that touch this set of nodes are included in the shell. In addition, the shell can be widened to include all elements with centroids that are within specified distances from the set of nodes (*x_width*, *y_width*, and *z_width*).

To allow for the definition of an active grid region with a regional topography of land surface, inactive elements are allowed above active elements. That is, the potentially saturated region of an unconfined flow simulation does not have to possess a flat upper surface. Inactive elements may not be defined for one-dimensional simulations (**GRID** data block, **-chemistry_dimensions**). If unconfined flow is simulated (**FREE_SURFACE_BC** data block), the specific storage is reset to zero (except for the cells containing the water table, where the specific storage is equal to the porosity) regardless of the value entered in the **MEDIA** data block.

Dispersivity is a mixing parameter with a magnitude that is dependent on the scale of observation distance over which solutes are transported in the flow system. A review of field data by Gelhar and others (1992) shows that, in the absence of more definitive information, the longitudinal dispersivity (**-longitidinal_dispersivity**) can be estimated to be on the order of 10 percent of the longitudinal transport distance, whereas horizontal transverse dispersivity (**-horizontal_dispersivity**) is on the order of 1 percent of the longitudinal transport distance, and vertical transverse dispersivity (**-vertical_dispersivity**) is on the order of 0.001 to 0.1 percent of the longitudinal transport distance.

Example Problems

The **MEDIA** data block is used in Chapter 6 example problems **1**, **2**, **3**, **4**, **5**, and **6**.

PRINT_FREQUENCY

This keyword data block is used to select the results to be written to output files and the frequencies at which the results are written. This data block and all identifiers are optional. The data for time "0 days" in the example below indicates the default settings. The effect of the default settings is to write data at the end of each simulation period for all identifiers with "end" or nonzero values, and not to write data for identifiers with values of zero or "false". Explicit definition of an identifier in this data block will override the default setting. The times specified in this data block indicate times during the simulation when print frequencies will be changed. Only print frequencies explicitly specified following a time specification are changed. A print frequency, default or explicitly specified, remains in effect during a simulation until a specified time when the frequency for the identifier is explicitly changed.

Example

Line numbers in the Example are used for identification purposes only and are referenced in the following Explanation. The line numbers do not indicate a required order for the keyword data block. Line numbers are not used in the flow and transport data file.

```
Line 0:     PRINT_FREQUENCY
Line 1:     0 days
Line 2:         -bc_flow_rates            0
Line 3:         -boundary_conditions      False
Line 4:         -components               0
Line 5:         -conductances             0
Line 6:         -end_of_period_default    True
Line 7:         -flow_balance             1       step
Line 8:         -force_chemistry_print    0
Line 9:         -HDF_chemistry            1       day
Line 10:        -HDF_heads                1       yr
Line 11:        -HDF_velocities           1       yr
Line 12:        -heads                    end
Line 13:        -progress_statistics      1       step
Line 14:        -restart_file             10      yr
Line 15:        -save_final_heads         False
Line 16:        -velocities               0
Line 17:        -wells                    end
Line 18:        -xyz_chemistry            0
Line 19:        -xyz_components           0
Line 20:        -xyz_heads                0
Line 21:        -xyz_velocities           0
Line 22:        -xyz_wells                0
Line 23:        -zone_flow                end
Line 24:        -zone_flow_heads          1       step
Line 25:        -zone_flow_tsv            end
Line 1a:    1 yr
Line 19a:       -xyz_chemistry            0.1     yr
```

Explanation

Line 0: **PRINT_FREQUENCY**

PRINT_FREQUENCY is the keyword for the data block; no other data are included on this line.

Line 1: *time* [*units*]

> *time*—Simulation time at which to change print frequencies. Any print frequencies defined in lines following this time specification, and preceding the next time specification or the end of the data block, will become effective at this time during the simulation. If any identifiers are included, this time will mark the beginning of a simulation period.

> *units*—*Units* can be "seconds", "minutes", "hours", "days", or "years" or an abbreviation of one of these units. Default units are defined by **-time** identifier in UNITS data block.

Line 2: **-bc_flow_rates** *frequency* [(*time unit* | **step**)]

> **-bc_flow_rates**—This identifier is used to control writing flow rates for all boundary-condition cells to the file *prefix***.bcf.txt**. By the default setting, no flow rates are written to the file *prefix***.bcf.txt**. Optionally, **bc_flow_rates** or **-bc_[flow_rates]**.

> *frequency*—Frequency at which data are written. Frequency can be an interval of time or a number of time steps between writing results to the file, or the string "end", which will cause writing at the end of each simulation period. If *frequency* is zero, no transient data will be written to the file.

> (*time unit* | **step**)—*Time unit* can be "seconds", "minutes", "hours", "days", or "years" or an abbreviation of one of these units. Alternatively, if **step** is entered, *frequency* specifies the number of time steps between writing results. Default units are defined by **-time** identifier in UNITS data block.

Line 3: **-boundary_conditions** [(**True** | **False**)]

> **-boundary_conditions**—This identifier is used to control writing heads, component concentrations, fluxes, and other boundary-condition information to the file *prefix***.probdef.txt**. These data are written at most once per simulation period. By the default setting, no boundary-condition data are written to the file *prefix***.probdef.txt**. Optionally, **boundary_conditions**, **boundary**, **bc**, **-b[c]**, or **-b[oundary_conditions]**.

> (**True** | **False**)—**True** causes writing of boundary-condition information to the file *prefix***.probdef.txt**. **False** suppresses writing. If neither **true** nor **false** is entered on the line, **true** is assumed. Optionally, **t[rue]** or **f[alse]**.

Line 4: **-components** *frequency* [(*time unit* | **step**)]

> **-components**—This identifier is used to control writing total chemical element (component) data for each cell to the file *prefix***.comps.txt**. By the default setting, no component data are written to the file *prefix***.comps.txt**. Optionally, **component**, **components**, or **-com[ponents]**.

> *frequency*—Same as Line 2.

> (*time unit* | **step**)—Same as Line 2.

Line 5: **-conductances** *frequency* [(*time unit* | **step**)]

> **-conductances**—This identifier is used to specify the frequency for writing static fluid and transient solute dispersive conductances for each cell face to the file *prefix***.kd.txt**. By the default setting, no static or transient conductances are written to the file *prefix***.kd.txt**. Optionally, **conductances**, **conductance**, or **-cond[uctances]**.

> *frequency*—Same as Line 2.

> (*time unit* | **step**)—Same as Line 2.

Line 6: **-end_of_period_default** [(**True** | **False**)]

-end_of_period_default—This identifier is used to control automatic writing of some data at the end of simulation periods. By default, data are written at the end of a simulation period for **flow_balance**, **HDF_chemistry**, **HDF_heads**, **HDF_velocities**, **heads**, **progress_statistics**, **wells**, and **zone_flow**. If the value for this identifier is false, then data will not be written to files for these options at the end of a simulation period. For example, it could be useful to omit the default printing at the end of each simulation period if the run includes hundreds of simulation periods. Optionally, **end_of_period_default** or **-en[d_of_period_default**.

(**True** | **False**)—**True** causes automatic writing of data at the end of simulation periods. **False** suppresses writing. If neither **true** nor **false** is entered on the line, **true** is assumed. Optionally, **t[rue]** or **f[alse]**.

Line 7: **-flow_balance** *frequency* [(*time unit* | **step**)]

-flow_balance—This identifier is used to specify the frequency for writing flow-balance information to the file *prefix***.bal.txt**. By the default setting, flow-balance information is written to the file *prefix***.bal.txt** at the end of each simulation period. Optionally, **flow_balance** or **-f[low_balance]**.

frequency—Same as Line 2.

(*time unit* | **step**)—Same as Line 2.

Line 8: **-force_chemistry_print** *frequency* [(*time unit* | **step**)]

-force_chemistry_print—This identifier is used to specify the frequency for writing detailed chemical descriptions of the composition of the solution and all reactants for each cell to the file *prefix***.chem.txt**. *Warning*: this file could exceed file-size limits of a 32-bit operating system because a long description of the chemistry in each cell is written for each selected time step. Writing this information may be useful for debugging, for small problems, or if the PRINT_LOCATIONS data block is used to limit the set of cells for which data are written. Data written to the file *prefix***.chem.txt** also can be limited by the options of the **PRINT** data block of the chemistry data file. By the default setting, no detailed chemical descriptions are written to the file *prefix***.chem.txt**. Optionally, **force_chemistry**, **force_chemistry_print**, or **-fo[rce_chemistry_print]**.

frequency—Same as Line 2.

(*time unit* | **step**)—Same as Line 2.

Line 9: **-HDF_chemistry** *frequency* [(*time unit* | **step**)]

-HDF_chemistry—This identifier is used to specify the frequency for writing chemistry data to the file *prefix***.h5**. Chemistry data to be written to the file *prefix***.h5** are defined in the **SELECTED_OUTPUT** and **USER_PUNCH** data blocks of the chemistry data file. By the default setting, chemistry data are written to the file *prefix***.h5** at the end of each simulation period. Optionally, **HDF_chemistry**, **HDF_concentration**, **HDF_concentrations**, **-HDF_c[oncentrations]**, or **-HDF_c[hemistry]**.

frequency—Same as Line 2.

(*time unit* | **step**)—Same as Line 2.

Line 10: **-HDF_heads** *frequency* [(*time unit* | **step**)]

-HDF_heads—This identifier is used to specify the frequency for writing heads to the file *prefix***.h5**. By the default setting, heads are written to the file *prefix***.h5** at the end of each simulation period. Optionally, **HDF_head**, **HDF_heads**, or **-HDF_h[eads]**.

frequency—Same as Line 2.

(*time unit* | **step**)—Same as Line 2.

Line 11: **-HDF_velocities** *frequency* [(*time unit* | **step**)]

> **-HDF_velocities**—This identifier is used to specify the frequency for writing X, Y, and Z velocities to the file *prefix*.**h5**. By the default setting, velocities are written to the file *prefix*.**h5** at the end of each simulation period. Optionally, **HDF_velocity**, **HDF_velocities**, **-HDF_v**[**elocities**], or **-HDF_v**[**elocity**].

> *frequency*—Same as Line 2.

> (*time unit* | **step**)—Same as Line 2.

Line 12: **-heads** *frequency* [(*time unit* | **step**)]

> **-heads**—This identifier is used to specify the frequency for writing potentiometric heads to the file *prefix*.**head.txt** and, for free-surface calculations, *prefix*.**wt.txt** (water-table file). By the default setting, heads are written to the file *prefix*.**head.txt** and *prefix*.**wt.txt** at the end of each simulation period. Data will be written to *prefix*.**wt.txt** only if a free-surface boundary condition is defined (FREE_SURFACE_BC). Optionally, **head**, **heads**, or **-h**[**eads**].

> *frequency*—Same as Line 2.

> (*time unit* | **step**)—Same as Line 2.

Line 13: **-progress_statistics** *frequency* [(*time unit* | **step**)]

> **-progress_statistics**—This identifier is used to specify the frequency for writing solver statistics, including solution-method information, number of iterations, and maximum changes in head and component concentrations (due to transport), to the file *prefix*.**log.txt** and to the screen. By the default setting, solver statistics are written to the file *prefix*.**log.txt** and screen at the end of each simulation period. Optionally, **progress_statistics**, **-pr**[**ogress_statistics**], **solver_statistics** or **-solv**[**er_statistics**].

> *frequency*—Same as Line 2.

> (*time unit* | **step**)—Same as Line 2.

Line 14: **-restart_file** *frequency* [(*time unit* | **step**)]

> **-restart_file**—This identifier is used to specify the frequency for writing all chemical compositions to the file *prefix*.**restart.tgz**. This file can be used to define starting conditions in subsequent runs. Data for only the most recent time of writing are stored in the file. When data are written, the previous version of *prefix*.**restart.tgz** is renamed to *prefix*.**restart.backup.tgz** and the current data are written to *prefix*.**restart.tgz**. The default *frequency* is 0 if no **-restart_file** identifier is defined; that is, no restart file is written. Line 10 specifies that the restart file will be written after every 10 years of simulation time. Optionally, **restart**, **restart_file**, or **-r**[**estart_file**].

> *frequency*—Same as Line 2.

> (*time unit* | **step**)—Same as Line 2.

Line 15: **-save_final_heads** [(**True** | **False**)]

> **-save_final_heads**—This identifier is used to control writing heads to the file *prefix*.**head.dat** at the end of the simulation. The file *prefix*.**head.dat** is an ASCII file than can be used for initial head conditions in subsequent simulations. Initial heads can be read from the file by using a zone that includes the entire grid region and "**-head file** *prefix*.**head.dat**" in HEAD_IC data block. By the default setting, no head data are written to the file *prefix*.**head.dat**. Optionally, **save_head**, **save_heads**, **save_final_heads**, **-sa**[**ve_heads**], or **-sa**[**ve_final_heads**].

(**True** | **False**)—**True** writes heads at the end of the simulation to the file *prefix*.**head.dat**; **false** suppresses writing. If neither **true** nor **false** is entered on the line, **true** is assumed. Optionally, **t**[**rue**] or **f**[**alse**].

Line 16: **-velocities** *frequency* [(*time unit* | **step**)]

> **-velocities**—This identifier is used to specify the frequency for writing interstitial velocities at cell boundaries and interpolated velocities at nodes to the file *prefix*.**vel.txt**. By the default setting, no velocities are written to the file *prefix*.**vel.txt**. Optionally, **velocities**, **velocity** **-v**[**elocities**], or **-v**[**elocity**].

> *frequency*—Same as Line 2.

> (*time unit* | **step**)—Same as Line 2.

Line 17: **-wells** *frequency* [(*time unit* | **step**)]

> **-wells**—This identifier is used to specify the frequency for writing transient well information, including fluid and solute flow rates, cumulative fluid and solute flow amounts, and solute concentrations, to the file *prefix*.**wel.txt**. Data are written in the order of the well sequence numbers. By the default setting, well information is written to the file *prefix*.**wel.txt** at the end of each simulation period. Optionally, **wells** or **-w**[**ells**].

> *frequency*—Same as Line 2.

> (*time unit* | **step**)—Same as Line 2.

Line 18: **-xyz_chemistry** *frequency* [(*time unit* | **step**)]

> **-xyz_chemistry**—This identifier is used to specify the frequency for writing selected chemical data to the file *prefix*.**chem.xyz.tsv**. The **SELECTED_OUTPUT** and **USER_PUNCH** data blocks of the chemistry data file are used to select data that are written to the file *prefix*.**chem.xyz.tsv**. Cells for which results are to be written can be restricted with the **PRINT_LOCATIONS** data block. By the default setting, no data are written to the file *prefix*.**chem.xyz.tsv**. Optionally, **concentrations**, **selected_output**, **selected_outputs**, **xyz_chemistry**, **-c**[**oncentrations**], **-se**[**lected_outputs**], or **-xyz_ch**[**emistry**].

> *frequency*—Same as Line 2.

> (*time unit* | **step**)—Same as Line 2.

Line 19: **-xyz_components** *frequency* [(*time unit* | **step**)]

> **-xyz_components**—This identifier is used to specify the frequency for writing component (chemical element) concentrations to the file *prefix*.**comps.xyz.tsv**. By the default setting, no component concentrations are written to the file *prefix*.**comps.xyz.tsv**. Optionally, **xyz_component**, **xyz_components**, or **-xyz_c**[**omponents**].

> *frequency*—Same as Line 2.

> (*time unit* | **step**)—Same as Line 2.

Line 20: **-xyz_heads** *frequency* [(*time unit* | **step**)]

> **-xyz_heads**—This identifier is used to specify the frequency for writing heads to the files *prefix*.**head.xyz.tsv** and *prefix*.**wt.xyz.tsv** (water-table file). By the default setting, no heads are written to the files *prefix*.**head.xyz.tsv** and *prefix*.**wt.xyz.tsv**. Data will be written to *prefix*.**wt.xyz.tsv** only if a free-surface boundary condition is defined (**FREE_SURFACE_BC**). Optionally, **xyz_head**, **map_head**, **-xyz_h**[**ead**], or **-map_h**[**ead**].

> *frequency*—Same as Line 2.

> (*time unit* | **step**)—Same as Line 2.

Line 21: **-xyz_velocities** *frequency* [(*time unit* | **step**)]

-**xyz_velocities**—This identifier is used to specify the frequency for writing interpolated velocities at cell nodes to the file *prefix*.**vel.xyz.tsv**. By the default setting, no velocities are written to the file *prefix*.**vel.xyz.tsv**. Optionally, **xyz_velocity**, **map_velocity**, -**xyz_v**[**elocity**], or -**map_v**[**elocity**].

frequency—Same as Line 2.

(*time unit* | **step**)—Same as Line 2.

Line 22: -**xyz_wells** *frequency* [(*time unit* | **step**)]

-**xyz_wells**—This identifier is used to specify the frequency for writing a time-series of concentrations for each well to the file *prefix*.**wel.xyz.tsv**. By the default setting, no concentrations are written to the file *prefix*.**wel.xyz.tsv**. Optionally, **xyz_well**, **xyz_wells**, **well_time_series**, -**xyz_w**[**ells**], -**well_**[**time_series**], **wells_time_series**, or -**wells_**[**time_series**].

frequency—Same as Line 2.

(*time unit* | **step**)—Same as Line 2.

Line 23: -**zone_flow** *frequency* [(*time unit* | **step**)]

-**zone_flow**—This identifier is used to specify the frequency for zone flow information to the file *prefix*.**zf.txt**, including water and solute flow into and out of the zone through the boundaries of each zone and through any boundary-condition cells within a zone. Zones for calculations of zone flows are defined with the ZONE_FLOW data block. By the default setting, zone flow information is written to the file *prefix*.**zf.txt** at the end of each simulation period. Optionally, **zone_flow**, **zone_flows**, **zone_flow_rates**, -**z**[**one_flow_rates**], or -**z**[**one_flows**].

frequency—Same as Line 2.

(*time unit* | **step**)—Same as Line 2.

Line 24: -**zone_flow_heads** *frequency* [(*time unit* | **step**)]

-**zone_flow_heads**—This identifier is used to specify the frequency for writing heads for zones defined in ZONE_FLOW data blocks. Data will be written for those zones defined with the ZONE_FLOW data block that have a file name defined with the -**write_heads_xyzt** identifier. The file will contain lines with X, Y, Z, time, and head for each print interval, where X, Y, Z ranges over all of the nodes in the zone. By the default setting, heads are written at the end of each simulation period. Optionally, **zone_flow_heads**, or -**zone_flow_h**[**eads**].

frequency—Same as Line 2.

(*time unit* | **step**)—Same as Line 2.

Line 25: -**zone_flow_tsv** *frequency* [(*time unit* | **step**)]

-**zone_flow_tsv**—This identifier is used to specify the frequency for zone flow information to the file *prefix*.**zf.tsv**, including water and solute flows into and out of the zone through the boundaries of each zone and through any boundary-condition cells within a zone. Zones for calculations of zone flows are defined with the ZONE_FLOW data block. By the default setting, zone flow information is written to the file *prefix*.**zf.tsv** at the end of each simulation period. Optionally, **zone_flow_tsv**, **tsv_zone_flow**, **tsv_zone_flows**, **tsv_zone_flow_rates**, -**zone_flow_t**[**sv**], -**tzv_z**[**one_flows**], or -**tsv_z**[**one_flow_rates**].

frequency—Same as Line 2.

(*time unit* | **step**)—Same as Line 2.

Notes

The **PRINT_FREQUENCY** data block controls writing of data during the simulation (times greater than zero), but it does not control printing data at the beginning of the simulation (time zero). The **PRINT_INITIAL** data block is used to control writing of information at initialization to output files, including initial conditions and media properties. Print frequencies may be changed at any time in the simulation by specifying the time at which the frequencies are to change (*time*), followed by the identifiers and frequencies that are to change. A new simulation period is begun for every time (*time*) that is specified.

User-specified units for the various print frequencies need not be the same. One **PRINT_FREQUENCY** data block may contain multiple time units and (or) the **step** unit. If time units are used for a print frequency and the simulation time does not fall on an even multiple of the print frequency, when the simulation time is within one time step of an even multiple of the print frequency, the time step will be decreased to reach the target simulation time at which printing is to occur. The time step will revert to the original time step until a smaller time step is needed to reach another target time for printing or the end of a simulation period.

A value of 0 (zero) for *frequency* suppresses writing of the specified data. By the default settings, print frequencies are set to "end" so that writing will occur at the end of each simulation period, except for frequencies defined by **-components**, **-conductances**, **-force_chemistry_print**, **-xyz_components**, **-xyz_heads**, **-xyz_velocities**, **-wells**, and **-xyz_wells**, which are set to 0 to suppress writing. Once set by default, or explicitly following a specified *time*, a print frequency remains in effect until the next *time* for which the print frequency is explicitly defined.

When using **steps** for a print frequency, writing to the files is determined by the number of time steps since the beginning of the simulation, not from the beginning of the current simulation period. For example, if a print frequency is specified to be every two steps beginning at time 0, every three steps beginning at time 5, and time step (**TIME_CONTROL**) is 1, the printing will occur at time steps 2 and 4 (evenly divisible by 2), 5 (last time step of the first simulation period), 6 and 9 (evenly divisible by 3), and 10 (last time step of the second simulation period). Similarly, when a frequency is specified in time units, printing will occur when the total time from the beginning of the simulation is evenly divisible by the specified frequency and at the end of each simulation period.

By default, if data are to be written at any nonzero frequency, they are written at the specified frequency and, in addition, at the end of each simulation period. It may be desirable to suppress printing at the end of each simulation period if a run contains many simulation periods. The **end_of_period_default** identifier can be used to suppress automatic printing at the end of time periods, while printing at the specified frequency remains unaffected.

When steady-state flow is simulated (**STEADY_FLOW true**), nonzero print frequencies for the identifiers **-head** (*prefix*.**head.txt** and *prefix*.**wt.txt**), **-flow_balance** (*prefix*.**bal.txt**), **-velocity** (*prefix*.**vel.txt**), **-HDF_heads** (*prefix*.**h5**), **-HDF_velocities** (*prefix*.**hdf.vel**), **-xyz_heads** (*prefix*.**head.xyz.tsv** and *prefix*.**wt.xyz.tsv**), and **-xyz_velocities** (*prefix*.**vel.xyz.tsv**) are treated in a special way. During the iterations to achieve steady-state flow, heads are written to the *prefix*.**head.txt** and *prefix*.**wt.xyz.tsv** file for every steady-state iteration; no data are written to these files during the transient part of the simulation. During the iterations to achieve steady-state flow, balances are written to the *prefix*.**bal.txt** file for every steady-state iteration; during the transient part of the simulation, the print frequency defined by **-flow_balance** is used to determine when data are written to the *prefix*.**bal.txt** file. For steady-flow simulations, velocities are written to the *prefix*.**vel.txt**, *prefix*.**h5**, and *prefix*.**vel.xyz.tsv** files only once if the *frequency* for the identifier corresponding to the file is nonzero for any simulation period within the flow and transport data file. Similarly, heads are written to their respective files only once. Printing of steady-flow heads and velocities also can be requested in the **PRINT_INITIAL** data block.

The files with names ending in **.txt** are data formatted to be printed or viewed on a screen. The orientation of the printout of the spatially distributed properties in the **.txt** files, either X–Y or X–Z planes, is controlled by the **-print_orientation** identifier in the **GRID** data block. The files *prefix*.**chem.xyz.tsv**, *prefix*.**comps.xyz.tsv**, *prefix*.**head.xyz.tsv**, *prefix*.**wt.xyz.tsv**, *prefix*.**vel.xyz.tsv**, *prefix*.**wel.xyz.tsv**, and *prefix*.**zf.tsv** are written as tab-separated values to facilitate importing into spreadsheets and postprocessing programs for graphical display.

The restart file, written at a frequency defined by the **-restart_file** option, contains a complete definition of the chemical compositions in each cell (solution, equilibrium phases, exchangers, surfaces, kinetic reactants, solid solutions, and gas phase). This file can be used to define initial conditions for subsequent PHAST runs by use of the **restart** option in CHEMISTRY_IC data block. The file is written in Gnu zip format (extension **.gz**), which is a binary, compressed format. The file can be converted to ASCII text by using the *gunzip* command. The zipped or unzipped format can be read with the **restart** option in CHEMISTRY_IC.

Example Problems

The **PRINT_FREQUENCY** data block is used in Chapter 6 example problems 1, 2, 3, 4, 5, and 6.

PRINT_INITIAL

This keyword data block is used to print the initial and static flow and transport data to various output files. The output controlled by this data block is useful for verifying that media properties, and initial and boundary conditions have been defined correctly. The example below indicates the default settings for each of the identifiers. The effect of the default settings is not to write data for identifiers with values of "false" and to write data for identifiers with values of "true". Options in this data block apply only to initial conditions and have no effect on printing data after transient time stepping begins.

Example

Line numbers in the Example are used for identification purposes only and are referenced in the following Explanation. The line numbers do not indicate a required order for the keyword data block. Line numbers are not used in the flow and transport data file.

```
Line 0:    PRINT_INITIAL
Line 1:         -boundary_conditions        false
Line 2:         -components                 false
Line 3:         -conductances               false
Line 4:         -echo_input                 true
Line 5:         -fluid_properties           true
Line 6:         -force_chemistry_print      false
Line 7:         -HDF_chemistry              true
Line 8:         -HDF_heads                  true
Line 9:         -HDF_media_properties       true
Line 10:        -HDF_steady_flow_velocities true
Line 11:        -heads                      true
Line 12:        -media_properties           false
Line 13:        -solution_method            true
Line 14:        -steady_flow_velocities     false
Line 15:        -wells                      true
Line 16:        -xyz_chemistry              false
Line 17:        -xyz_components             false
Line 18:        -xyz_heads                  false
Line 19:        -xyz_steady_flow_velocities false
Line 20:        -xyz_wells                  false
```

Explanation

Line 0: **PRINT_INITIAL**

   PRINT_INITIAL is the keyword for the data block; no other data are included on this line.

Line 1: -**boundary_conditions** [(**True** | **False**)]

   -**boundary_conditions**—This identifier is used to control writing initial boundary-condition information to the file *prefix*.**probdef.txt**, including data for specified-head, flux, leaky, river, drain, and well boundary conditions and the definitions for all solutions related to boundary conditions. By the default setting, no initial boundary-condition information is written to the file *prefix*.**probdef.txt**. Optionally, **boundary_conditions**, **boundary**, **bc**, -**b**[**oundary_conditions**], or -**b**[**c**].

   (**True** | **False**)—**True** writes data to the file. **False** suppresses writing. If neither **true** nor **false** is entered on the line, **true** is assumed. Optionally, **t**[**rue**] or **f**[**alse**].

Line 2: **-components** [(**True** | **False**)]

> **-components**—This identifier is used to control writing initial indices and mixing fractions for solutions, equilibrium phases, exchangers, surfaces, kinetic reactants, solid solutions, and gas phases that define initial conditions for the simulation and initial component concentrations to the file *prefix*.**comps.txt**. By the default setting, no initial indices, mixing fractions, or component concentrations are written to the file *prefix*.**comps.txt**. Optionally, **component**, **components**, or **-c**[**omponents**].

> (**True** | **False**)—See line 1.

Line 3: **-conductances** [(**True** | **False**)]

> **-conductances**—This identifier is used to control writing static fluid-conductance factors and initial-condition conductances to the file *prefix*.**kd.txt**. By the default setting, no conductance factors or initial-condition conductances are written to the file *prefix*.**kd.txt**. Optionally, **conductance**, **conductances**, or **-con**[**ductances**].

> (**True** | **False**)—See line 1.

Line 4: **-echo_input** [(**True** | **False**)]

> **-echo_input**—This identifier is used to control writing of lines from the flow and transport data file to the file *prefix*.**log.txt** as they are processed. The option takes effect as soon as it is encountered in the flow and transport data file. (Writing lines from the chemistry data file to the file *prefix*.**log.txt** is controlled by **-echo_input** in the **PRINT** data block of the chemistry data file.) By the default setting, lines from the flow and transport data file are written to the file *prefix*.**log.txt**. Optionally, **echo_input** or **-e**[**cho_input**].

> (**True** | **False**)—See line 1.

Line 5: **-fluid_properties** [(**True** | **False**)]

> **-fluid_properties**—This identifier is used to control writing fluid property data to the file *prefix*.**probdef.txt**, including compressibility, molecular diffusivity, viscosity, and density. By the default setting, fluid properties are written to the file *prefix*.**probdef.txt**. Optionally, **fluid_properties**, **fluid**, or **-f**[**luid_properties**].

> (**True** | **False**)—See line 1.

Line 6: **-force_chemistry_print** [(**True** | **False**)]

> **-force_chemistry_print**—This identifier is used to control writing detailed chemical descriptions of the composition of the solution and all reactants for each cell to the file *prefix*.**chem.txt**. *Warning*: this file could exceed file-size limits on computers with operating systems using 32-bit addresses because it will produce a long description of the chemistry for each cell. Writing this information may be useful for debugging, for small problems, or if the cells for which writing results are restricted by cell selections made in the PRINT_LOCATIONS data block. Data written to the file can be restricted by options within the **PRINT** data block of the chemistry data file. By the default setting, no chemical descriptions of solution and reactant compositions are written to the file *prefix*.**chem.txt**. Optionally, **force_chemistry**, **force_chemistry_print**, or **-fo**[**rce_chemistry_print**].

> (**True** | **False**)—See line 1.

Line 7: **-HDF_chemistry** [(**True** | **False**)]

> **-HDF_chemistry**—This identifier is used to control writing chemistry data at time zero to the file *prefix*.**h5**. Data to be written are defined in **SELECTED_OUTPUT** and **USER_PUNCH** data blocks of the chemistry data file. By the default setting, chemistry data at time zero are written to the file *prefix*.**h5**. Optionally, **HDF_chemistry**,

HDF_concentration, HDF_concentrations, -HDF_c[hemistry], or
-HDF_c[oncentrations].
(**True** | **False**)—See line 1.
Line 8: -HDF_heads [(True | False)]
-**HDF_heads**—This identifier is used to control writing initial heads to the file *prefix*.**h5**. By the default setting, initial heads are written to the file *prefix*.**h5**. Optionally, **HDF_head**, **HDF_heads**, or -**HDF_h[eads]**.
(**True** | **False**)—See line 1.
Line 9: -HDF_media_properties [(True | False)]
-**HDF_media_properties**—This identifier is used to control writing media properties—hydraulic conductivities, porosity, specific storage, and dispersivities—to the file *prefix*.**h5**. The spatial distribution of these properties can then be checked for accuracy with Model Viewer (see Appendix A). By the default setting, media properties are written to the file *prefix*.**h5**. Optionally, **HDF_media**, **HDF_media_properties**, or -**HDF_m[edia_properties]**.
(**True** | **False**)—See line 1.
Line 10: -HDF_steady_flow_velocities [(True | False)]
-**HDF_steady_flow_velocities**—This identifier is used to control writing steady-flow velocities to the file *prefix*.**h5**. This option has meaning only if steady-state flow is specified in the **STEADY_FLOW** data block. By the default setting, steady-flow velocities are written to the file *prefix*.**h5**. Optionally, **HDF_steady_flow_velocity**, **HDF_steady_flow_velocities**, -**HDF_s[teady_flow_velocity]**, -**HDF_s[teady_flow_velocities]**, **HDF_ss_velocity**, **HDF_ss_velocities**, -**HDF_s[s_velocity]**, or -**HDF_s[s_velocities]**.
(**True** | **False**)—See line 1.
Line 11: -heads [(True | False)]
-**heads**—This identifier is used to control writing initial heads to the file *prefix*.**head.txt**. By the default setting, initial heads are written to the file *prefix*.**head.txt**. Optionally, **head**, **heads**, or -**h[eads]**.
(**True** | **False**)—See line 1.
Line 12: -media_properties [(True | False)]
-**media_properties**—This identifier is used to control writing problem definition data for all the media properties for porous-media zones, including element zone definitions, porosities, hydraulic conductivities, specific storages, and dispersivities to the file *prefix*.**probdef.txt**. By the default setting, no media properties data are written to the file *prefix*.**probdef.txt**. Optionally, **media_properties**, **media**, **medium**, -**m[edia_properties]**, or -**m[edium]**.
(**True** | **False**)—See line 1.
Line 13: -solution_method [(True | False)]
-**solution_method**—This identifier is used to control writing input data related to the solution method for flow and transport equations to the file *prefix*.**probdef.txt**, including type of solver used and relevant solver parameters. By the default setting solution-method data are written to the file *prefix*.**probdef.txt**. Optionally, **solution_method**, **method**, -**s[olution_method]**, or -**met[hod]**.
(**True** | **False**)—See line 1.
Line 14: -steady_flow_velocities [(True | False)]
-**steady_flow_velocities**—This identifier is used to control writing steady-flow velocities to the file *prefix*.**vel.txt**. This option has meaning only if steady-state flow is specified in the

STEADY_FLOW data block. By the default setting, no steady-flow velocities are written to the file *prefix*.**vel.txt**. Optionally, **steady_flow_velocity, steady_flow_velocities, -st[eady_flow_velocity], -st[eady_flow_velocities], ss_velocity, ss_velocities, -ss[_velocity], or -ss[_velocities]**.

(**True | False**)—See line 1.

Line 15: **-wells [(True | False)]**

-wells—This identifier is used to control writing static well information, including location, diameter, screened intervals, and well indices to the file *prefix*.**wel.txt**. By the default setting, static well information is written to the file *prefix*.**wel.txt**. Optionally, **wells**, or **-w[ells]**.

(**True | False**)—See line 1.

Line 16: **-xyz_chemistry [(True | False)]**

-xyz_chemistry—This identifier is used to control writing initial chemistry data to the file *prefix*.**chem.xyz.tsv**. The **SELECTED_OUTPUT** and **USER_PUNCH** data blocks of the chemistry data file are used to select data that are written to the file *prefix*.**chem.xyz.tsv**. Cells for which results are to be written can be restricted with the PRINT_LOCATIONS data block. By the default setting, no initial chemistry data are written to the file *prefix*.**chem.xyz.tsv**. Optionally, **xyz_chemistry** or **-xyz_ch[emistry]**.

(**True | False**)—See line 1.

Line 17: **-xyz_components [(True | False)]**

-xyz_components—This identifier is used to control writing initial dissolved-component concentrations to the file *prefix*.**comps.xyz.tsv**. By the default setting, no initial component concentrations are written to the file *prefix*.**comps.xyz.tsv**. Optionally, **xyz_component, xyz_components**, or **-xyz_c[omponents]**.

(**True | False**)—See line 1.

Line 18: **-xyz_heads [(True | False)]**

-xyz_heads—This identifier is used to control writing initial heads to the file *prefix*.**head.xyz.tsv**. By the default setting, no initial heads are written to the file *prefix*.**head.xyz.tsv**. Optionally, **xyz_head, xyz_heads**, or **-xyz_h[eads]**.

(**True | False**)—See line 1.

Line 19: **-xyz_steady_flow_velocities [(True | False)]**

-xyz_steady_flow_velocities—This identifier is used to control writing velocities from the steady-flow calculation to the file *prefix*.**vel.xyz.tsv**. This option has meaning only if steady-state flow is specified in the STEADY_FLOW data block. By the default setting, no steady-flow velocities are written to the file *prefix*.**vel.xyz.tsv**. Optionally, **xyz_steady_flow_velocity, xyz_steady_flow_velocities, -xyz_s[teady_flow_velocity], -xyz_s[teady_flow_velocities], xyz_ss_velocity, xyz_ss_velocities, -xyz_s[s_velocity]**, or **-xyz_s[s_velocities]**.

(**True | False**)—See line 1.

Line 20: **-xyz_wells [(True | False)]**

-xyz_wells—This identifier is used to control writing initial concentrations at wells to the file *prefix*.**wel.xyz.tsv**. By the default setting, no initial concentrations at wells are written to the file *prefix*.**wel.xyz.tsv**. Optionally, **xyz_well, xyz_wells**, or **-xyz_w[ells]**.

(**True | False**)—See line 1.

Notes

Default settings at the beginning of a simulation are **true** for **-echo_input**, **-fluid_properties**, **-HDF_chemistry**, **-HDF_heads**, **-HDF_media_properties**, **-HDF_steady_flow_velocities**, **-heads**, **-solution_method**, and **-wells**; default settings at the beginning of a simulation are false for all other identifiers. The **PRINT_INITIAL** options are used only once, prior to any transient calculations. The files with names ending in **.txt** contain data formatted to be printed or viewed on a screen. The orientation of the printout of the spatially distributed properties in the *prefix***.txt** files, either X–Y or X–Z planes, is controlled by the **-print_orientation** identifier in the **GRID** data block. The files *prefix***.chem.xyz.tsv**, *prefix***.comps.xyz.tsv**, *prefix***.head.xyz.tsv**, *prefix***.vel.xyz.tsv**, *prefix***.wel.xyz.tsv**, and *prefix***.zf.tsv** are written in tab-separated-values format to facilitate importing into spreadsheets and postprocessing programs for graphical display. The files with names containing **.xyz.** have node-by-node output.

Example Problems

The **PRINT_INITIAL** data block is used in Chapter 6 example problems 3, 4, 5, and 6.

PRINT_LOCATIONS

This keyword data block is used to limit printing of results to the chemistry output file *prefix*.**chem.txt** and (or) the spreadsheet file *prefix*.**chem.xyz.tsv** to a subset of cells within the grid region. This data block is optional, and in its absence, results are printed for all active nodes. The identifiers -**chemistry** and -**xyz_chemistry** in the PRINT_FREQUENCY data block control the frequency of printing to the files. The **PRINT** data block of the chemistry data file specifies the data to be written to the file *prefix*.**chem.txt**. The **SELECTED_OUTPUT** and **USER_PUNCH** data blocks in the chemistry data file are used to specify the data to be included in the file *prefix*.**chem.xyz.tsv**. The **PRINT_LOCATIONS** data block contains only static data, which do not change over the course of the transport calculations.

Example

Line numbers in the Example are used for identification purposes only and are referenced in the following Explanation. The line numbers do not indicate a required order for the keyword data block. Line numbers are not used in the flow and transport data file.

```
Line  0:    PRINT_LOCATIONS
Line  1:         -chemistry
Line  2:             -sample X 2
Line  3:             -box          5     5     5     10    10    10
Line  4:                 -print        0
Line  3a:            -wedge        5     5     5     10    10    10    Y2 grid
Line  4a:                -print        1
Line  5:         -xyz_chemistry
Line  2a:            -sample Y 2
Line  3b:            -box          0     0     0     5     5     5
Line  4b:                -print        1
```

Explanation

Line 0: **PRINT_LOCATIONS**

> **PRINT_LOCATIONS** is the keyword for the data block; no other data are included on this line. Optionally, **PRINT_LOCATION**.

Line 1: -**chemistry**

> -**chemistry**—Zone definitions following this identifier are used to select nodes for which results are written to the file *prefix*.**chem.txt**. Optionally, **chemistry** or -**c**[**hemistry**].

Line 2: -**sample** (**X** | **Y** | **Z**) *sample frequency*

> -**sample**—This identifier is used to specify a subgrid for writing results to the file *prefix*.**chem.txt** (when -**sample** follows -**chemistry**) or the file *prefix*.**chem.xyz.tsv** (when -**sample** follows -**xyz_chemistry**). Optionally, **sample**, **sample_grid**, **thin**, **thin_grid**, -**s**[**ample_grid**] or -**t**[**hin_grid**].

> (**X** | **Y** | **Z**)—The coordinate direction for which a subset of grid nodes will be selected.

> *sample frequency*—Writing to the specified file will occur for the first and last nodes in the coordinate direction (provided these nodes are active); between the first and last nodes, results will be printed at node intervals of *sample frequency*. If *sample frequency* is 2, then data will be written for every other node. To preserve the edges of the grid region, the nodes are selected by working to the interior of the grid region from each end of the grid in the specified coordinate direction.

Line 3: *zone*

zone—A zone definition as defined in section 4.5. Description of Input for Zones. Lines 3 and 3b define box zones; line 3a defines a wedge zone.

Line 4: **-print** *property*

-print—This identifier is used to control writing results for cells selected by the zone to the file *prefix*.**chem.txt** (when **-print** follows **-chemistry**) or the file *prefix*.**chem.xyz.tsv** (when **-print** follows **-xyz_chemistry**). Optionally, **print** or **-p[rint]**.

property—Property input as defined in section 4.6. Description of Input for Properties. Property values are indicators of whether printing will or will not occur for cells. If the interpolated value for a node is less than or equal to 0, the data will not be printed for the cell; if the interpolated value is greater than zero, data will be printed for the cell. Line 4 defines no printing for cells with nodes inside the zone; lines 4a and 4b define printing for cells with nodes inside the zone.

Line 5: **-xyz_chemistry**

-xyz_chemistry—Zone definitions following this identifier are used to select nodes for which results are written to the file *prefix*.**chem.xyz.tsv**. Optionally, **xyz_chemistry** or **-x[yz_chemistry]**.

Notes

Frequency of printing to the files *prefix*.**chem.txt** and *prefix*.**chem.xyz.tsv** is controlled by the respective identifiers **-chemistry** and **-xyz_chemistry** in the **PRINT_FREQUENCY** data block. By default, when printing to one of these files is enabled, printing will occur for all active nodes (**MEDIA** data block, identifier **-active**) for the dimensions for which chemistry is calculated (**GRID** data block, identifier **-chemistry_dimensions**). The **PRINT_LOCATIONS** data block may be used to limit the nodes selected for printing by specifying **-print 0** for some nodes. It may be convenient to define **-print 0** for all nodes in the grid region, and then define **-print 1** for selected nodes. The number of nodes selected for writing chemical data to the file *prefix*.**chem.txt** should be kept to a minimum to avoid generating an output that could exceed file-size limits of the computer operating system.

If both **-sample_grid** and **-print** identifiers are defined, the **-print** definitions supersede the **-sample_grid** definitions. The **-sample_grid** definitions are applied to the cells first, after which the zone definitions (with **-print**) are applied. If a node is included in more than one zone, the value specified for **-print** in the last definition will apply. In the example above, the cell chemistry information to be printed to the *prefix*.**chem.txt** file is determined as follows (assuming the grid region extends from 0 to 10 in each direction): first, printing and not printing are alternated for cells in the X direction; then all cells in the top half of the domain are set not to print; and finally half of the cells (a wedge) in the top half are set to print. The result is that every other cell in the X direction in the bottom half of the grid region and a wedge of cells in the top half of the grid region will have results printed to the chemistry output file. Similarly for the *prefix*.**chem.xyz.tsv** file, the example specifies printing for every other plane of cells perpendicular to the Y axis in the top half of the grid region and printing for every cell in the bottom half of the grid region.

Example Problems

The **PRINT_LOCATIONS** data block is used in Chapter 6 example problem **4**.

RIVER

This keyword data block is used to define boundary conditions that describe a single river. This data block is optional and only is needed if a river boundary condition is included in the simulation. A river is defined by a series of X–Y points, which define line segments that locate the centerline of the river across the land surface. For each river point, the river width, head, riverbed-leakance parameters, and, for reactive-transport simulations, an associated-solution composition must be defined, either by explicit definitions or by interpolation from other river points. Leakage into and out of the aquifer through the riverbed is calculated from these parameters. Head and solution composition may vary over the course of the simulation. For a vertical stack of cells, any river leakage enters or leaves the cell that contains the water table. Multiple **RIVER** data blocks are used to define multiple rivers in the grid region.

Example

Line numbers in the Example are used for identification purposes only and are referenced in the following Explanation. The line numbers do not indicate a required order for the keyword data block. Line numbers are not used in the flow and transport data file.

```
Line 0:     RIVER 1 Rubicon River
Line 1:         -xy_coordinate_system    grid
Line 2:         -z_coordinate_system     grid
Line 3:         -point  155.0      3633. # X, Y location of first river point
Line 4:             -head
Line 5:                           0     day   275.
Line 5a:                          1.5   day   276.
Line 6:             -solution
Line 7:                           0     4
Line 7a:                          1.8   5
Line 8:             -width                125.
Line 9:             -bed_hydraulic_conductivity      1.5e-2
Line 10:            -bed_thickness        1.6
Line 11:            -depth                3.5
Line 3a:        -point  165.0      3663.  # X, Y location of second river point
Line 3b:        -point  175.0      3603.  # X, Y location of third river point
Line 4a:            -head
Line 5b:                           0     274.
Line 5c:                          2.1    274.5.
Line 6a:            -solution
Line 7b:                           0     4
Line 8a:            -width                150.
Line 9a:            -bed_hydraulic_conductivity      1.5e-2
Line 10a:           -bed_thickness        2.
Line 12:            -bottom               271.5
```

Explanation

Line 0: **RIVER** *number* [*description*]

> **RIVER** is the keyword for the data block.
>
> *number*—Positive number to designate this river. Default is 1.
>
> *description*—Optional character field that identifies the river.

Line 1: **-xy_coordinate_system** (**grid** | **map**)

-xy_coordinate_system—Coordinate system for X–Y points that define the river. If **-xy_coordinate_system** is not defined for a river, the default is **grid**. Optionally, **xy_coordinate_system** or **-xy[_coordinate_system]**.

(grid | map)—**Grid** indicates that the X–Y points are defined in the grid coordinate system with units defined by **-horizontal_grid** in the **UNITS** data block. **Map** indicates that the X–Y points are defined in the map coordinate system with units defined by **-map_horizontal** in the **UNITS** data block. The transformation from map to grid coordinates is defined in the **GRID** data block.

Line 2: **-z_coordinate_system (grid | map)**

-z_coordinate_system—Coordinate system for Z coordinates of river bottom when defined with **-bottom** identifier. If **-z_coordinate_system** is not defined for a river, the default is **grid**. Optionally, **z_coordinate_system** or **-z_[coordinate_system]**

(grid | map)—**Grid** indicates that the X–Y points are defined in the grid coordinate system with units defined by **-horizontal_grid** in the **UNITS** data block. **Map** indicates that the X–Y points are defined in the map coordinate system with units defined by **-map_horizontal** in the **UNITS** data block. The transformation from map to grid coordinates is defined in the **GRID** data block.

Line 3: **-point** *x, y*

-point—The X–Y location of a river point is defined. Line 2 may be repeated as many times as needed to define the entire length of the river. At least two points must be defined. The series of points in sequence define the river by a series of line segments. Optionally, **point**, **-p[oint]**, **node**, or **-n[ode]**.

x—X coordinate location of a river point. Units, L, are defined by the **-horizontal_grid** (**grid** coordinate system) or **-map_horizontal** (**map** coordinate system) identifier in the **UNITS** data block.

y—Y coordinate location of a river point. Units, L, are defined by the **-horizontal_grid** (**grid** coordinate system) or **-map_horizontal** (**map** coordinate system) identifier in the **UNITS** data block.

Line 4: **-head**

-head—This identifier is used to specify the head at the river point. Units, L, are defined by the **-head** identifier in the **UNITS** data block. The head is defined on line 5. A time series of heads may be defined by using multiple line 5s. The first *time* in the series must be zero. Optionally, **head**, or **-he[ad]**.

Line 5: *time [units] head*

time—Simulation time (T) at which the head definition (*head*) will take effect. Units may be defined explicitly with *units*; default units are defined by **-time** identifier in **UNITS** data block.

units—Units for *time* can be "seconds", "minutes", "hours", "days", or "years" or an abbreviation of one of these units.

head—Head at the river point starting at *time*.

Line 6: **-solution**

-solution—This identifier is used to specify the solution index number for the X–Y river point. Solution index numbers correspond to solution compositions defined in the chemistry data file. The composition of water at the river point is defined in line 7. A time series of solution compositions may be defined by using multiple line 7s. The first *time* in the series must be zero. Optionally, **solution**, **associated_solution**, **-s[olution]**, or **-a[ssociated_solution]**.

Line 7: *time* [*units*] *solution*

time—Simulation time (T) at which the solution composition will take effect. Units may be defined explicitly with *units*; default units are defined by **-time** identifier in **UNITS** data block.

units—Units for *time* can be "seconds", "minutes", "hours", "days", or "years" or an abbreviation of one of these units.

solution—Solution index number for the river point starting at *time*.

Line 8: **-width** *width*

-width—This identifier is used to specify the width of the river at the X–Y river point. Units, L, are defined by the **-river_width** identifier in the **UNITS** data block. Optionally, **width**, or **-w[idth]**.

width—Width of the river at the river point.

Line 9: **-bed_hydraulic_conductivity** *bed_hydraulic_conductivity*

-bed_hydraulic_conductivity—This identifier is used to specify the hydraulic conductivity for the riverbed at the X–Y river point. Units, L/T, are defined by the **-river_bed_hydraulic_conductivity** identifier in the **UNITS** data block. Optionally, **bed_hydraulic_conductivity**, **bed_k**, **hydraulic_conductivity**, **k**, **-bed_h[ydraulic_conductivity]**, **-bed_k**, **-hy[draulic_conductivity]**, or **-k**.

bed_hydraulic_conductivity—Hydraulic conductivity of the riverbed at the river point.

Line 10: **-bed_thickness** *bed_thickness*

-bed_thickness—This identifier is used to specify the thickness of the riverbed at the X–Y river point. Units, L, are defined by the **-river_bed_thickness** identifier in the **UNITS** data block. Optionally, **bed_thickness**, **thickness**, **-bed_t[hickness]**, or **-t[hickness]**.

bed_thickness—Thickness of the riverbed at the river point.

Line 11: **-depth** *depth*

-depth—This identifier is used to specify the depth of the river at the X–Y river point and is alternative to the **-bottom** identifier. The depth of the river is subtracted from the initial head defined for the river point to define the elevation of the bottom of the river at that point. Units, L, are defined by the **-river_depth** identifier in the **UNITS** data block. Optionally, **depth**, or **-dep[th]**.

depth—Depth of the river at the river point.

Line 12: **-bottom** *bottom*

-bottom—This identifier is used to specify the elevation of the top of the riverbed at the X–Y river point and is alternative to the **-depth** identifier. Units, L, are defined by the **-z_coordinate_system** to be either grid coordinates, with units defined by **-vertical_grid** identifier in the **UNITS** data block or map coordinate units, with units defined by **-map_vertical** identifier in the **UNITS** data block. Optionally, **bottom**, **z**, **river_bottom**, **-bo[ttom]**, **-z**, or **-r[iver_bottom]**.

bottom—Elevation of the top of the riverbed at the river point.

Notes

River boundary conditions accept water from the aquifer when the head in the aquifer is greater than the head in the river and contribute water to the aquifer when the head in the aquifer is less than the head in the river. The maximum flow into the aquifer is limited to that attained when the head in the aquifer is equal to the bottom of the riverbed; this maximum flow is applied whenever the head in the aquifer is at or below the bottom of the riverbed. Multiple rivers are defined by using multiple **RIVER** data blocks, where each river is uniquely identified by the integer following the **RIVER** keyword.

The coordinate system for the X–Y location of the river points may be either of the two available coordinate systems, grid or map, as defined by the **-xy_coordinate_system** identifier. The coordinate system for the elevation (**-z** identifier) may be grid or map as defined by the **-z_coordinate_system** identifier. If the elevation is defined by the **-depth** identifier, the units of depth are defined by the **-river_depth** identifier of the **UNITS** data block. In this case, the elevation of the top of the riverbed in grid units is calculated by converting the initial head to grid units and subtracting the depth converted to grid units. Ultimately, all data are converted to the grid coordinate system and then converted to SI units for the simulation calculations.

River points can be defined in upstream or downstream order. A sequential set of two points within a **RIVER** data block define a line segment. The head, width, and riverbed elevation, hydraulic conductivity, and thickness at each point are used to quantify river leakage to or from the aquifer. The elevation of the river bottom can be defined in two ways: it can be defined explicitly with **-river_bottom** identifier, or it can be defined by a river depth (**-depth**) that will be subtracted from the head (**-head**) that applies at time zero. For reactive-transport simulation, a solution composition is associated with each point that defines the composition of water leaking from the river into the aquifer.

Parameters may be defined explicitly for a river point, or they may be interpolated from other river points. The head, width, and riverbed elevation or depth, hydraulic conductivity, thickness, and solution (for reactive transport simulations) must be defined for the first and last points of the river. For a point between the first and last points, a parameter can be defined explicitly with its identifier (**-head**, for example), or it can be interpolated from the nearest river points upstream and downstream where the parameter has been explicitly defined. Linear interpolation is performed by the ratio of the distance of the interpolation point from one of the explicitly defined points and the total distance along the river line segments that lie between the two explicitly defined points. Solution compositions at river points are determined by explicit definition (**-solution** identifier) or by mixing explicitly defined solutions at other river points by mixing fractions determined by linear interpolation.

Figure 4.9 shows an example of points that define a river and its tributary and the initial set of polygons that are processed to define the river in each cell that the polygons intersect (see section D.5.5. River Boundary for details). The line segment (fig. 4.9, line

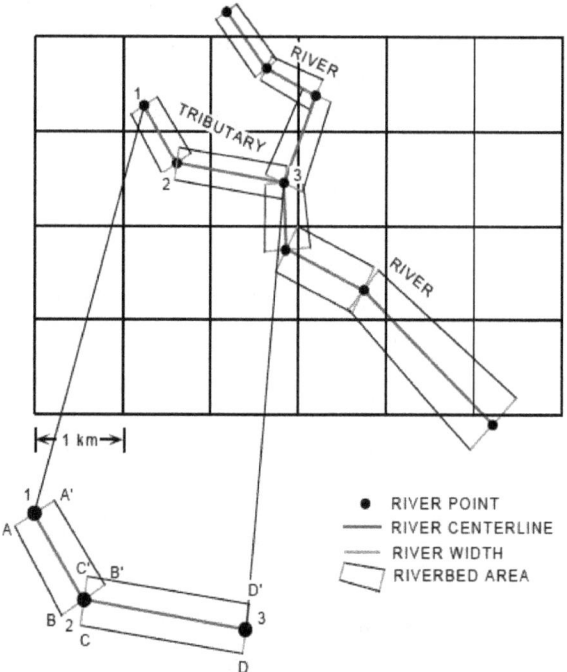

Figure 4.9. Discretization of river segments for a river and its tributary. Initial riverbed areas are shown before fill-in of gaps and removal of overlapping areas.

segment 1–2) joining two sequential river points forms the centerline of a trapezoid. At each of the two river points, the width of the river is defined by a line segment perpendicular to the centerline (fig. 4.9, lines A–A' and B–B'). A trapezoid is formed by connecting the endpoints of the two river-width line segments (fig. 4.9, lines A–B and A'–B'). Several steps are involved in processing the data at a series of river points to arrive at the parameter values that describe the river in each grid cell (see section D.5.5. River Boundary). The trapezoids are modified to add areas between adjacent trapezoids and to remove duplicated areas. These modified polygons are intersected with the grid cells, and properties of the river areas in each grid cell are interpolated from the properties at the river points.

Complete explicit or implicit definition of all properties for all river points is required for time zero. Implicit definition is by interpolation. A time-sequence of values can be defined for the head (**-head**) and the solution composition (**-solution**) for a river point. All interpolation is repeated at each time that the head or solution composition change at any river point. At the beginning of each simulation period, terms that incorporate both the transient head and solution composition and the static area and riverbed definition for each cell are calculated for the flow and transport equations.

A river boundary condition always applies to the positive Z face of a cell. River, flux, and leaky boundary conditions may be defined for the positive Z face of a cell. If both a river and a specified-head boundary condition are defined for a cell, the cell is a specified-head boundary cell and the river boundary condition is ignored. Rivers cannot be used in confined flow simulations. See section D.5.8. Boundary-Condition Compatibility for more details.

Example Problems

The **RIVER** data block is used in Chapter 6 example problem 4.

SOLUTE_TRANSPORT

This keyword data block is a switch used to include or exclude transport and reaction calculations. If **SOLUTE_TRANSPORT** is set to false, any keyword data blocks or identifiers related to transport or chemistry in the flow and transport data file are ignored. Molecular diffusivity is the only parameter that can be defined in this data block, and the diffusivity is pertinent only to transport calculations. The diffusivity is a scalar value that is uniform over the entire grid region; it is not spatially variable. The **SOLUTE_TRANSPORT** data block is mandatory for all simulations and contains only static data.

Example

Line numbers in the Example are used for identification purposes only and are referenced in the following Explanation. The line numbers do not indicate a required order for the keyword data block. Line numbers are not used in the flow and transport data file.

```
Line 0:    SOLUTE_TRANSPORT          true
Line 1:        -diffusivity          1e-9        # m^2/s
```

Explanation

Line 0: **SOLUTE_TRANSPORT [(True | False)]**

> **SOLUTE_TRANSPORT**—This keyword is used to switch the program from a reactive-transport simulation to a flow-only simulation. By default, a simulation includes flow, transport, and reaction calculations.

> **[(True | False)]**—A value of **true** indicates that transport and reaction calculations are made. A value of **false** indicates that only flow is simulated and no transport or reaction calculations are to be done. If neither **true** nor **false** is entered on the line, **true** is assumed. Optionally, **t[rue]** or **f[alse]**.

Line 1: **-diffusivity** *diffusivity*

> **-diffusivity**—This identifier is used to specify molecular diffusivity of all chemical solutes. The diffusivity only applies when **SOLUTE_TRANSPORT** is defined to be **true**. Default is 10^{-9} square meter per second (m^2/s). Optionally, **diffusivity** or **-d[iffusivity]**.

> *diffusivity*—Molecular diffusivity, m^2/s.

Notes

The diffusivity is used in simulations only if the transport equations are solved, that is, when **SOLUTE_TRANSPORT** is true. The units for diffusivity are square meters per second (m^2/s); no option for alternative units is available.

Example Problems

The **SOLUTE_TRANSPORT** data block is used in Chapter 6 example problems **1**, **2**, **3**, **4**, **5**, and **6**.

SOLUTION_METHOD

This keyword data block is used to select the linear equation solver that solves the finite-difference flow and transport equations and to set parameters related to the solver and the formulation of the finite-difference equations. The two solvers available are a D4 Gaussian elimination direct solver and a restarted generalized conjugate-gradient (ORTHOMIN) iterative solver (section D.7.2. Linear-Equation Solvers for Flow and Transport Finite-Difference Equations). If this data block is not included, the iterative solver with default parameter values is used. The **SOLUTION_METHOD** data block contains only static data.

The choice of equation solver depends on the memory storage requirement and computational workload for a specific simulation. Both factors are related to the size of the problem being simulated and to the rate of convergence of the iterative method. The importance of each factor depends on the computer system used. Thus, it is difficult to give a definitive formula for the selection of the solver. Small problems of a few hundred nodes are most efficiently solved with the direct solver. The iterative solver (example 1) usually is more efficient for simulations with several thousand nodes or more. The direct solver (example 2) is faster for small problems of 1,000 nodes or less). The PHAST simulator writes out the storage requirements for the selected solver to the *prefix*.**log.txt** file.

Example 1

Line numbers in the Example are used for identification purposes only and are referenced in the following Explanation. The line numbers do not indicate a required order for the keyword data block. Line numbers are not used in the flow and transport data file.

```
Line 0:    SOLUTION_METHOD
Line 1:         -iterative_solver        true
Line 2:         -tolerance               1e-12
Line 3:         -save_directions         20
Line 4:         -maximum_iterations      500
Line 5:         -space_differencing      0.0
Line 6:         -time_differencing       1.0
Line 7:         -cross_dispersion        false
Line 8:         -rebalance_fraction      0.5
Line 9:         -rebalance_by_cell       false
```

Explanation 1

Line 0: **SOLUTION_METHOD**
> **SOLUTION_METHOD** is the keyword for the data block; no other data are included on this line.

Line 1: **-iterative_solver [(True | False)]**
> **-iterative_solver**—This identifier is used to specify whether the iterative solver will be used for the flow and transport equations. By default, the program will use the iterative solver. Optionally, **iterative_solver**, **iterative**, or **-iterativ[e_solver]**.
>> **(True | False)**—**True**, the iterative solver is used, **False**, the direct solver is used. If neither **true** nor **false** is entered on the line, **true** is assumed. Optionally, **t[rue]** or **f[alse]**.

Line 2: **-tolerance** *tolerance*

-tolerance—This identifier is used to specify a tolerance for the iterative solver. The iterative solver has converged to a numerical solution when the scaled residual is smaller than the tolerance. Default is 1×10^{-12} (unitless). Optionally, **tolerance**, or **-to[lerance]**.

tolerance—Convergence criterion for the iterative solver.

Line 3: **-save_directions** *number*

-save_directions—This identifier is used to specify the number of search directions that are saved between restarts of the iterative solver. Default is 20. Optionally, **save_directions**, **save**, or **-sa[ve_directions]**.

number—Number of saved search directions.

Line 4: **-maximum_iterations** *number*

-maximum_iterations—This identifier is used to specify the maximum number of iterations that are allowed for the iterative solver. Default is 500. Optionally, **maximum_iterations**, **maximum**, **iterations**, **-m[aximum_iterations]**, or **-iterati[ons]**.

number—Maximum number of iterations for the iterative solver.

Line 5: **-space_differencing** *weight*

-space_differencing—This identifier is used to specify the weighting used in spatial differencing for the advective term in the transport equation. Default is 0.0, upstream weighting. Optionally, **space_differencing**, **space**, or **-sp[ace_differencing]**.

weight—The weighting can range from 0.0 (upstream-in-space differencing) to 0.5 (centered-in-space differencing).

Line 6: **-time_differencing** *weight*

-time_differencing—This identifier is used to specify the weighting used in time differencing for the flow and transport equations during transient calculations. [This weighting factor does not apply to steady-state flow equations where fully implicit weighting (*weight* = 1.0) is used.] Default is 1.0, fully implicit. Optionally, **time_differencing**, **time**, or **-ti[me_differencing]**.

weight—Value can range from 0.5 (centered-in-time differencing) to 1.0 (fully implicit, backwards-in-time differencing).

Line 7: **-cross_dispersion** [(**True** | **False**)]

-cross_dispersion—This identifier is used to include cross-dispersion terms in the transport equations. By default, cross-dispersion terms are not included. Optionally, **cross_dispersion**, or **-c[ross_dispersion]**.

(**True** | **False**)—**True**, cross-dispersion terms are included in transport equations; **false**, cross-dispersion terms are not included in transport equations. If neither **true** nor **false** is entered on the line, **true** is assumed. Optionally, **t[rue]** or **f[alse]**.

Line 8: **-rebalance_fraction** *fraction*

-rebalance_fraction—This identifier affects the way cells are distributed to multiple processes when running the parallel version of PHAST. After each time step, PHAST evaluates how to optimally rebalance the cells among the processes for the chemical calculation. It estimates a number of cells to transfer from one process to the next. The factor *fraction* is multiplied times the estimated number of cells to obtain the actual number of cells transferred. Optionally, **rebalance_fraction**, or **-r[ebalance_fraction]**.

fraction—The fraction of the estimated number of cells transferred to the next process at the end of a time step (parallel version only). Default is 0.5.

Line 9: **-rebalance_by_cell** [(**True** | **False**)]

-**rebalance_by_cell**—This identifier affects the way cells are distributed to multiple processes when running the parallel version of PHAST. After each time step, PHAST evaluates how to optimally rebalance the cells among the processes for the chemical calculation. The default rebalancing method (**false**) uses an average time per cell for each process in the rebalancing calculation. When -**rebalance_by_cell** is **true**, the time required to solve chemistry in each cell is measured and the individual cell timings are used to calculate the redistribution of cells among processes. Optionally, **rebalance_by_cell**, or -**rebalance_b[y_cell]**.

(**True | False**)—**True**, individual cell timings are used to calculate the redistribution of cells among processes; **false**, average cell timings for each process are used to calculate the redistribution of cells among processes. If neither **true** nor **false** is entered on the line, **true** is assumed. Default is **false**. Optionally, **t[rue]** or **f[alse]**.

Notes 1

By default, the iterative solver is used, which is a restarted ORTHOMIN method that solves the Schur complement, reduced matrix, which is preconditioned by triangular factorization. See section D.7.2. Linear-Equation Solvers for Flow and Transport Finite-Difference Equations for details on the implementation of the iterative solver. The identifiers, -**tolerance**, -**save_directions**, and -**maximum_iterations** apply only to the iterative solver. *Tolerance* is the convergence criterion for the iterative solver; if the Euclidean norm of the relative residual vector is less than *tolerance*, the iterative solution to the linear equations has converged. It is not feasible to determine this tolerance parameter on the basis of the problem specifications and the desired accuracy of the solution. A value for *tolerance* can be determined empirically by examining the magnitude of the changes in heads and concentrations with successively smaller values of *tolerance*. Experience with test problems has shown that a tolerance of 10^{-6} to 10^{-12} is necessary to obtain 3- or 4-digit agreement with the direct solver.

The solver can be sensitive to the number of search directions (-**save_directions**) that are retained between restarts of the ORTHOMIN solver. The default value of 20 works satisfactorily for most problems. If solver convergence requires more than about 100 iterations, the number of retained search directions should be increased. However, the number of search directions should not be set to a very large number because, as the number of search directions is increased, the memory needed to save the vectors of the search directions also increases.

The limit on the number of iterations (-**maximum_iterations**) prevents runaway conditions when the convergence rate is very slow. Iteration counts are written to the *prefix*.**log.txt** file as controlled by -**progress_statistics** of the **PRINT_FREQUENCY** data block. For some simulations, the user may need to double or triple the default maximum of 500 iterations. However, more than a few hundred iterations indicates that adjustments probably need to be made in the spatial or temporal discretization or that there is an incompatibility in the problem formulation.

Default weighting is upstream-in-space (-**space_differencing** *0.0*) and backward-in-time (fully implicit, -**time_differencing** *1.0*). For steady-flow simulations, the time differencing scheme for the flow equation is always fully implicit. For transient-flow simulations with reactive transport, the same time weighting (-**time_differencing**) is used for both the flow and the transport equations. The flow equation does not have an advective term and thus does not use a spatial differencing factor. For additional information on the effects of weighting in time and space on oscillations and numerical dispersion, see sections D.1.2. Spatial Discretization, D.1.3. Temporal Discretization, and D.8. Accuracy from Spatial and Temporal Discretization.

If cross-dispersion terms (see section D.3. Property Functions and Transport Coefficients) are used (-**cross_dispersion** true), the solution to the transport equations may result in negative concentrations, which will be set to zero in the chemical calculations. However, the negative concentrations were balanced by off-setting excess positive concentrations. therefore, setting negative concentrations to zero will increase the mass of a solute in the system and increase global mass-balance errors because the excess positive concentrations remain.

If the parallel version of PHAST is used, the program tries to optimize the distribution of cells among the available processes to minimize CPU time. After each time step, the distribution of cells among the processes is changed by transferring cells among the processes. An optimum number of cells to be transferred from one process to another is calculated. However, it is possible that the rebalancing of cells can cause oscillations, where too many cells are transferred after one time step, which causes cells to be transferred back after the next time step. The rebalance fraction (**-rebalance_fraction**) allows only a fraction of the estimated number of cells to be transferred from one process to another, which may damp the oscillations in cell transfers. The default rebalancing strategy uses average times per cell for each process to calculate the redistribution of cells among processes. An alternative strategy (**-rebalance_by_cell true**) uses actual timings for each cell to calculate the redistribution of cells. Empirically, there has been no clear indication that one strategy is better than the other.

Example 2

Line numbers in the Example are used for identification purposes only and are referenced in the following Explanation. The line numbers do not indicate a required order for the keyword data block. Line numbers are not used in the flow and transport data file.

```
Line 0:     SOLUTION_METHOD
Line 1:         -direct_solver          true
Line 2:         -space_differencing     0.5
Line 3:         -time_differencing      0.5
Line 4:         -cross_dispersion       False
Line 5:         -rebalance_fraction     0.5
Line 6:         -rebalance_by_cell      false
```

Explanation 2

Line 0: **SOLUTION_METHOD**

> **SOLUTION_METHOD** is the keyword for the data block; no other data are included on this line.

Line 1: **-direct_solver** [(**True** | **False**)]

> **-direct_solver**—This identifier is used to specify whether the direct solver will be used to solve the flow and transport equations. This option is equivalent to the **-iterative_solver** option with opposite meaning for true and false. By default, the program will use the iterative solver. Optionally, **direct_solver**, **direct**, or **-d[irect_solver]**.

> (**True** | **False**)—**True**, the direct solver is used; **false**, the iterative solver is used. If neither **true** nor **false** is entered on the line, **true** is assumed. Optionally, **t[rue]** or **f[alse]**.

Line 2: **-space_differencing** *weight*

> Same as example 1, line 5.

Line 3: **-time_differencing** *weight*

> Same as example 1, line 6.

Line 4: **-cross_dispersion** [(**True** | **False**)]

> Same as example 1, line 7.

Line 5: **-rebalance_fraction** *fraction*
   Same as example 1, line 8.
Line 6: **-rebalance_by_cell** [(**True** | **False**)]
   Same as example 1, line 9.

Notes 2

By default, the iterative solver is used, but the direct solver is faster for small problems (as many as 1,000 nodes). The direct solver does not have the iteration-termination (convergence-tolerance) errors of the iterative solver and, unlike the iterative solver, the direct solver has no additional parameters that need to be defined.

Example Problems

The **SOLUTION_METHOD** data block is used in Chapter 6 example problems 1, 2, 3, 4, 5, and 6.

SPECIFIED_HEAD_BC

This keyword data block is used to define specified-head boundary conditions. This keyword data block is optional and only is needed if a specified-head boundary condition is included in the simulation. For flow-only simulations, only the heads are required for each specified-head boundary-condition node. For reactive-transport simulations, solution composition also is required for each specified-head boundary-condition node. The solution composition for a specified-head cell may be fixed—solution composition is constant over a simulation period—or the composition may be associated with the flux of water into the grid region, in which case the solution composition for a specified-head cell may vary over a simulation period. The head and solution composition may vary independently over the course of the simulation.

Example

Line numbers in the Example are used for identification purposes only and are referenced in the following Explanation. The line numbers do not indicate a required order for the keyword data block. Line numbers are not used in the flow and transport data file.

```
Line 0:     SPECIFIED_HEAD_BC
Line 1:          -box           0     0     0     10    2     10
Line 2:              -exterior_cells_only      X
Line 3:              -head
Line 4:                  0     10
Line 4a:                 10    X     101.5 0     103.  10
Line 5:              -fixed_solution
Line 6:                  0     6
Line 6a:                 100   mixture 2     4     xyz   grid  mixf.dat
Line 6b:                 200   points     grid
Line 7:                      0.0   5.0   0.0   1
Line 7a:                     10.0  5.0   0.0   2
Line 8:                      end_points
Line 1a:         -box           0     8     0     10    10    10
Line 2a:             -exterior_cells_only      X
Line 3a:             -head
Line 4b:                 0     xyzt       grid  head.xyzt
Line 9:              -associated_solution
Line 6c:                 0     6
```

Explanation

Line 0: **SPECIFIED_HEAD_BC**

   SPECIFIED_HEAD_BC is the keyword for the data block; no other data are included on this line.

Line 1: *zone*

   zone—A zone definition as defined in section 4.5. Description of Input for Zones. Lines 1 and 1a define box zones.

Line 2: **-exterior_cells_only (X | Y | Z | all)**

   -exterior_cells_only—This identifier is used to restrict the selection of specified-head cells to those cells in the zone that have cell faces on the exterior of the active grid region. By default, all cells within the zone are selected to have the specified-head boundary condition. By using the **-exterior_cells_only** identifier, the selection can be restricted to cells with exterior X, Y, or Z faces or to cells with exterior faces of any orientation. Optionally, **exterior_cells_only** or **-e[xterior_cells_only]**.

(**X** | **Y** | **Z** | **all**)—Restricts the selection of specified-head boundary-condition cells to those cells within the zone that have faces on the exterior of the active grid region of the specified orientation (**X**, **Y**, or **Z**) or to any cell within the zone that has a face on the exterior of the active grid region (**all**).

Line 3: **-head**

-head—This identifier is used to specify a time series of head for the specified-head boundary cells. The head is defined by the property definition on line 4. A time series of head properties may be defined by using multiple line 4s. The first *time* in the series must be zero. Optionally, **head** or **-he[ad]**.

Line 4: *time* [*units*] *property*

time—Simulation time (T) at which the head property definition (*property*) will take effect. Units may be defined explicitly with *units*; default units are defined by **-time** identifier in **UNITS** data block.

units—Units for *time* can be "seconds", "minutes", "hours", "days", or "years" or an abbreviation of one of these units.

property—Property input as defined in section 4.6. Description of Input for Properties. Property values are the heads for the specified-head boundary-condition cells selected by the zone and the **-exterior_cells_only** identifier. Line 4 defines constant head for the specified-head cells of the zone starting at time 0. Line 4a defines a linearly varying head in the X direction for the specified-head cells of the zone starting at time 10. Line 4b defines a time series of head conditions for the specified-head cells in the zone that are extracted from the file *head.xyzt*. The file contains heads at scattered points at one or more time planes. Units of head, L, are defined by the **-head** identifier in the **UNITS** data block.

Line 5: **-fixed_solution**

-fixed_solution—This identifier is used to specify a time series of solution composition for the cells of the specified-head boundary. By using the **-fixed_solution** identifier, the composition of each specified-head cell within the zone is constant from one time in the time series until the next time in the series. The solution composition time series is defined by the property definitions on the subsequent lines. The first *time* in the series must be zero. Optionally, **associated_solution**, **-a[ssociated_solution]**, **solution**, or **-s[olution]**.

Line 6: *time* [*units*] [**mixture** *i j*] *property*

time—Simulation time (T) at which the fixed-solution composition property definition (*property*) will take effect. Units may be defined explicitly with *units*; default units are defined by **-time** identifier in **UNITS** data block.

units—Units for *time* can be "seconds", "minutes", "hours", "days", or "years" or an abbreviation of one of these units.

mixture *i j*—Optionally, solution compositions can be defined as a mixture of solution compositions *i* and *j* in the chemistry data file. For a **mixture**, the *property* defines mixing fraction, *f*, of solution composition *i*; mixing fraction for solution composition *j* is *1 − f*.

property—Property input as defined in section 4.6. Description of Input for Properties. Property values are index numbers or mixing fractions (when **mixture** *i j* is used) that define the solution composition of the specified-head boundary-condition cells. Lines 6 and 6c define spatially uniform solution compositions. Line 6a defines mixtures of solutions 2 and 4, where the mixing fractions at scattered points are read from a file; values of mixing fraction of solution 2 will be interpolated at the nodes of the specified-head cells within the zone by closest point interpolation. Lines 6b through 8 define solution composition at a set of points, and solution compositions at cell nodes are interpolated by closest point interpolation.

Line 7: *x, y, z value*

> *x*—X coordinate of a point.
>
> *y*—Y coordinate of a point.
>
> *z*—Z coordinate of a point.
>
> *value*—Solution index number for the specified point.

Line 8: **end_points**

> **end_points**—The character string **end_points** must be included on the next line following the last scattered data point definition of the **points** property option.

Line 9: **-associated_solution**

> **-associated_solution**—This identifier is used to specify a time series of solution composition for the flux of water into the active grid region through the specified-head boundary cells in the zone. The solution composition time series is defined by the property definitions on line 6c. The first *time* in the series must be zero. Optionally, **associated_solution**, **-a[ssociated_solution]**, **solution**, or **-s[olution]**.

Notes

Two identifiers, **-fixed_solution** and **-associated_solution**, can be used to define chemical compositions for specified-head boundaries. With the **-fixed_solution** identifier, the concentrations of components in a boundary cell are defined by a time series of solution compositions, regardless of reactions or flow into or out of the cell. With the **-associated_solution** identifier, the concentrations of the components in the boundary cell may vary as determined by flow into the cell of water with the associated-solution composition, reaction, and transport. For the associated-solution option, if flow is out of a specified-head cell during a time step, the concentrations in the water that exits the cell are equal to the concentrations in the cell. If flow enters a specified-head cell during a time step, the concentrations in the water that enters the cell are equal to the concentrations in the associated solution. Specified-head boundary conditions are described in detail in section D.5.1. Specified-Head Boundary with Associated-Solution Composition and D.5.2. Specified-Head Boundary with Specified-Solution Composition.

Multiple zones may be used within **SPECIFIED_HEAD_BC** data blocks to define boundary conditions within the grid region. Different boundary conditions for a single cell (node) may be defined multiple times as part of different zone definitions in different keyword data blocks. The last boundary condition defined for a cell determines the type of boundary conditions that apply to the cell. If the last boundary condition defined is a specified-head boundary, then the cell is exclusively a specified-head boundary. If the last definition is flux or leaky, then the cell is not a specified-head cell. No other boundary conditions apply to a specified-head cell, including leaky, flux, drain, and river boundary conditions. See section D.5.8. Boundary-Condition Compatibility for more details.

Example Problems

The **SPECIFIED_HEAD_BC** data block is used in Chapter 6 example problems **1**, **2**, **3**, **4**, **5**, and **6**.

STEADY_FLOW

This keyword data block is used to specify whether flow conditions are steady state and to define criteria for the determination that a steady-state flow condition is attained. If the simulation is defined to have steady-state flow, then an initial flow-only calculation is performed to determine the steady-state velocities, and these velocities are used in the transport equations throughout the simulation. Steady-flow is calculated by time stepping with the finite-difference flow equations until (1) changes in head and (2) flow balance are within specified tolerances. The data in this data block apply only to the initial calculation to obtain steady-state flow.

Example

Line numbers in the Example are used for identification purposes only and are referenced in the following Explanation. The line numbers do not indicate a required order for the keyword data block. Line numbers are not used in the flow and transport data file.

```
Line 0:    STEADY_FLOW true
Line 1:        -head_tolerance            1e-5
Line 2:        -flow_balance_tolerance    0.001
Line 3:        -minimum_time_step         1       s
Line 4:        -maximum_time_step         1000    day
Line 5:        -head_change_target        100.
Line 6:        -iterations                100
Line 7:        -growth_factor             2.0
```

Explanation

Line 0: **STEADY_FLOW [(True | False)]**

STEADY_FLOW—This keyword is used to specify whether or not flow conditions are steady state. By default, the simulation is assumed to be transient flow.

(True | False)—**True**, an initial calculation is performed to calculate steady-state velocities, **False**, transient velocities will be calculated at each time step of the simulation. If neither **true** nor **false** is entered on the line, **true** is assumed. Optionally, **t[rue]** or **f[alse]**.

Line 1: **-head_tolerance** *tolerance*

-head_tolerance—This identifier is used to specify a head tolerance for determining when flow is at steady state. Head changes for a time step must be less than this tolerance for the flow system to be at steady state. Default is 10^{-5} (units are specified in the **UNITS** data block, **-head** identifier). Optionally, **head_tolerance**, **head_tol**, or **-h[ead_tolerance]**.

tolerance—Tolerance for head changes. Units, L, are defined by **-head** identifier in the **UNITS** data block.

Line 2: **-flow_balance_tolerance** *tolerance*

-flow_balance_tolerance—This identifier is used to specify a relative flow-balance tolerance for determining when flow is at steady state. The fractional flow balance must be less than this tolerance for the flow system to be at steady state. The fractional flow balance is the difference between the inflow and outflow rates divided by the average of the inflow and outflow rates for the current time step. Default is 0.001 (unitless). Optionally, **flow_tol**, **flow_tolerance**, **flow_balance_tol**, **flow_balance_tolerance**, **f[low_tolerance]**, or **-f[low_balance_tolerance]**.

tolerance—Fractional flow-balance tolerance. Units are dimensionless.

Line 3: **-minimum_time_step** *time_step* [*units*]

-minimum_time_step—This identifier is used to specify the initial time step used in calculating steady-state flow and also is the minimum time step allowed for automatic time stepping in the steady-state flow calculation. Default is the value for **-time_step** defined in **TIME_CONTROL** divided by 1,000. Optionally, **minimum**, **minimum_time**, **minimum_time_step**, or **mi[nimum_time_step]**.

time_step—Initial and minimum time step for steady-flow calculation. Units, T, may be defined explicitly with *units*; default units are defined by **-time** identifier in **UNITS** data block.

units—Units for time step can be "seconds", "minutes", "hours", "days", or "years" or an abbreviation of one of these units.

Line 4: **-maximum_time_step** *time_step* [*units*]

-maximum_time_step—This identifier is used to specify the maximum time step used in the steady-state flow calculation. Default is the value for **-time_step** defined in **TIME_CONTROL** times 1,000. Optionally, **maximum**, **maximum_time**, **maximum_time_step**, or **ma[ximum_time_step]**.

time_step—Maximum time step for steady-state flow calculation. Units, T, may be defined explicitly with *units*; default units are defined by **-time** identifier in **UNITS** data block.

units—Units for time step can be "seconds", "minutes", "hours", "days", or "years" or an abbreviation of one of these units.

Line 5: **-head_change_target** *target*

-head_change_target—This identifier is used to specify the target head change for a single time step in the steady-flow calculation. The time step will be decreased or increased during the steady-flow calculation to try to achieve head changes of *target* over each time step. Default is 0.3 times the thickness of the grid region. Optionally, **head_change**, **head_change_target**, **head_target**, **-head_c[hange_target]**, or **-head_ta[rget]**.

target—Target head change. Units, L, are defined by **-head** identifier in the **UNITS** data block.

Line 6: **-iterations** *maximum_iterations*

-iterations—This identifier is used to specify the maximum number of iterations (time steps) that will be used in attempting to attain a steady-flow conditions. Default is 100 iterations. Optionally, **iterations** or **-i[terations]**.

maximum_iterations—Maximum number of iterations (time steps) used in attempting to attain a steady-flow condition.

Line 7: **-growth_factor** *factor*

-growth_factor—This identifier is used to specify a factor that limits the increase in time step from one iteration to the next. If *t* is the time step at one steady-state iteration, the maximum value for the time step at the next iteration is *factor* times *t*. Optionally, **iterations** or **-i[terations]**.

factor—Factor that limits the increase in time step from one iteration to the next. *Factor* should be greater than 1.0. Default is 2.0.

Notes

STEADY_FLOW can be specified with or without transport calculations (**SOLUTE_TRANSPORT**). If **STEADY_FLOW** is true, an additional flow-only calculation is performed at the beginning of the PHAST simulation to determine the steady-state head field and associated steady-state velocity field. The steady-state calculation ensures that heads are constant within the tolerance specified by **-head_tolerance** and global flow balance is satisfied within a relative tolerance specified by **-flow_balance_tolerance**. See section D.1.4. Automatic Time-Step Algorithm for Steady-State Flow Simulation for details. For steady-flow simulations, the steady-state velocity field is used in all transport calculations. The **STEADY_FLOW true** option saves CPU time by eliminating the solution of the groundwater flow equation at each time step.

The difference equations for unconfined flow (**FREE_SURFACE_BC true**) contain explicit terms, which are calculated on the basis of conditions from the previous time step: for example, fraction of cell saturated, fraction of well screen receiving water, and the vertical location of the cell receiving river flux. These explicit terms may cause instability (for example, oscillations) in the numerical method, such that the method does not converge to a steady-state flow condition. The only remedy available is to use smaller time steps so that the explicit terms change more slowly. To ensure a sufficiently small time step, it may be necessary to start with a small time step (**-minimum_time_step**) and limit the maximum time step (**-maximum_time_step**). If time steps are small, it also may be necessary to increase the number of iterations (**-iterations**) to allow the method to slowly approach the steady-state solution. The growth factor (**-growth_factor**) can also be used to slow down the rate of increase of the time step in the automatic time-step algorithm.

The automatic time-step algorithm will fail if the maximum change in head after the second steady-state time step exceeds the target change in head (**-head_change_target** *target*). For the first two steady-state time steps the minimum time step (**-minimum_time_step** *time_step*) is used. Therefore, the simulator fails because it cannot reduce the time step to less than the minimum, but the head changes are still too large. The remedies to this problem are to increase the maximum acceptable change in head (**-head_change_target** *target*), reduce the minimum allowable time step (**-minimum_time_step** *time_step*), or adjust the initial head conditions to be closer to the steady-state head conditions.

Example Problems

The **STEADY_FLOW** data block is used in Chapter 6 example problems 1, 2, 3, 4, and 5.

TIME_CONTROL

This keyword data block is used to define the time steps for the simulation, times at which simulation periods end, and a time at which to start the calculations. The time step that applies at time zero must be defined, but a new time step may be defined to apply beginning at any time in the simulation. Times at which to end simulation periods may be explicitly defined in this data block; however, the number of simulation periods also depends on the time-series definitions in **FLUX_BC**, **LEAKY_BC**, **PRINT_FREQUENCY**, **RIVER**, **SPECIFIED_HEAD_BC**, and **WELL** data blocks. The last time defines the end of the simulation. The time specified for the end of each simulation period is measured from time zero. A start time other than zero may also be defined. The **TIME_CONTROL** data block is mandatory for all simulations.

Example

Line numbers in the Example are used for identification purposes only and are referenced in the following Explanation. The line numbers do not indicate a required order for the keyword data block. Line numbers are not used in the flow and transport data file.

```
Line  0:   TIME_CONTROL
Line  1:       -time_step
Line  2:              0           1      day
Line  2a:            10     day    5      day
Line  3:       -time_end
Line  4:              1     yr
Line  4a:             2     yr
Line  5:       -time_start        0.5 yr
```

Explanation

Line 0: **TIME_CONTROL**

    TIME_CONTROL is the keyword for the data block; no other data are included on this line.

Line 1: **-time_step**

    -time_step—This identifier is used to specify that time-step data will be entered. Optionally, **step**, **time_step**, **delta**, **delta_time**, **-s[tep]**, **-time_s[tep]**, or **-d[elta_time]**.

Line 2: *time* [*units$_1$*] *time_step* [*units$_2$*]

    time—Simulation time (T) at which the new time step will be used. The first *time* in a series of line 2s must be zero. Units may be defined explicitly with *units1*; default units are defined by **-time** identifier in **UNITS** data block.

    units$_1$—Units for *time* can be "seconds", "minutes", "hours", "days", or "years" or an abbreviation of one of these units.

    time_step—Time-step length (T), units may be defined explicitly with *units2*; default units are defined by **-time** identifier in **UNITS** data block.

    units$_2$—Units for *time_step* can be "seconds", "minutes", "hours", "days", or "years" or an abbreviation of one of these units.

Line 3: **-time_end**

    -time_end—This identifier is used to specify a series of times at which simulation periods end. The last time in the series marks the end of the simulation. All times are measured from time zero. The number of simulation periods also depends on the time-series definitions in **FLUX_BC, LEAKY_BC, PRINT_FREQUENCY, RIVER, SPECIFIED_HEAD_BC,**

and WELL data blocks. Optionally, **time_change**, **change_time**, **end**, **end_time**, **time_end**, **-time_c[hange]**, **-c[hange_time]**, **-e[nd_time]**, or **-time_e[nd]**.

Line 4: *time_end* [*units*]

time_end—Time at which a simulation period ends. The last line 4 indicates the time of the end of the simulation. *Time* (T) units may be defined explicitly with *units*; default units are defined by **-time** identifier in UNITS data block.

units—Units for *time* can be "seconds", "minutes", "hours", "days", or "years" or an abbreviation of one of these units.

Line 5: **-time_start** *time_start* [*units*]

-time_start—This identifier is used to specify a time at which to begin the simulation. Optionally, **time_initial**, **initial_time**, **start**, **start_time**, **time_start**, **-time_i[nitial]**, **-i[nitial_time]**, **-s[tart_time]**, or **-time_s[tart]**.

time_start—Time at which the first simulation period starts. *Time_start* (T) units may be defined explicitly with *units*; default units are defined by **-time** identifier in UNITS data block.

units—Units for *time* can be "seconds", "minutes", "hours", "days", or "years" or an abbreviation of one of these units.

Notes

A **TIME_CONTROL** data block, including **-time_step** and **-time_end** identifiers, is mandatory for every input file. If a time unit is specified for a time variable, then the specified unit overrides the default unit (UNITS data block, **-time** identifier). The units specified for each variable need not be the same. The first *time* in the **-time_step** series must be zero. Every *time_end* marks the end of a simulation period; the final *time_end* marks the end of the simulation. Every time less than the final *time_end* that is included in a time series for *time_step* or in any time series in FLUX_BC, LEAKY_BC, PRINT_FREQUENCY, RIVER, SPECIFIED_HEAD_BC, and WELL data blocks will mark the end of one simulation period and the beginning of another.

The time to start a simulation is zero by default. A start time greater than zero may be defined with the **-time_start** identifier. All time-series definitions for boundary conditions and time step must begin at time zero. If *time_start* is greater than zero, then, even though no flow or transport is simulated, time-series definitions are applied in sequence until the start time is reached, at which time, flow, transport, and reaction calculations are begun.

The time step, in addition to grid discretization, is critical for obtaining a sufficiently accurate numerical solution to the flow and transport equations. Criteria for estimating appropriate time steps for different finite-difference weightings are discussed in section D.1.3. Temporal Discretization.

The difference equations for unconfined flow (FREE_SURFACE_BC **true**) contain explicit terms, which are calculated on the basis of conditions from the previous time step: for example, fraction of cell saturated, fraction of well screen receiving water, and the vertical location of the cell receiving river flux. These explicit terms may cause instability (for example, oscillations) in the numerical method. The only remedy available is to use smaller time steps (*time_step*) so that the explicit terms change more slowly.

Default values for the automatic time-step algorithm for achieving steady state are based on the value for *time_step* in the **TIME_CONTROL** data block. The default value for the minimum time step for automatic time-stepping (STEADY_FLOW) is 0.001 times the *time_step* value. Similarly, the default value for the maximum time step for automatic time-stepping (STEADY_FLOW) is 1,000 times the *time_step* value.

Example Problems

The **TIME_CONTROL** data block is used in Chapter 6 example problems **1, 2, 3, 4, 5,** and **6**.

TITLE

This keyword data block is used to define two title lines that are written to each of the output files, except the **.xyz.tsv** and *prefix*.**h5** files.

Example

Line numbers in the Example are used for identification purposes only and are referenced in the following Explanation. Line numbers are not used in the flow and transport data file.

```
Line 0:   TITLE
Line 1:   line 1
Line 1a:  line 2
```

Explanation

Line 0: **TITLE**

   TITLE is the keyword for the data block.

Line 1: *title*

   title—The title may continue on as many lines as desired; however, only the first two lines are written to output files. Lines are read and saved as part of the title until a keyword begins a line or until the end of the file.

Notes

The **TITLE** data block is used to identify the simulation in the output files, except the **.xyz.tsv** and *prefix*.**h5** files. Be careful not to begin a line of the title with a keyword because that marks the end of the **TITLE** data block. If more than one title keyword data block is included, only the lines from the last **TITLE** data block will appear in the output files.

Example Problems

The **TITLE** data block is used in Chapter 6 example problems **1**, **2**, **3**, **4**, **5**, and **6**.

UNITS

This keyword data block is used to specify the units of measure for the input data. Internal to PHAST, all flow and transport calculations use SI units. The **UNITS** data block provides the information necessary to convert the input data to SI units. Output units are always SI except for time, which is output in the unit defined by the **-time** identifier of this keyword data block. Units for time (T) can be "seconds", "minutes", "hours", "days", or "years" or an abbreviation of one of these units. Units for distance (L) can be U.S. customary ("inches", "feet", or "miles"), metric ("millimeters", "centimeters", "meters", or "kilometers") or an abbreviation of one of these units. The **UNITS** data block contains only static data.

Example

Line numbers in the Example are used for identification purposes only and are referenced in the following Explanation. The line numbers do not indicate a required order for the keyword data block. Line numbers are not used in the flow and transport data file.

```
Line 0:    UNITS
Line 1:         -time                               years
Line 2:         -horizontal_grid                    km
Line 3:         -vertical_grid                      ft
Line 4:         -map_horizontal                     km
Line 5:         -map_vertical                       ft
Line 6:         -head                               ft
Line 7:         -hydraulic_conductivity             m/d
Line 8:         -specific_storage                   1/ft
Line 9:         -dispersivity                       m
Line 10:        -flux                               m/s
Line 11:        -leaky_hydraulic_conductivity       m/s
Line 12:        -leaky_thickness                    km
Line 13:        -well_diameter                      in
Line 14:        -well_flow_rate                     gpm
Line 15:        -well_depth                         m
Line 16:        -river_bed_hydraulic_conductivity   m/s
Line 17:        -river_bed_thickness                m
Line 18:        -river_width                        m
Line 19:        -river_depth                        m
Line 20:        -drain_hydraulic_conductivity       m/s
Line 21:        -drain_thickness                    m
Line 22:        -drain_width                        m
Line 23:        -equilibrium_phases                 ROCK
Line 24:        -exchange                           ROCK
Line 25:        -surface                            WATER
Line 26:        -solid_solutions                    WATER
Line 27:        -kinetics                           WATER
Line 28:        -gas_phase                          WATER
```

Explanation

Line 0: **UNITS**

UNITS is the keyword for the data block; no other data are included on this line.

Line 1: **-time** *time_units*

-time—This identifier is used to specify the default unit of time for input and output data. Optionally, **time** or **-t[ime]**.

time_units—Units for time (T).

Line 2: **-horizontal_grid** *units*

-horizontal_grid—This identifier is used to specify the units of distance for input data related to the X and Y grid coordinate directions, which include horizontal distances in the GRID data block and all zone definitions that are specified in grid coordinates. Optionally, **horizontal_grid**, **horizontal**, or **-ho[rizontal_grid]**.

units—Input units of measure for horizontal grid coordinate system (L).

Line 3: **-vertical_grid** *units*

-vertical_grid—This identifier is used to specify the units of distance for input data related to the Z grid coordinate direction, which include vertical distances in the GRID data block and all zone definitions that are specified in grid coordinates. Optionally, **vertical_grid**, **vertical**, or **-v[ertical_grid]**.

units—Input units for vertical grid coordinate system (L).

Line 4: **-map_horizontal** *units*

-map_horizontal—This identifier is used to specify the units of distance for input data related to the X and Y map coordinate directions, which include horizontal distances in all zone definitions that are specified in map coordinates. Optionally, **map_horizontal** or **-map_h[orizontal]**.

units—Input units of measure for horizontal map coordinate system (L).

Line 5: **-map_vertical** *units*

-map_vertical—This identifier is used to specify the units of distance for input data related to the Z map coordinate direction, which include vertical distances in all zone definitions that are specified in map coordinates. Optionally, **map_vertical** or **-map_v[ertical]**.

units—Input units for vertical map coordinate system (L).

Line 6: **-head** *units*

-head—This identifier is used to specify the units of head for input data in all keyword data blocks. Optionally, **head** or **-he[ad]**.

units—Input units for head (L).

Line 7: **-hydraulic_conductivity** *units*

-hydraulic_conductivity—This identifier is used to specify the units of hydraulic conductivity for input data in the MEDIA data block. Optionally, **hydraulic_conductivity**, **K**, **-hy[draulic_conductivity]**, or **-K**.

units—Input units for hydraulic conductivity (L/T). Units must include the divide symbol, "/" and must contain no spaces.

Line 8: **-specific_storage** *units*

-specific_storage—This identifier is used to specify the units of specific storage for input data in the MEDIA data block. Optionally, **specific_storage**, **storage**, **-s[pecific_storage]**, or **-s[torage]**.

units—Input units for specific storage (1/L). Units must include the "1/" and must contain no spaces.

Line 9: **-dispersivity** *units*

-dispersivity—This identifier is used to specify the units of dispersivity for input data in the MEDIA data block. Optionally, **dispersivity**, **alpha**, **-d[ispersivity]**, or **-a[lpha]**.

units—Input units for dispersivity (L).

Line 10: **-flux** *units*

> **-flux**—This identifier is used to specify the units of fluid flux for input data in **FLUX_BC** data blocks. Optionally, **flux** or **-f[lux]**.
>
> *units*—Input units for fluid flux (L/T). Units must include the divide symbol, "/" and must contain no spaces.

Line 11: **-leaky_hydraulic_conductivity** *units*

> **-leaky_hydraulic_conductivity**—This identifier is used to specify the units of hydraulic conductivity for input data in **LEAKY_BC** data blocks. Optionally, **leaky_hydraulic_conductivity**, **leaky_K**, **-leaky_h[ydraulic_conductivity]**, or **-leaky_K**.
>
> *units*—Input units for hydraulic conductivity (L/T) for **LEAKY_BC** data blocks. Units must include the divide symbol, "/" and must contain no spaces.

Line 12: **-leaky_thickness** *units*

> **-leaky_thickness**—This identifier is used to specify the units of thickness for the leaky boundary for input data in **LEAKY_BC** data blocks. Optionally, **leaky_thickness**, **thickness**, **-leaky_t[hickness]**, or **-t[hickness]**.
>
> *units*—Input units for thickness of the leaky boundary (L).

Line 13: **-well_diameter** *units*

> **-well_diameter**—This identifier is used to specify the units of the diameter or radius of wells for input data in **WELL** data blocks. Optionally, **well_diameter** or **-well_d[iameter]**.
>
> *units*—Input units for well diameter or radius, L.

Line 14: **-well_flow_rate** *units*

> **-well_flow_rate**—This identifier is used to specify the units of flow rates for input data in **WELL** data blocks. Optionally, **well_flow_rate**, **well_pumpage**, **-well_f[low_rate]**, or **-well_p[umpage]**.
>
> *units*—Input units for flow rate (L^3/T). Flow-rate units can be U.S. customary (for example, "gal/min", "gpm", "ft^3/s", or "ft3/s"), metric (for example, "liter/minute", "meters^3/day", or "meters3/day"), or an abbreviation of one of these units.

Line 15: **-well_depth** *units*

> **-well_depth**—This identifier is used to specify the units of the depth below the land-surface datum of open intervals when using the **-depth** identifier of the **WELL** data block. Optionally, **well_depth** or **-well_de[pth]**.
>
> *units*—Input units for depths of open intervals of the well (L).

Line 16: **-river_bed_hydraulic_conductivity** *units*

> **-river_bed_hydraulic_conductivity**—This identifier is used to specify the units of hydraulic conductivity for input data in **RIVER** data blocks. Optionally, **river_bed_hydraulic_conductivity**, **river_bed_k**, **river_k**, **-river_bed_h[ydraulic_conductivity]**, **-river_bed_k**, or **-river_k**.
>
> *units*—Input units for hydraulic conductivity (L/T) of riverbeds. Units must include the divide symbol, "/" and must contain no spaces.

Line 17: **-river_bed_thickness** *units*

> **-river_bed_thickness**—This identifier is used to specify the units of thickness of the riverbed for input data in **RIVER** data blocks. Optionally, **river_bed_thickness**, **river_thickness**, **-river_bed_t[hickness]**, or **-river_t[hickness]**.
>
> *units*—Input units for thickness of the riverbed (L).

Line 18: **-river_width** *units*

-river_width—This identifier is used to specify the units of width of the river for input data in RIVER data blocks. Optionally, **river_width** or **-river_w[idth]**.

units—Input units for width of the river (L).

Line 19: **-river_depth** *units*

-river_depth—This identifier is used to specify the units of the depth of the river bottom below the initial head (specified or interpolated) of the river. The units apply to data defined with the **-depth** identifier of the RIVER data block. Optionally, **river_depth** or **-river_de[pth]**.

units—Input units for depths of the river bottom (L).

Line 20: **-drain_hydraulic_conductivity** *units*

-drain_hydraulic_conductivity—This identifier is used to specify the units of hydraulic conductivity for input data in DRAIN data blocks. Optionally, **drain_hydraulic_conductivity, drain_bed_k, drain_k, -drain_h[ydraulic_conductivity], -drain_bed_k,** or **-drain_k**.

units—Input units for hydraulic conductivity (L/T) of the layer surrounding drains. Units must include the divide symbol, "/" and must contain no spaces.

Line 21: **-drain_thickness** *units*

-drain_thickness—This identifier is used to specify the units of thickness of the layer surrounding the drain for input data in DRAIN data blocks. Optionally, **drain_thickness, drain_bed_thickness, -drain_t[hickness],** or **-drain_bed_t[hickness]**.

units—Input units for thickness of the layer surrounding the drain (L).

Line 22: **-drain_width** *units*

-drain_width—This identifier is used to specify the units of width of the layer surrounding the drain for input data in DRAIN data blocks. Optionally, **drain_width** or **-drain_w[idth]**.

units—Input units for width of the layer surrounding the drain (L).

Line 23: **-equilibrium_phases** (**WATER | ROCK**)

-equilibrium_phases—This identifier is used to specify whether the number of moles of minerals in EQUILIBRIUM_PHASES data blocks are interpreted as moles per liter of water or moles per liter of rock when defining PHAST initial conditions. Optionally, **equilibrium_phases** or **-eq[uilibrium_phasesh]**.

(**WATER | ROCK**)—Water indicates moles of each equilibrium phase are mol/L water—the units used in PHAST. Rock indicates moles of each equilibrium phase are converted from mol/L rock to mol/L water as initial conditions are distributed to each cell.

Line 24: **-exchange** (**WATER | ROCK**)

-exchange—This identifier is used to specify whether the number of moles of exchangers in EXCHANGE data blocks are interpreted as moles per liter of water or moles per liter of rock when defining PHAST initial conditions. Optionally, **exchange** or **-e[xchange]**.

(**WATER | ROCK**)—Water indicates moles of each exchanger are mol/L water—the units used in PHAST. Rock indicates moles of each exchanger are converted from mol/L rock to mol/L water as initial conditions are distributed to each cell.

Line 25: **-surface** (**WATER | ROCK**)

-surface—This identifier is used to specify whether the number of moles of surfaces in SURFACE data blocks are interpreted as moles per liter of water or moles per liter of rock when defining PHAST initial conditions. Optionally, **surface** or **-su[rface]**.

(**WATER** | **ROCK**)—Water indicates moles of each surface are mol/L water—the units used in PHAST. Rock indicates moles of each surface are converted from mol/L rock to mol/L water as initial conditions are distributed to each cell.

Line 26: **-solid_solutions** (**WATER** | **ROCK**)

-solid_solutions—This identifier is used to specify whether the number of moles of solid solutions in **SOLID_SOLUTIONS** data blocks are interpreted as moles per liter of water or moles per liter of rock when defining PHAST initial conditions. Optionally, **solid_solutions** or **-so[lid_solutions]**.

(**WATER** | **ROCK**)—Water indicates moles of each solid solution are mol/L water—the units used in PHAST. Rock indicates moles of each solid solution are converted from mol/L rock to mol/L water as initial conditions are distributed to each cell.

Line 27: **-kinetics** (**WATER** | **ROCK**)

-kinetics—This identifier is used to specify whether the number of moles of kinetic reactants in **KINETICS** data blocks are interpreted as moles per liter of water or moles per liter of rock when defining PHAST initial conditions. Optionally, **kinetics** or **-k[inetics]**.

(**WATER** | **ROCK**)—Water indicates moles of each kinetic reactant are mol/L water—the units used in PHAST. Rock indicates moles of each kinetic reactant are converted from mol/L rock to mol/L water as initial conditions are distributed to each cell.

Line 28: **-gas_phase** (**WATER** | **ROCK**)

-gas_phase—This identifier is used to specify whether the number of moles of gas-phase components in **GAS_PHASE** data blocks are interpreted as moles per liter of water or moles per liter of rock when defining PHAST initial conditions. Optionally, **gas_phase** or **-g[as_phase]**.

(**WATER** | **ROCK**)—Water indicates moles of each gas-phase component are mol/L water—the units used in PHAST. Rock indicates moles of each gas-phase component are converted from mol/L rock to mol/L water as initial conditions are distributed to each cell.

Notes

The **UNITS** data block must be defined for all flow and transport simulations. The definitions in this data block specify the units for input data for all keyword data blocks in the flow and transport data file. Note that all definitions apply to the input data. All output data are in SI units, except time. The unit of time defined with the **-time** identifier is used for both input and output data. Units for time (T) can be "seconds", "minutes", "hours", "days", or "years" or an abbreviation of one of these units. Units for distance (L) can be U.S. customary ("inches", "feet", or "miles"), metric ("millimeters", "centimeters", "meters", or "kilometers") or an abbreviation of one of these units. Units need not be defined for features that are not used in a simulation. For example, units applying to rivers need not be defined if no rivers are used in the simulation.

The identifiers **-equilibrium_phases**, **-exchange**, **-surface**, **-solid_solutions**, **-kinetics**, and **-gas_phase** can be used to specify whether moles of solid reactants are interpreted as moles per liter of water or moles per liter of rock. By default, the number of moles of solid reactants are interpreted as per liter of water (**WATER** option). However, when the domain has a nonuniform distribution of porosity, it is difficult to define the number of moles of solid reactants per liter of water—the units used by PHAST—because the varying porosity must be taken into account. By using the **ROCK** option, a constant composition relative to rock (mol/L rock) can be defined and the factor $(1-\phi)/\phi$ is used to scale the number of moles of solid reactants as the initial conditions are distributed to the cells of the domain, where ϕ is the porosity for the cell. Note that an initial PHREEQC calculation is made at the beginning of a PHAST run. During this calculation all reaction calculations interpret the concentration of solid reactants as simply moles. (When a solution is reacted with the solids, the solution contains a mass of water, which is usually 1 kg, but not necessarily.) It is only when initial conditions are subsequently distributed before the transport calculations that solid reactants are scaled.

Four Basic functions that can be used in **USER_PUNCH** (chemistry input file) are provided to be able to accumulate the number of moles in a PHAST finite-difference cell. CELL_VOLUME is the total volume of the current cell, in liters. CELL_PORE_VOLUME is the total void space in the cell, in liters (no adjustment for compressibility is made). CELL_SATURATION is the fraction of the void space filled with water, unitless. CELL_SATURATION is 1.0 for confined flow, and between 0 and 1.0, inclusive, for unconfined flow. CELL_POROSITY is the ratio of void to rock volume, unitless, and is equal to CELL_PORE_VOLUME / CELL_VOLUME. The values of these functions are set to the appropriate values for each cell as the chemistry in the cell is computed.

For confined flow, the concentrations of dissolved or solid constituents multiplied by the pore volume is equal to the total number of moles in the finite-difference cell, for example, TOT("Cl")*CELL_PORE_VOLUME, or EQUI("Calcite")*CELL_PORE_VOLUME. For a water-table cell in unconfined flow, the solid reactants are split into an unreactive pool representing the unsaturated part of the cell and a reactive pool representing the saturated part of the cell. The formula TOT("Cl")*CELL_SATURATION*CELL_PORE_VOLUME gives the number of moles of dissolved chloride in the current finite-difference cell. The formula EQUI("Calcite")*CELL_SATURATION*CELL_PORE_VOLUME gives the moles of calcite in the reactive pool of the current finite-difference cell. It is a deficiency in PHAST that no functions exist to determine the unreactive pool of solid reactants in a cell. Note that in steady-state unconfined flow, the unreactive pool of solid reactants for a water-table cell cannot change, whereas, as the water table rises and falls for transient unconfined flow, the unreactive pool will vary.

Example Problems

The **UNITS** data block is used in Chapter 6 example problems 1, 2, 3, 4, 5, and 6.

WELL

This keyword data block is used to define an injection or pumping well. This data block is optional and only is needed if a well boundary condition is included in the simulation. The location, diameter, open intervals, pumping or injection rate, and solution composition must be defined for the well. The pumping or injection rate and the chemical composition of injected water may vary over the course of a simulation. Multiple **WELL** data blocks are used to define all of the wells in the grid region.

Example

Line numbers in the Example are used for identification purposes only and are referenced in the following Explanation. The line numbers do not indicate a required order for the keyword data block. Line numbers are not used in the flow and transport data file.

```
Line 0:      WELL 122 Metropolis Injection Well 122
Line 1:          -xy_coordinate_system      map
Line 2:          -z_coordinate_system       map
Line 3:          -location    1766.         2356.
Line 4:          -injection_rate
Line 5:              0                0.0
Line 5a:             1        yr     4.5
Line 6:          -solution
Line 7:              0                16
Line 7a:            0.5       yr     17
Line 8:          -diameter              12
Line 9:          -elevation        101.   107.
Line 9a:         -elevation        143.   153.
Line 9b:         -elevation        175.   183.
Line 0a:     WELL 165 Metropolis Supply Well 165
Line 1a:         -xy_coordinate_system      map
Line 2a:         -z_coordinate_system       grid
Line 3a:         -location    1833.         2320.
Line 10:         -pumping_rate
Line 11:             0                4.5
Line 11a:            2                3.5
Line 12:         -land_surface_datum    193.
Line 13:         -radius                12
Line 14:         -depth             220.   250.
Line 14a:        -depth             100.   105.
Line 15:         -allocate_by_head_and_mobility      true
```

Explanation

Line 0: **WELL** *number* [*description*]

> **WELL** is the keyword for the data block.
>
> *number*—Positive number that identifies this well. Default is 1.
>
> *description*—Optional character field that identifies the well.

Line 1: **-xy_coordinate_system (grid | map)**

> **-xy_coordinate_system**—Coordinate system for X–Y location of the well. If
> **-xy_coordinate_system** is not defined for a well, the default is **grid**. Optionally,
> **xy_coordinate_system** or **-xy[_coordinate_system]**.

(**grid** | **map**)—**Grid** indicates that the X–Y location is defined in the grid coordinate system with units defined by -**horizontal_grid** in the UNITS data block. **Map** indicates that the X–Y location is defined in the map coordinate system with units defined by -**map_horizontal** in the UNITS data block. The transformation from grid to map coordinates is defined in the GRID data block.

Line 2: -**z_coordinate_system** (**grid** | **map**)

   -**z_coordinate_system**—Coordinate system for screen intervals defined with the -**elevation** identifier and for the land-surface datum (-**land_surface_datum** identifier). If -**z_coordinate_system** is not defined for a well, the default is **grid**. Optionally, **z_coordinate_system** or -**z_[coordinate_system]**.

   (**grid** | **map**)—**Grid** indicates that the Z coordinates for the -**elevation** identifier and land-surface datum are defined in the grid coordinate system, with units defined by -**vertical_grid** in the UNITS data block. **Map** indicates that the Z coordinates defined for the -**elevation** identifier and land-surface datum are in the map coordinate system, with units defined by -**map_vertical** in the UNITS data block. The transformation from map to grid coordinates is defined in the GRID data block.

Line 3: -**location** *x, y*

   -**location**—X–Y location of the well. Units, L, are defined by the -**horizontal_grid** in the UNITS data block when -**xy_coordinate_system grid**, or -**map_horizontal** in the UNITS data block when -**xy_coordinate_system map**. Optionally, **location**, **xy_location**, -**lo[cation]**, or -**xy_l[ocation]**.

   x—X location of the well.

   y—Y location of the well.

Line 4: -**injection_rate**

   -**injection_rate**—This identifier is used to specify the rate of fluid injection into the well. The injection rate is defined on line 5. A time series of injection rates may be defined by using multiple line 5s. The first *time* in the series must be zero. Optionally, **injection_rate**, **injection**, or -**i[njection_rate]**.

Line 5: *time* [*units*] *injection_rate*

   time—Simulation time (T) at which the injection rate will take effect. Units may be defined explicitly with *units*; default units are defined by -**time** identifier in UNITS data block.

   units—Units for *time* can be "seconds", "minutes", "hours", "days", or "years" or an abbreviation of one of these units.

   injection_rate—Rate of fluid injection into the well starting at *time*. Units, L^3/T, are defined by the -**well_flow_rate** identifier in the UNITS data block.

Line 6: -**solution**

   -**solution**—This identifier is used to specify a solution index number that defines the composition of water injected into the well. The identifier is required for each injection well. Solution index numbers correspond to solution compositions defined in the chemistry data file. The composition of injected water is defined on line 7. A time series of solution indices may be defined by using multiple line 7s. The first *time* in the series must be zero. Optionally, **solution**, **associated_solution**, -**s[olution]**, or -**a[ssociated_solution]**.

Line 7: *time* [*units*] *solution*

   time—Simulation time (T) at which the solution composition will take effect. Units may be defined explicitly with *units*; default units are defined by -**time** identifier in UNITS data block.

units—Units for *time* can be "seconds", "minutes", "hours", "days", or "years" or an abbreviation of one of these units.

solution—Solution index number for injected water composition starting at *time*.

Line 8: **-diameter** *diameter*

-diameter—This identifier is used to specify the diameter of the well. Optionally, **diameter**, or **-di[ameter]**.

diameter—Diameter of the well. Units, L, are defined by the **-well_diameter** identifier in the **UNITS** data block.

Line 9: **-elevation** *elevation$_1$, elevation$_2$*

-elevation—This identifier is used to specify a top and bottom elevation for an open interval in the well. The top and bottom elevations may be entered in either order. Multiple **-elevation** identifiers are used to define all open intervals for the well; each identifier defines one open interval. Optionally, **elevation**, **elevations**, or **-e[levations]**.

elevation$_1$—Elevation of start of open interval in the well. Units, L, are defined by either the **-vertical_grid** identifier or the **-map_vertical** identifier in the **UNITS** data block, depending on the value of **-z_coordinate_system** (Line 2).

elevation$_2$—Elevation of end of open interval in the well. Units, L, are defined by either the **-vertical_grid** identifier or the **-map_vertical** identifier in the **UNITS** data block, depending on the value of **-z_coordinate_system** (Line 2).

Line 10: **-pumping_rate**

-pumping_rate—This identifier is used to specify the pumping rate for the well. The pumping rate is defined on line 11. A time series of pumping rates may be defined by using multiple line 11s. Optionally, **pumping_rate**, **pumping**, **pumpage**, **-pu[mping_rate]**, or **-pu[mpage]**.

Line 11: *time [units] pumping_rate*

time—Simulation time (T) at which the pumping rate will take effect. Units may be defined explicitly with *units*; default units are defined by **-time** identifier in **UNITS** data block.

units—Units for *time* can be "seconds", "minutes", "hours", "days", or "years" or an abbreviation of one of these units.

pumping_rate—Pumping rate (L^3/T) starting at *time*. Units are defined by the **-well_flow_rate** identifier in the **UNITS** data block.

Line 12: **-land_surface_datum** *land_surface_datum*

-land_surface_datum—This identifier is used to specify the elevation of the well datum. This data item is required only if open intervals are defined with the **-depth** identifier. The coordinate system for the land-surface datum is defined by the **-z_coordinate system** identifier (Line 2). Optionally, **land_surface_datum**, **lsd**, **-l[and_surface_datum]**, or **-l[sd]**.

land_surface_datum—Elevation of land surface at the well. Units, L, are defined by either the **-vertical_grid** identifier or the **-map_vertical** identifier in the **UNITS** data block, depending on the value of **-z_coordinate_system** (Line 2).

Line 13: **-radius** *radius*

-radius—This identifier is used to specify the radius of the well. Optionally, **radius**, or **-r[adius]**.

radius—Radius of the well. Units, L, are defined by the **-well_diameter** identifier in the **UNITS** data block.

Line 14: **-depth** *depth₁, depth₂*

    -depth—This identifier is used to specify the top and bottom of an open interval as depths below the well land-surface datum. The two depths may be entered in either order. The **-land_surface_datum** identifier must be used to specify the elevation of the well land-surface datum. Multiple **-depth** identifiers are used to define all open intervals for the well. Optionally, **depth**, **depths**, or **-de[pths]**.

    depth₁—Depth of top of open interval in the well. Units, L, are defined by the **-well_depth** identifier in the UNITS data block.

    depth₂—Depth of bottom of open interval in the well. Units, L, are defined by the **-well_depth** identifier in the UNITS data block.

Line 15: **-allocate_by_head_and_mobility** [(**True** | **False**)]

    -allocate_by_head_and_mobility—This identifier is used to specify the method by which water is distributed to the aquifer from the well or received from the aquifer to the well. The well-bore flow rate can be allocated by the product of mobility and head difference or by mobility only (Kipp, 1987, p. 34 and p. 122). By default, allocation is by mobility only. Optionally, **allocation_by_head_and_mobility**, **allocate_by_head_and_mobility**, **head_and_mobility**, **-al[location_by_head_and_mobility]**, **-al[location_by_head_and_mobility]**, or **-h[ead_and_mobility]**.

    (**True** | **False**)—**True** allocates flow by mobility and head difference. **False** allocates flow by mobility only. If neither true nor false is entered on the line, true is assumed. Optionally, **t[rue]** or **f[alse]**.

Notes

Multiple **WELL** data blocks are defined by using multiple **WELL** data blocks, where each well is uniquely identified by the integer following the **WELL** keyword. The well location, geometry, and flow rate must be defined for each well, including X–Y coordinate location, diameter or radius, injection or pumping rate, open intervals, and solution composition. Only the flow rate (**-injection_rate** or **-pumping_rate**) and the solution composition (**-solution**) for a well may vary during a simulation. Open intervals for a well could include screened intervals, perforated intervals, and open-hole intervals.

The coordinate system for the X–Y location of a well may be either of the two available coordinate systems, grid or map, as defined by the **-xy_coordinate_system** identifier. The coordinate system for the well-screen elevations (**-elevation** identifier) and land-surface datum (**-land_surface_datum** identifier) may be grid or map as defined by the **-z_coordinate_system** identifier.

The elevation of open intervals in a well can be defined in two ways: (1) the elevations of the top and bottom of an interval can be defined explicitly with the **-elevation** identifier, or (2) the well land-surface datum (elevation of a measuring point) can be defined with the **-land_surface_datum** identifier and the depths to the top and bottom of each open interval can be defined with the **-depth** identifier. All data are converted to grid units for calculations of screened intervals. Ultimately, all data are converted to SI units for the simulation calculations.

If the well is an injection well, the flow rate is defined with the **-injection_rate** identifier as a positive number, and an associated-solution composition (**-solution** identifier) must be specified for reactive-transport simulations. If the well is a pumping well, the flow rate is defined with the **-pumping_rate** identifier as a positive number. A well can be changed from an injection well to a pumping well (or the reverse) in the time-series definition for injection or pumping rate by changing the sign of the rate.

Wells in corner cells or side cells are assumed to be located exactly at the corner or at the mid-side of the cell, respectively. The well flow rate into or out of the active grid region is specified by the user. The program does not adjust the well flow rate to account for symmetry at boundaries. Thus, if the active grid region is one-quarter or one-half of a symmetric region (see Example 2 for an example of the use of symmetry) then the user must reduce the well flow rate by a factor of one-quarter or one-half, respectively, to account for the actual flow into the active grid region.

By default, allocation of well-bore flow to the open intervals is by a formula that depends only on mobility. Alternatively, the **-allocate_by_head_and_mobility** identifier can be used to allocate well-bore flow by a formula that uses the product of mobility and head difference between the well and the aquifer (Kipp, 1987, p. 34 and p. 122). For numerical stability, wells should have at least a small flow rate. More information on wells is provided in section D.4. Well-Source Conditions.

Example Problems

The **WELL** data block is used in Chapter 6 example problem **4**.

ZONE_FLOW

This keyword data block is used to define a zone for which water and solute flows will be calculated. The calculated water and solute (component) flows include the flow in and out of the zone through active cell faces and summations of flows in and out through each boundary-condition type within the zone: specified head, leaky, flux, river, drain, and well. This data block is optional and only is needed if an accounting of water and solute flows for a subregion of the grid is desired. Multiple ZONE_FLOW data blocks are used to define all of the subregions for which accounting is desired. Multiple zone-flow regions may be aggregated with the **-combination** identifier.

Example

Line numbers in the Example are used for identification purposes only and are referenced in the following Explanation. The line numbers do not indicate a required order for the keyword data block. Line numbers are not used in the flow and transport data file.

```
Line 0:    ZONE_FLOW 1 Flow into and out upper half of domain
Line 1:        -box         0      0      5      10     10     10
Line 2:        -write_heads_xyzt    subdomain.heads.xyzt
Line 0a:   ZONE_FLOW 2 Pond 7
Line 1a:       -prism
Line 3a:           -perimeter   shape       map    pond7.shp
Line 4a:           -bottom      shape       map    pond_bottom7.shp   3
Line 0b:   ZONE_FLOW 3 Sum of Pond 7 and Pond 5
Line 5:        -combination     2
Line 1b:       -prism
Line 3b:           -perimeter   shape       map    pond5.shp
Line 4b:           -bottom      shape       map    pond_bottom5.shp   3
```

Explanation

Line 0: **ZONE_FLOW** *number,* [*description*]

    ZONE_FLOW is the keyword for the data block.

    number—Positive number that identifies this zone. Default is 1.

    description—Optional character field that identifies the zone.

Line 1: *zone*

    zone—A zone definition as defined in section 4.5. Description of Input for Zones. Line 1 defines a box zone; lines 1a, 3a, and 4a and 1b, 3b, and 4b define prism zones.

Line 2: **-write_heads_xyzt** *file_name*

    -write_heads_xyzt—This identifier causes heads for the zone to be written to the specified file at a print frequency defined by the **-zone_flow_heads** identifier of the PRINT_FREQUENCY data block. Optionally, **write_heads_xyzt**, or **-w**[**rite_heads_xyzt**].

    file_name—A file name to which the time series of heads will be written. This file is suitable for use with an XYZT property definition.

Line 3: *zone continued*

    zone continued—Continuation of zone definition, see line 1.

Line 4: *zone continued*

    zone continued—Continuation of zone definition, see line 1.

Line 5: **-combination** *list_of_zone_flows*

> **-combination**—This identifier is used to include the cells defined by another **ZONE_FLOW** data block in the current **ZONE_FLOW** data block. Optionally, **combination**, or **-c[ombination]**.
>
> *list_of_zone_flows*—A list of one or more index numbers corresponding to **ZONE_FLOW** data blocks. Index numbers are white space delimited.

Notes

The **ZONE_FLOW** data block is used to provide water and solute flows in and out of subregions of the grid. An appropriate box zone definition could be used, for example, to calculate the flux of solute across a plane intersecting the active grid region. The zones may be arbitrarily shaped as allowed by zone definitions (section 4.5. Description of Input for Zones). Multiple accounting zones may be defined by use of multiple **ZONE_FLOW** data blocks, where each flow-rate zone is uniquely identified by the integer following the **ZONE_FLOW** keyword. Flows for multiple zones may be aggregated by use of the **-combination** identifier; however, the resulting flows may not be the numeric sum of the flows for the two separate zones if the zones overlap. A new zone is calculated that includes all unique cells from the two (or more) zones and flow rates are calculated and printed for the combined zone.

For each zone defined by a **ZONE_FLOW** data block, water and solute flow data will be written to the *prefix*.**zf.txt** and *prefix*.**zf.tsv** files at the frequency specified by the **-zone_flows** and **-zone_flow_tsv** identifiers in the **PRINT_FREQUENCY** data block, respectively. The *prefix*.**zf.txt** file contains the water and solute flow information for each print cycle in a form readable in a text editor. The *prefix*.**zf.tsv** file contains rows of data, one row for water and each component for each print cycle. Each row contains the time, the zone number, the component name, and the flow rates in kilograms per time unit in and out for the boundary of the zone and in and out of each boundary condition within the zone.

Example Problems

The **ZONE_FLOW** data block is used in Chapter 6 example problems **5** and **6**.

Chapter 5. Output Files

PHAST has many options to control writing data to output files. Many of the output files are in paragraph or tabular form to be displayed or printed. Some files are columnar data or binary, and are intended for post-processing by plotting, statistical, and visualization programs. All of the output files are written in ASCII text format except for the HDF and restart files, which are in compressed binary formats. Programs are included in the PHAST distribution package to visualize and extract data from the HDF file. The restart file can be used to provide initial chemical conditions for PHAST simulations.

5.1. Content of Output Files

Each output file is listed in this section with a description of the types of data that are written to that file. In addition, the keyword data blocks and identifiers that control the content and print frequency are listed. Files are listed alphabetically by suffix.

prefix.**bal.txt**—Fluid and chemical-component global mass balances for the time step, cumulative balances, and balances listed by boundary-condition type are written to this file. This ASCII file is intended to be viewed with a text editor.

The frequency of writing data to the file is specified with the **-flow_balance** identifier in the PRINT_FREQUENCY data block. By default, the flow-balance information is written at the end of each simulation period.

prefix.**bcf.txt**—Fluid and solute flow rates through each boundary-condition cell are listed in this file by boundary-condition type. Data in the file are formatted as planes of nodes, either X–Y or X–Z, as defined by **-print_orientation** in the GRID data block. This ASCII file is intended to be viewed with a text editor.

The frequency of writing data to the file is specified with the **-bc_flow_rates** identifier in the PRINT_FREQUENCY data block. By default, boundary-condition flow rates are not written to the file.

prefix.**chem.txt**—Solution concentrations, distribution of aqueous species, saturation indices, and compositions of equilibrium-phase assemblages, exchangers, surfaces, kinetic reactants, solid solutions, and gas phases are written to this file while processing the chemistry data file at the beginning of a PHAST simulation. Data also may be printed for cells during the subsequent reactive-transport part of the simulation. By default, these data are not written during the reactive-transport simulation because this file could be very large. However, for diagnostic purposes, for small problems, or for selected locations, it is possible to enable printing during the reactive-transport simulation. This ASCII file is intended to be viewed with a text editor.

Writing to the file can be enabled and the frequency of writing specified by using the **-force_chemistry_print** identifier in the PRINT_FREQUENCY data block. By default, data for all cells will be written to this file, if printing is enabled with the **-force_chemistry_print** identifier. The PRINT_LOCATIONS data block can be used to restrict the set of cells for which data will be written. In addition, the PRINT and USER_PRINT data blocks in the chemistry data file can be used to choose the types of data that are written to the file.

prefix.**chem.xyz.tsv**—Selected initial-condition and transient chemical data for solutions, equilibrium-phases, exchangers, surfaces, kinetic reactants, solid solutions, and gas phases are written to this file for all active nodes in natural order (see section 4.2. Spatial Data). Basic-language programs can be used to calculate other chemical quantities that are written to this file. Data in the file are formatted in columns with one row per node; X–Y– Z coordinates, time, and a cell-saturation indicator begin each row. This tab-separated-values ASCII file is intended to be used for postprocessing simulation results by user-supplied spreadsheet, statistical, plotting, and visualization programs.

Initial-condition data are written to the file at the beginning of the simulation as defined by the **-xyz_chemistry** identifier in **PRINT_INITIAL** data block. By default, no initial-condition chemical data are written to the file. The frequency for writing transient chemical data to the file is specified with the **-xyz_chemistry** identifier in the **PRINT_FREQUENCY** data block. By default, no transient chemical data are written to the file. The **PRINT_LOCATIONS** data block in the flow and transport data file can be used to specify the cells for which data will be printed. By default, chemical data will be printed for all cells in the active grid region.

The chemical data to be written to the *prefix*.**chem.xyz.tsv** file are defined in the **SELECTED_OUTPUT** data block of the chemistry data file and by Basic-language programs in the **USER_PUNCH** data block of the chemistry data file.

prefix.**comps.txt**—Initial-condition chemical information and transient total dissolved component concentrations are written to this file. Initial-condition information includes the indices and mixing fractions for solutions, equilibrium phases, exchangers, surfaces, kinetic reactants, solid solutions, and gas phases that define chemical conditions at the beginning of the simulation. Component concentrations may be written to the file during the reactive-transport simulation. Data in the file are formatted as planes of nodes, either X–Y or X–Z, as defined by **-print_orientation** in the **GRID** data block. This ASCII file is intended to be viewed with a text editor.

Initial-condition chemical information is written at the beginning of the simulation as specified by the **-components** identifier in the **PRINT_INITIAL** data block. By default, initial-condition information is not written to the file. The frequency of writing transient data during the reactive-transport simulation is specified with the **-components** identifier in the **PRINT_FREQUENCY** data block. By default, transient component concentrations are not written to the file.

prefix.**comps.xyz.tsv**—Initial-condition and transient concentrations for each component (chemical element) are written to this file for all nodes in natural order (see section 4.2. Spatial Data). Data in the file are formatted in columns with one row per node; X–Y–Z coordinates, time, and a cell-saturation indicator begin each row. This tab-separated-values ASCII file is intended to be used for postprocessing simulation results by user-supplied spreadsheet, statistical, plotting, and visualization programs.

Initial-condition component data are written at the beginning of the simulation as defined by the **-xyz_components** identifier in **PRINT_INITIAL** data block. By default, initial component data are not written to the file. The frequency for writing transient chemical component data to the file is specified with the **-xyz_components** identifier in the **PRINT_FREQUENCY** data block. By default, transient chemical data are not written to the file.

prefix.**h5**—Grid and boundary-condition information, media properties, heads, nodal velocities, and selected chemical concentration data are written to this HDF (hierarchical data format) file. Grid node locations and the identity of nodes with each boundary-condition type are written to this file. Boundary-condition types include specified head, flux, leaky, river, drain, and well. Media properties, including hydraulic conductivities, porosity, storage coefficient, and dispersivities, can be written to this file. In addition, heads, nodal velocities, and chemical data can be written to this file at varying frequencies. Basic-language programs can be used to calculate chemical quantities that are written to this file. This binary file is intended to be used with the data visualization program, Model Viewer (Appendix A) and the data extraction program PHASTHDF (Appendix B).

Writing of media properties is controlled by the **-HDF_media_properties** identifier in the **PRINT_INITIAL** data block. By default, media properties are written to the file. Writing of initial-condition heads and chemical data is controlled by the identifiers **-HDF_heads** and **-HDF_chemistry** in the **PRINT_INITIAL** data block. By default, initial-condition heads and chemical data are written to the file. If steady-state flow is simulated, the steady-state velocities at nodes (section D.11. Nodal Velocity Calculation) can be written to the HDF file by using the **-HDF_steady_flow_velocities** identifier of the PRINT_INITIAL data block. By default, steady-state flow velocities are not written to the file. During the simulation, heads, nodal velocities, and chemical data are written at selected frequencies as defined by the **-HDF_heads**, **-HDF_velocities**, and **-HDF_chemistry** identifiers in the PRINT_FREQUENCY data block. By default, the heads, velocities, and chemistry data are written to the HDF file at the end of each simulation period. If steady-state flow is simulated, the heads and velocities will be written only once.

The chemical data to be written to the HDF file are defined in the **SELECTED_OUTPUT** data block of the chemistry data file and by Basic-language programs in the **USER_PUNCH** data block of the chemistry data file.

prefix.**head.dat**—This ASCII file is intended to be used for initial head conditions in subsequent simulations. At the end of the simulation, heads for each node in the grid region are written to this file. The data are saved in tab-separated-values ASCII file with the X–Y–Z location and head for each active grid node; the file contains one row per grid node. These heads can be used as the initial head condition in subsequent simulations by using the **xyz** option to define values for the **-head** identifier in the HEAD_IC data block. The data from the *prefix*.**head.dat** file can be used as initial conditions even if the new simulation has a different grid. The node locations saved in the file are treated as scattered X–Y–Z data and heads are interpolated to a new grid by closest point interpolation.

Writing to this file is controlled by the **-save_final_heads** identifier of the PRINT_FREQUENCY data block. If a file named *prefix*.**head.dat** exists, it is overwritten. By default, final heads are not written to the file.

prefix.**head.txt**—Initial and transient heads for each node are written to this file. If steady-state flow is simulated, heads are written only once during the simulation. Data in the file are formatted as planes of nodes, either X–Y or X–Z, as defined by **-print_orientation** in the GRID data block. This ASCII file is intended to be viewed with a text editor.

Initial heads are written at the beginning of the simulation as defined by the **-heads** identifier in PRINT_INITIAL data block. By default, initial heads are written to this file. The frequency of writing transient heads is specified by the **-heads** identifier in the PRINT_FREQUENCY data block. By default, heads are written to the file at the end of each simulation period.

prefix.**head.xyz.tsv**—Initial-condition and transient potentiometric heads for all active nodes are written to this file in natural order (see section 4.2. Spatial Data). If steady-state flow is simulated, heads are written at most once during the simulation. Data in the file are formatted in columns with one row per node; X–Y–Z coordinates, time, and a cell-saturation indicator begin each row. This tab-separated-values ASCII file is intended to be used for postprocessing simulation results by user-supplied spreadsheet, statistical, plotting, and visualization programs.

Initial-condition heads are written at the beginning of the simulation as defined by the **-xyz_heads** identifier in PRINT_INITIAL data block. By default, initial heads are not written to the file. The frequency for writing transient heads to the file is specified with the **-xyz_heads** identifier in the PRINT_FREQUENCY data block. By default, transient heads are not written to the file.

prefix.**kd.txt**—Static fluid-conductance factors, fluid conductances (Kipp, 1987, p. 104), and transient solute dispersive conductances (Kipp, 1987, p. 110) are written to this file for positive X, Y, and Z cell faces for each cell. Data in the file are formatted as planes of nodes, either X–Y or X–Z, as defined by **-print_orientation** in the GRID data block. This ASCII file is intended to be viewed with a text editor.

Static fluid-conductance factors are written at the beginning of the simulation as specified by the **-conductances** identifier in PRINT_INITIAL data block. By default, conductance factors are not written to the file. The frequency of writing transient conductances is specified by **-conductances** identifier in the PRINT_FREQUENCY data block. By default, transient conductances are not written to the file.

prefix.**log.txt**—This file is used to document the progress of the simulation. The *prefix*.**trans.dat** and *prefix*.**chem.dat** input data files are echoed to this file along with any error and warning messages resulting from processing these input files. Where possible, error messages immediately follow the echoed line from the input file that contains the error. Additional log messages indicate the progress of the simulation through simulation periods and time steps. Error and warning messages generated during the simulation calculations are written to this file. This ASCII file is intended to be viewed with a text editor.

The identifiers **-echo_input** in **PRINT_INITIAL** data block and in the **PRINT** data block of the chemistry data file can be used to disable echoing input data to the log file; by default, echoing is enabled. The **-progress_statistics** identifier in **PRINT_FREQUENCY** data block can be used to control the time-step frequency for detailed printing of progress statistics to the log file. By default, progress statistics are written at the end of each simulation period.

prefix.**probdef.txt**—Static and initial-condition flow and transport information that relates to the problem definition for the simulation is written to this file. The *prefix*.**probdef.txt** file is used to verify that the input data properly define the simulation. Information written to the file includes array sizes, grid definition, permeability distribution, porosity distribution, dispersivity distribution, fluid properties, initial head distribution, indices for initial conditions in each cell (solutions, equilibrium phases, exchangers, surfaces, kinetic reactants, solid solutions, and gas phases), initial component concentrations in each cell, specific storage distribution, static and transient boundary-condition information, cell volumes, and component concentrations for fixed and associated solutions for each boundary-condition cell. Transient boundary-condition information may be written to the file during the simulation. Data in the file that are spatially distributed are formatted as planes of nodes, either X–Y or X–Z, as defined by **-print_orientation** in the **GRID** data block. This ASCII file is intended to be viewed with a text editor.

The identifiers **-boundary_conditions**, **-fluid_properties**, and **-media_properties** in the **PRINT_INITIAL** data block can be used to select problem definition data to be printed in this file. By default, only the fluid properties are written to the file. Writing of transient boundary-condition information is specified by using the **-boundary_conditions** identifier in the **PRINT_FREQUENCY** data block. By default, transient boundary-condition information is not written to the file.

prefix.**restart.gz**—Chemical composition of solutions and reactants at all nodes in the active grid region are written to this file plus a list that contains the location of each node and the presence or absence of each type of reactant at the node. This file is intended to be used to specify initial chemical conditions in subsequent model runs. The data are written in compressed *gzip* format, which can be translated to an ASCII file with the *gunzip* utility. In ASCII format the file can be inspected and potentially edited, although the file is likely to be large. The file can be used to specify initial conditions for one or more types of reactants by using the **restart** option for identifiers in the **CHEMISTRY_IC** data block. Either the *gzip* or the ASCII version of the file can be read with a **restart** option. The data from the restart file can be used as initial conditions even if the new simulation has a different grid. The nodes of the restart file are treated as scattered X–Y–Z data and interpolated to a new grid by closest point interpolation.

The frequency for writing the restart file is specified with the **-restart** identifier in the **PRINT_FREQUENCY** data block. By default, data are not written to the file. If a file with the same name as the restart file (*prefix*.**restart.gz**) exists when the file is to be written, the old file is renamed to *prefix*.**restart.backup.gz**. If a file named *prefix*.**restart.backup.gz** exists, it is overwritten.

prefix.**vel.txt**—Interstitial velocity-vector components across cell faces and velocity-vector components at grid nodes (see section D.11. Nodal Velocity Calculation) are written to this file. If steady-state flow is simulated, velocity-vector components are written only once during the simulation. Data in the file are formatted as planes of nodes, either X–Y or X–Z, as defined by **-print_orientation** in the **GRID** data block. This ASCII file is intended to be viewed with a text editor.

If steady-state flow is simulated (**STEADY_FLOW** data block), the **-steady_flow_velocities** identifier in **PRINT_INITIAL** data block can be used to control printing of velocity-vector components to the file. By default, steady-state velocities are not written to the file. For transient-flow simulations, the frequency for writing transient velocity data is specified by the **-velocities** identifier in the **PRINT_FREQUENCY** data block. By default, transient velocities are not written to the file.

prefix.**vel.xyz.tsv**—Steady-state or transient velocity-vector components interpolated to grid nodes (see section D.11. Nodal Velocity Calculation) are written to this file for all nodes in natural order (see section 4.2. Spatial Data). If steady-state flow is simulated, velocity-vector components are written at most once during the simulation. Data in the file are formatted in columns with one row per node; X–Y– Z coordinates, time, and a cell-saturation indicator begin each row. This tab-separated-values ASCII file is intended to be used for postprocessing simulation results by user-supplied spreadsheet, statistical, plotting, and visualization programs.

If steady-state flow is simulated (**STEADY_FLOW** data block), the **-xyz_steady_flow_velocities** identifier in **PRINT_INITIAL** data block can be used to control printing of velocity-vector components to the file. By default, steady-state velocities are not written to the file. For transient-flow simulations, the frequency for writing transient velocity data is specified by the **-xyz_velocities** identifier in the **PRINT_FREQUENCY** data block. By default, transient velocities are not written to the file.

prefix.**wel.txt**—Static and transient well information is written to this file, including well location, well identification number, fluid and solute flow rates, cumulative fluid and solute flow amounts, solute concentrations, and injection and production rates per node for each well. This ASCII file is intended to be viewed with a text editor.

Static data are written at the beginning of the simulation as defined by the **-wells** identifier in **PRINT_INITIAL** data block. By default, static well data are written to the file. The frequency for writing transient well data to the file is specified by the **-wells** identifier in the **PRINT_FREQUENCY** data block. By default, transient well data are written to the file at the end of each simulation period.

prefix.**wel.xyz.tsv**—Initial and transient concentration data for wells are written to this file, including component concentrations, pH, and alkalinity. Data in the file are formatted in columns with one row per well; columns with X–Y well coordinates, vertical datum, time, and well number begin each row. This tab-separated-values ASCII file is intended to be used for postprocessing simulation results by user-supplied spreadsheet, statistical, plotting, and visualization programs.

The **-xyz_wells** identifier in **PRINT_INITIAL** data block is used to control printing of initial component concentrations for wells to the file. By default, initial component concentrations are not written to the file. The frequency for writing transient component concentrations to the file is specified with the **-xyz_wells** identifier in the **PRINT_FREQUENCY** data block. By default, transient component concentrations are not written to the file.

prefix.**wt.txt**—Steady-state or transient water-table heads are written to this file. If steady-state flow is simulated, heads are written only once during the simulation. Data in the file are formatted as an X–Y grid of water levels. This ASCII file is intended to be viewed with a text editor.

The frequency of writing transient water-table heads is specified by the **-heads** identifier in the **PRINT_FREQUENCY** data block. By default, heads are written to the file at the end of each simulation period for unconfined flow simulations.

prefix.**wt.xyz.tsv**—Steady-state or transient water-table heads are written to this file in natural order (see section 4.2. Spatial Data) for the X–Y grid. If steady-state flow is simulated, heads are written at most once during the simulation. Data in the file are formatted in columns with one row per X–Y location; X–Y coordinates, time, and a water-table head are written for each location. This tab-separated-values ASCII file is intended to be used for postprocessing simulation results by user-supplied spreadsheet, statistical, plotting, and visualization programs.

The frequency for writing transient water-table heads to the file is specified with the **-xyz_heads** identifier in the **PRINT_FREQUENCY** data block. By default, transient water-table heads are not written to the file.

*prefix***.zf.tsv**—Water and solute flow into and out of the boundaries of specified zones and water and solute flows through any boundary-condition cells within zones are written to this file. The zones are defined with the **ZONE_FLOW** data block. Data in the file are formatted in columns with rows for water and each constituent for each zone; time, zone number, and identifier ("water" or component name) begin each row followed by flow rates in kilograms per time unit for flows in and out of the boundary of the zone and in and out of each boundary-condition type within the zone. This tab-separated-values ASCII file is intended to be used for postprocessing simulation results by user-supplied spreadsheet, statistical, plotting, and visualization programs.

The frequency for writing zone flow information to the file is specified with the **-zone_flow_tsv** identifier in the **PRINT_FREQUENCY** data block. By default, zone flow information is not written to the file.

*prefix***.zf.txt**—Water and solute flow into and out of the boundaries of specified zones and water and solute flows through any boundary-condition cells within zones are written to this file. The zones are defined with the **ZONE_FLOW** data block. This ASCII file is intended to be viewed with a text editor.

The frequency for writing zone flow information to the file is specified with the **-zone_flow** identifier in the **PRINT_FREQUENCY** data block. By default, zone flow information is not written to the file.

user defined **xyzt** *file*—A time series of heads for a zone are written to a file that is named by the user with the **-write_heads_xyzt** identifier in the **ZONE_FLOW** data block. This ASCII file is intended to be used to define a time series of head boundary conditions (typically for specified-head boundaries) that is used in subsequent submodel simulations.

The frequency for writing zone heads is specified with the **-zone_flow_heads** identifier in the **PRINT_FREQUENCY** data block. By default, zone flow information is written at the end of simulation periods.

selected_output—Selected chemical data for solutions, equilibrium phases, exchangers, surfaces, kinetic reactants, solid solutions, and gas phases are written to this file, but only during the processing of the chemistry data file from the initial call to PHREEQC at the beginning of the simulation. This is a tab-separated-values ASCII file that is rarely used because it does not contain any results from the reactive-transport part of the simulation.

The default name of the file is *selected_output*, but the file name may be defined by the user with the **-file** identifier in the **SELECTED_OUTPUT** data block of the chemistry data file. Data items to be written are selected in the **SELECTED_OUTPUT** data block of the chemistry data file and by Basic-language statements defined in the **USER_PUNCH** data block of the chemistry data file.

5.2. Selection of Data for Chemical Output Files

Because of the many chemical components and species in reactive-transport modeling, it is unwieldy to write all of the chemical data to the *prefix***.chem.xyz.tsv** and *prefix***.h5** files. It is necessary to select the data that are written to these files with the **SELECTED_OUTPUT** and **USER_PUNCH** data blocks in the chemistry data file (*prefix***.chem.dat**). If more than one **SELECTED_OUTPUT** or **USER_PUNCH** data blocks are included in the chemistry data file, the last data block of each type determines the chemical data to be written during the reactive-transport simulation to both the *prefix***.chem.xyz.tsv** and *prefix***.h5** files.

The **SELECTED_OUTPUT** data block has identifiers that select printing of specific quantities, such as the total moles of an element in solution, the activity of an aqueous species, the saturation index of a mineral, and several other quantities. The quantities are selected by lists that are specified in the data block. The **USER_PUNCH** data block allows mathematical manipulation of chemical quantities and printing of data items through Basic-language statements. Through the use of these two data blocks, the large amount of output data produced by a simulation can be limited to the most important values, and files can be limited to manageable sizes.

5.3. Output Files for Postprocessing

The file *prefix*.**h5** contains potentiometric head, velocity components, and concentrations of chemical species for the active grid region at user-specified time levels. These data can be extracted to produce files from which statistics, contour maps, isosurface plots, or vector maps of velocity fields can be generated with user-supplied software. The interactive utility program, PHASTHDF, is used to extract subsets of data from the HDF file. It is possible to extract data for specified subregions of the active grid region and for specified simulation times. PHASTHDF generates ASCII files with the same format as *prefix*.**xyz.tsv** files. The capabilities and use of PHASTHDF are described in Appendix B.

The Model Viewer visualization software (Hsieh and Winston, 2002) is included in the distribution of PHAST. The software is used to visualize three-dimensional media-property, concentration, head, and velocity fields from PHAST simulations that are stored in the *prefix*.**h5** file. The capabilities and use of Model Viewer are described in Appendix A.

An alternative way to obtain data in a form suitable for postprocessing is to write data to files *prefix*.**head.xyz.tsv**, *prefix*.**chem.xyz.tsv**, *prefix*.**comps.xyz.tsv**, and *prefix*.**vel.xyz.tsv**, which have columns of data with one row for each active node. The columns of these files are tab delimited and the first columns are X, Y, and Z coordinate locations for the node, the time, and a 1 or 0 to indicate whether the cell for the node is at least partially saturated or dry. The *prefix*.**wt.xyz.tsv** file is a tab-delimited file written at the same frequency as the *prefix*.**head.xyz.tsv** and contains a water-table head for each active stack of nodes for each time of printing. The tab-separated-values format for these files allows importation into spreadsheet, statistical, plotting, or visualization programs. To limit the file size of the *prefix*.**chem.xyz.tsv** file, a subset of nodes can be specified with the **PRINT_LOCATIONS** data block, and only data from nodes in the subset will be written. No similar facility exists for the other tab-separated-values files.

Concentration data for chemical components, pH, and alkalinity at wells can be written at selected times in columnar format to the file *prefix*.**wel.xyz.tsv**. The first columns of the file are the X, Y coordinate location of the well, the well datum, the time, and the well number. The data that are output apply to the well datum, which is the elevation of the top of the uppermost open interval in the well (Kipp, 1987). For a pumping well, the water at the well datum is a mixture of water from all of the production zones of the well, weighted by the flow rates for the zones.

It is often useful to obtain the flows of water and solutes through specific parts of the grid region or through specific boundaries. The *prefix*.**zf.tsv** file provides this information in a tab-separated-values format for a set of zones that are defined with the **ZONE_FLOW** data block. The flows for an individual component for a given zone and time are found on one line of the file.

5.4. Diagnostic Output Files

Diagnostic messages, initial conditions, and problem definition information are written to files to be able to verify problem definition and to debug simulation failures. To identify what is happening when things go wrong, the PHAST simulator writes diagnostic output to the file *prefix*.**log.txt**. All warnings and errors are written to this file, including those detected while processing the input files and those that occur during the flow, transport, and reaction calculations. The *prefix*.**log.txt** file also contains simulation progress and information on the solution of the finite-difference equations for each time step. The equation-solution information includes iteration counts for the solver (if applicable) and maximum changes in the dependent variables written out at an interval requested by the user (**PRINT_FREQUENCY** data block, **-progress_statistics**).

The *prefix*.**h5** file contains the locations of all boundary conditions and, optionally, media properties (**PRINT_INITIAL** data block, **-HDF_media_properties**), which can be visualized with the program Model Viewer. Visualization of boundary conditions and media properties for a flow-only run is the first step in any simulation. The run must be successful to be able to visualize results in the file *prefix*.**h5**, so it may be necessary to run a transient simulation with start time equaling end time to have a successful run. Results of a run of zero time should reproduce the expected initial head distribution. It is also possible to visualize velocities from the *prefix*.**h5** file. Once boundary conditions, media properties, and initial heads are reasonable, it should be possible to make a complete flow simulation. The next step is to make a solute transport simulation for zero time and visualize initial chemical compositions. The final step is to make complete flow, solute transport, and reaction simulations.

Additional checking can be done through the **.txt** files containing problem-definition information. Velocities in the *prefix*.**vel.txt** file can be checked to ensure that the magnitudes are reasonable. The first sets of data in the *prefix*.**head.txt** and *prefix*.**comps.txt** files contain the initial-condition distributions of the variables for potentiometric head and component concentrations, respectively, if requested by the user (**PRINT_INITIAL** data block, **-heads** and **-components**). These files can be used in addition to Model Viewer and the *prefix*.**h5** file to verify that the initial conditions have been defined correctly. The *prefix*.**probdef.txt** file contains information related to the problem definition, including locations of nodes, fluid properties and distributions of porous-media properties, initial conditions, and boundary conditions.

A listing of boundary-condition types and excluded cells for the simulation grid can be written to the *prefix*.**probdef.txt** file if requested (**PRINT_INITIAL** data block, **-boundary_conditions**). The array is printed in X–Y or X–Z planes of nodes (depending on the value of **-print_orientation** in the **GRID** data block) and can be used to locate errors in the location and type of boundary conditions. Character strings of up to seven characters identify all the types of boundary conditions that apply at a cell. The characters in the string have the following meanings: Sa, Specified head, associated-solution boundary condition; Ss, Specified head, specified (or fixed) solution boundary condition; F, Flux boundary condition; L, Leaky boundary condition; R, River boundary condition; D, Drain boundary condition; W, well; and X, excluded cell. The cell is identified by index numbers i, j, and k, which are the node indices of the cell in the X, Y, and Z coordinate directions. Further static and transient information about specified-head, flux, leaky, river, drain, and well conditions is written to the *prefix*.**probdef.txt** file, which is indexed by the i, j, and k indices.

The *prefix*.**kd.txt** file contains conductance factors for flow, flow conductances, and solute-dispersive conductances for transport, if requested (**PRINT_INITIAL** data block, **-conductances** identifier; **PRINT_FREQUENCY** data block, **-conductances** identifier). The solute-dispersive conductances are functions of the velocity field and, thus, may be transient. This file is used primarily for diagnostic purposes in the later stages of tracking down a problem.

Chapter 6. Examples

A large number of example problems have been simulated to (1) confirm the numerical accuracy of the PHAST program for simulating coupled flow, transport, and chemical reactions by comparing results with analytical solutions, hand calculations, or published numerical results and (2) test and demonstrate most of the capabilities of the program. Seven of these verification examples are listed with no further documentation; PHAST simulations successfully reproduced the results of these relatively simple problems. However, six examples are presented in detail in the following sections for verification and tutorial purposes. The first two examples are one- and three-dimensional simulations of reactive solutes for which analytical solutions exist. The third example is a problem from the literature that includes kinetic biological processes. The fourth example is a simplified field-scale problem that uses most of the available boundary conditions and all of the commonly used chemical processes (except kinetics)—ion exchange, surface complexation, and mineral equilibria. The fifth and sixth examples demonstrate many of the features added in PHAST Version 2, including the use of spatial data from ArcInfo, the drain boundary condition, and aggregation of flows over arbitrary zones. The flow and transport data files and the chemistry data files for all examples are included in tables (and in the *examples* directory in the distribution of the program) and selected results are shown. These example files can be used as templates for other reactive-transport simulations. Run times for the 6 examples using Intel processors purchased in 2009 are shown in table 6.1

Table 6.1. Run times for example problems.

Example	Description	Run time, in seconds
1	Pulse source, sorption, and decay	3
2	Chain of four kinetically decaying reactants	85
3	Aerobic consumption of a substrate with biomass growth	9
4	Central Oklahoma naturally occurring arsenic	108
5A	Cape Cod steady-state flow	37
5B	Cape Cod transient flow	111
6	Cape Cod nitrogen transport and reactions	
	Serial version	2,751
	Parallel version, four processors	1,417

6.1. Verification Examples

The following examples were simulated with PHAST to verify that the program performs calculations correctly. Results from PHAST were compared with either published tables of numerical results, numerical output from computer evaluations of published analytical solutions, or published graphical results. All the verification examples were simulated by PHAST with satisfactory agreement with the published results; no further discussion of these examples is included.

Groundwater flow examples:

- Unconfined flow in one dimension with precipitation recharge. Analytical solution is presented in Bear (1972, p. 366).

- Single-well injection into two-dimensional rectangular region, confined flow. Finite-difference numerical solution is presented in Wang and Anderson (1982, p. 79).

- Well pumping from one layer of a two-layer leaky aquifer. Analytical solution is presented in Cheng and Morohunfola (1993, p. 791).

Solute-transport examples:

- Solute transport in one-dimensional column with advection and dispersion. Analytical solution is presented in Lapidus and Amundson (1952, p. 984).

- Single-well injection of tracer into two-dimensional rectangular region. Method-of-characteristics numerical solution is presented in Konikow and others (1996, p. 49).

Reactive-transport examples:

- Solute transport with decay in one-dimensional column. Method-of-characteristics numerical solution is presented in Konikow and others (1996, p. 45).

- Aerobic biodegradation combined with nitrate-reducing metabolism in a one-dimensional column. Numerical solution is presented in Kindred and Celia (1989, section 4.4, p. 1155). This problem is included in the set of example data sets distributed with PHAST. However, this example differs from the example in section 6.4. Example 3: Aerobic Consumption of a Substrate with Biomass Growth because the growth of biomass is not considered.

6.2. Example 1: Pulse Source of Chemical Constituent that Undergoes Sorption and Decay

An analytical solution for the concentration of a constituent as a function of time and distance following a pulse injection into a one-dimensional column is given by de Marsily (1986, p. 268). The constituent simultaneously undergoes linear decay and linear sorption. This example compares PHAST simulation results with

Table 6.2. Chemical and physical parameters for example 1.

[cm, centimeter; s, second; m, meter; mol, mole; kgw, kilogram of water]

Parameter	Value
Column length, cm	12.0
Interstitial velocity, cm/s	.1
Porosity, unitless	.1
Longitudinal dispersivity, cm	.1
Sorption constant, K_d, (mol/kgw)/(mol/kgw)	1.0
Decay rate constant, 1/s	.01

the analytical solution for an injection with the chemical and physical parameters given in table 6.2. A solution containing the constituent is injected into the column for 60 seconds; a solution without the constituent is then injected for an additional 60 seconds.

In this example, the prefix "ex1" is used for the input file names. The chemistry for the simulation is defined in the *ex1*.**chem.dat** file and the flow and transport characteristics of the simulation are defined in the *ex1*.**trans.dat** file. The third input file that is needed for a reactive-transport simulation is the thermodynamic database file, which is named *phast.dat* by default and, for this example, is the same as the *phreeqc.dat* thermodynamic database file distributed with PHREEQC.

The chemistry data file for this example is given in table 6.3. Typically, the database file *phast.dat* is used for the definition of most reactions and thermodynamic data. However, because of the idealized nature of this example problem, it is convenient to define all of the chemical characteristics of the chemical system within the chemistry data file (except for water and its associated species, which are found in the thermodynamic database file). The first data block, **SOLUTION_MASTER_SPECIES**, defines a new "element", A, which will be used as the reactive constituent. The **SOLUTION_SPECIES** data block defines a single uncharged[1] aqueous species for the element.

The **SURFACE_MASTER_SPECIES** data block defines a surface that will sorb constituent A. The **SURFACE_SPECIES** data block defines two surface complexes; the sites of the surface will be either an uncomplexed site, denoted by "Surf", or a site complexed with constituent A, denoted "SurfA". The empty **SOLUTION** *1* data block defines pure water by default. This pure water solution will be used to fill the column initially and also to flush the column for the last 60 seconds of the simulation. **SOLUTION** *2* defines a solution containing 1 mmol/kgw (millimole per kilogram of water) of constituent A. This solution will be injected into the column for the first 60 seconds of the simulation. **SURFACE** *1* defines the name of a surface sorption site and the number of sorption sites available to react.

PHREEQC does not have an explicit mechanism for simulating linear sorption. However, it is possible to define the number of surface sites and equilibrium constants for surface species in such a way that linear sorption is simulated accurately. Consider the reaction and log form of the mass-action equation for the sorption of A on surface Surf:

$$Surf + A = SurfA \text{ with } Log(K) = Log(SurfA) - Log(A) - Log(Surf),\qquad(6.1)$$

where $SurfA$, $Surf$, and A are activities. For an uncharged aqueous species (A), activity in dilute solutions is nearly equal to molality. For surface species, activity is equal to mole fraction. If the number of sites is large relative to the amount of A that sorbs, 10^{10} moles for example, the mole fraction $Surf$ is very nearly 1 and the mole fraction $SurfA$ is the number of moles of $SurfA$ divided by 10^{10}. (Note that for charged surface species, by default, an electrostatic term is included in the mass-action equation; **-no_edl** in the **SURFACE** *1* data block eliminates that term and also precludes the need to define the surface-area parameters for the surface.) For this example, K_d is 1, $\dfrac{\dfrac{moles(SurfA)}{kgwater}}{(A)} = 1$, where (A) is molality. Substituting and rewriting equation 6.1 gives

---

[1]If the constituent were an ion, additional issues would be raised. **KINETICS** removes the element and water is split to form dissolved O_2 or H_2 and hydronium or hydroxide ions to maintain charge balance. Thus, unwanted pH and redox effects will be obtained if a major cation undergoes significant decay. It is advisable to remove an oppositely charged element or replace the ionic element with another like ionic element so that charge balance is maintained.

Table 6.3. Chemistry data file for example 1.

```
TITLE
.         Example 1.--Pulse of solute undergoing sorption and decay
.         Analytical solution from Quantitative Hydrogeology, de Marsily
#         Kd = 1 mol sorbed/mol dissolved
#         K  = 0.01 1/s   Decay constant
SOLUTION_MASTER_SPECIES
          A        A        0        A        1
SOLUTION_SPECIES
          A = A
          log_k    0
SURFACE_MASTER_SPECIES
          Surf     Surf
SURFACE_SPECIES
          Surf = Surf
                  log_k 0
          Surf + A = SurfA
                  log_k   -10
SOLUTION 1
END
SOLUTION 2
          A          1.0 mmol/kgw
END
SURFACE 1
          -no_edl
          Surf     1e10
END
KINETICS 1
          A_decay
          -formula A 1.0
          -m 0
RATES
          A_decay
          -start
10 rate = -(TOT("A") + MOL("SurfA")) * .01
20 moles = rate * TIME
30 SAVE moles
          -end
END
SELECTED_OUTPUT
          -file ex1.dummy.sel
          -reset false
USER_PUNCH
          -headings A SurfA SurfA/A
10 PUNCH TOT("A")*1000
20 PUNCH MOL("SurfA")*1000
30 IF TOT("A") > 0 THEN PUNCH MOL("SurfA")/TOT("A") ELSE PUNCH -1
END
```

$$Log(K) = Log\left(\frac{moles(SurfA)}{kgwater}\right) + Log(kgwater) - Log(10^{10}) - Log(A) - 0 \text{, and} \tag{6.2}$$

$$Log(K) + 10 - Log(kgwater) = Log\left(\frac{\dfrac{moles(SurfA)}{kgwater}}{(A)}\right) = Log(K_d) = 0. \tag{6.3}$$

The mass of water in each cell is fixed at 1 kg (kilogram), so the term $Log(kgwater)$ is zero. Therefore, $Log(K)$ must be -10 to obtain the required K_d. These definitions are implemented in
SURFACE_SPECIES and **SURFACE** *1* data blocks.

The **KINETICS** *1* data block defines the name of a rate expression, "A_decay", that removes constituent A from solution. The **RATES** data block defines the rate of the kinetic reaction "A_decay". The rate expression defines a rate that is proportional to the total number of moles of A in the system (dissolved plus sorbed). Because the chemical calculation is performed for one kilogram of water, the rate has the same proportionality to concentration.

The **SELECTED_OUTPUT** and **USER_PUNCH** data blocks define selected data to be written to output files. The **SELECTED_OUTPUT** data block is used to control writing selected-output data to as many as three files: *ex1.dummy.sel* (as defined with **-file** in the **SELECTED_OUTPUT** data block in this example), *ex1*.**chem.xyz.tsv**, and *ex1*.**h5**. In the **SELECTED_OUTPUT** data block, **-reset** *false* eliminates all default writing to the selected-output files. In this example, all data to be written to selected-output files are defined in the **USER_PUNCH** data block. To understand which data go to each of the three selected output files, it is necessary to know the sequence of calculations in PHAST. PHAST runs the chemistry data file as a PHREEQC simulation before it begins to simulate flow and transport. During this initial PHREEQC simulation, any results specified by **SELECTED_OUTPUT** and **USER_PUNCH** data blocks will be written to the selected-output file as defined by the **-file** identifier (*ex1.dummy.sel*). However, in this example, these data blocks are located at the end of the file after an **END** keyword, and no chemical simulations are defined after this point. Therefore, the file *ex1.dummy.sel* will be empty. After the PHREEQC simulation, PHAST begins the flow and transport simulation. During the flow and transport simulation, PHAST writes all selected output to the file *ex1*.**chem.xyz.tsv** and to the file *ex1*.**h5**. The identifiers **-xyz_chemistry** and **-HDF_chemistry** in the **PRINT_FREQUENCY** data block of the flow and transport data file are used to specify the frequencies at which data are written to these two files.

The flow and transport data file (table 6.4) defines all system geometry, spatial discretization, characteristics of the porous media, boundary conditions, initial conditions, time steps, and duration of the simulation. The flow and transport data file is linked to the chemistry data file by the index numbers of solutions, surface, and kinetics that are used as initial conditions (**CHEMISTRY_IC** in the flow and transport data file) and the index numbers of solutions used as boundary conditions (**SPECIFIED_HEAD_BC** data block in the flow and transport data file).

The first data block in the flow and transport file is **TITLE**, which simply defines two lines to be written to the output files to identify the simulation. The **SOLUTE_TRANSPORT** data block is used to specify that this is not only a flow simulation, but transport and reactions are to be simulated as well. The flow field is defined to be steady state for the duration of the simulation with the **STEADY_FLOW** data block. A steady-state velocity field will be calculated once from the boundary conditions. The velocity field will then be constant for all time steps of the transport calculation. The **UNITS** data block is required to set the units of measure for all of the input quantities. On output, all units are in SI, including seconds for time, as specified with the **-time** identifier.

The **GRID** data block defines the locations of nodes that discretize the model domain. The grid is 12 cm (centimeters) (X) by 1 cm (Y), by 1 cm (Z), with two nodes in the Y and Z directions and 61 nodes equally spaced in the X direction. The grid region begins at the first node (at X=0 cm) and ends at the last node (at X=12 cm). The flow and transport calculations are always performed on a three-dimensional grid. However, for symmetric systems (both flow and chemistry must be symmetric), the chemistry can be run on a subset of nodes, either a line (one dimensional) or plane (two dimensional), and the results copied to the remaining symmetric parts of the grid region. The identifier **-chemistry_dimensions X** indicates that chemical reactions will be calculated for one line of nodes in the X direction and then copied to the other three lines of nodes in the X direction. The **-print_orientation** identifier specifies that X–Y planes of data will be written to the files with names beginning with *ex1* and ending with **.txt** that contain spatially distributed data.

The **MEDIA** data block specifies the properties of the porous medium. In this example, a single zone, inclusive of the entire grid region, is used to define uniform porous-medium properties throughout the grid region. Multiple zones could be used to represent heterogeneity of porous-media properties within the grid region. An interstitial velocity of 0.1 cm/s (centimeter per second) is desired for this example simulation. From Darcy's law (equation D.3), if the head gradient is defined to be 1/12 by the boundary conditions

Table 6.4. Flow and transport data file for example 1.

```
TITLE
.            Example 1.--Pulse of solute undergoing sorption and decay
.            Analytical solution from Quantitative Hydrogeology, de Marsily
#            1D region        12 cm
#            Velocity         0.1 cm/s
#            Porosity         0.1
#            Dispersivity      1 cm      (isotropic)
#            60 s pulse, 60 s chaser
SOLUTE_TRANSPORT                          true
STEADY_FLOW                               true
UNITS
             -time                         s
             -horizontal                   cm
             -vertical                     cm
             -head                         cm
             -hydraulic_cond               cm/s
             -specific_storage             1/cm
             -dispersivity                 cm
GRID
             -uniform X  0 12              61
             -uniform Y  0  1              2
             -uniform Z  0  1              2
             -chemistry_dimensions         X
             -print_orientation            XY
MEDIA
             -box    0 0 0 12 1 1
                     -Kx                  0.12
                     -Ky                  0.12
                     -Kz                  0.12
                     -porosity            0.1
                     -storage             0
                     -longitudinal_dispersivity  0.1
                     -horizontal_dispersivity    0.1
                     -vertical_dispersivity      0.1
FREE_SURFACE_BC false
SPECIFIED_HEAD_BC
             -box    0 0 0 0 1 1
                     -head               0          1
                     -associated_solution 0         2
                                         60          1
             -box   12 0 0 12 1 1
                     -head               0          0.0
                     -associated         0          1
HEAD_IC
             -box    0 0 0 12 1 1
                     -head    X 1.0 0.0 0.0 12.0
CHEMISTRY_IC
             -box    0 0 0 12 1 1
                     -solution            1
                     -surface             1
                     -kinetics            1
SOLUTION_METHOD
             -direct
             -space               0.5
             -time                0.5
TIME_CONTROL
             -time_step           0          0.4 s
             -time_end            120 s
PRINT_FREQUENCY
             0
             -HDF_chemistry       10 s
             -xyz_chemistry        0 s
             60
             -velocity            60 s
             -xyz_chemistry       60 s
             -force_chemistry_print 60 s
END
```

(**SPECIFIED_HEAD_BC** data block) and the porosity is 0.1, then a hydraulic conductivity of 0.12 cm/s will produce the desired interstitial velocity of 0.1 cm/s.

The example simulates flow in a confined aquifer (column), which is specified by inactivating the free-surface boundary condition (**FREE_SURFACE_BC** *false*). The flow is prescribed by specified heads at both ends of the column to achieve the desired head gradient. The **SPECIFIED_HEAD_BC** data block defines two zones. At the plane of nodes at X=0 cm, the head is specified to be 1 cm and beginning at time 0 seconds, any inflowing solution will have the composition defined by solution 2. At 60 seconds, the composition of water entering at X=0 cm changes to solution 1. At the plane of nodes at X=12 cm, the head is specified to be 0 cm and any inflowing solution will have the composition defined by solution 1 for the entire simulation. Recall that solution 2 is defined in the chemistry data file to contain constituent A, whereas solution 1 is defined to be pure water. For confined flow, specified-head values may be relative to any elevation datum, without regard to the Z datum of the coordinate system. (For unconfined flow, heads must be defined relative to the Z datum of the coordinate system.)

The initial head is defined to be linear in the X direction, ranging from 1 cm at X=0 cm to 0 cm at X=12 cm by the **HEAD_IC** data block. This initial head distribution is exactly the steady-state head distribution for this system, but any initial head distribution should result in the same steady-state head distribution. The **CHEMISTRY_IC** data block is used to define the initial composition of water throughout the grid region and the types of chemistry that are active in each cell. In this example, the grid region initially contains pure water (solution 1), and every cell has surface and kinetic chemical reactions. The index numbers for the identifiers **-solution**, **-surface**, and **-kinetics** refer to the index numbers of solutions, surfaces, and kinetic reactants defined in the chemistry data file (table 6.3).

The **SOLUTION_METHOD** data block specifies that the flow and transport equations are solved by the direct method, which is appropriate for the small number of nodes in this problem. The finite-difference approximations to the flow and transport differential equations use centered-in-space and centered-in-time weighting (**SOLUTION_METHOD** data block, **-space** and **-time** identifiers). This weighting results in second-order truncation errors, which contain no numerical dispersion, but may introduce oscillations in the numerical solutions for head and concentration. In this example, the Peclet number is 2 (equation D.5), which indicates that there will be no oscillations due to the spatial discretization (see section D.1.2. Spatial Discretization). The **TIME_CONTROL** data block specifies the time step (**-time_step**) of 0.4 seconds will be used from 0 seconds until the end of the simulation. The time at the end of the simulation (**-time_end**) is specified to be 120 seconds. For this example, the oscillation criterion (equation D.8) resulting from temporal discretization is $\Delta t = 0.4 < \dfrac{\Delta x^2}{v\alpha} = \dfrac{0.2^2}{0.1(0.1)} = 4$, which indicates that there should be no oscillations (see section D.1.3. Temporal Discretization).

The **PRINT_FREQUENCY** data block specifies the frequency of writing data to the output files. Starting at 0 seconds, the default print frequencies will be used for all data, except that chemistry data will be written to the HDF file after every 10 seconds of simulation (**-HDF_chemistry** 10 s), and no data will be written to the *ex1*.chem.xyz.tsv file (**-xyz_chemistry** 0 s). At 60 seconds, print frequencies are specified to be 60 seconds for the *ex1*.vel.txt (**-velocity**), *ex1*.chem.xyz.tsv (**-xyz_chemistry**), and *ex1*.chem.txt (**-force_chemistry_print**) data files (**PRINT_FREQUENCY** data block). The print frequency for the HDF file is not redefined at 60 seconds, so the print frequency of 10 seconds remains in effect until the end of the simulation. The **END** statement marks the end of the definitions for the flow and transport data file.

The time-series definitions for the **SPECIFIED_HEAD_BC** and **PRINT_FREQUENCY** data blocks specify a change at time 60 seconds. Thus, the total number of simulation periods is two, from 0 seconds to 60 seconds, and from 60 seconds to the end of the simulation at 120 seconds (**-time_end** in the **TIME_CONTROL** data block).

The results of the simulation are written to several output files. The velocity field, which is written to the *ex1*.vel.txt file and also can be viewed in Model Viewer from data written in the *ex1*.h5 file, is uniform at 0.001 m/s (meters per second) or 0.1 cm/s (centimeters per second). The head field is listed in *ex1*.**head.txt**

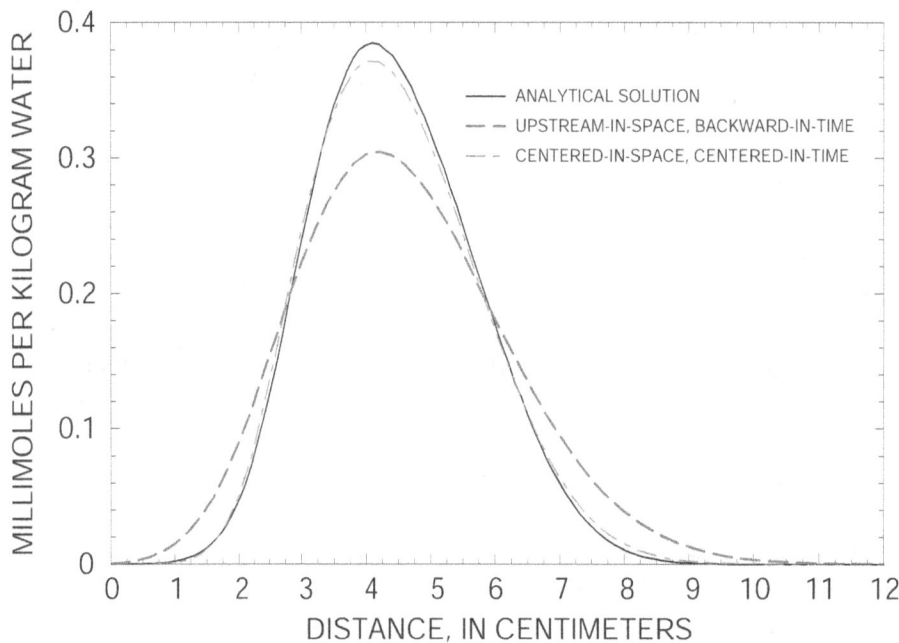

Figure 6.1. Example 1, analytical solution for concentration at 120 seconds of a pulse of solute that undergoes linear sorption and linear decay compared to numerical solutions using (1) centered-in-space and centered-in-time weighting and (2) upstream-in-space and backward-in-time weighting.

and also stored in the *ex1*.**h5** file. Descriptions of the chemistry in each cell are written to *ex1*.**chem.txt**. Selected output (as defined in the chemistry data file), including the concentration of A and SurfA, is written to *ex1*.**chem.xyz.tsv** and also to the *ex1*.**h5** file. The concentration of A from the PHAST simulation as a function of distance at 120 seconds is plotted in figure 6.1 along with the analytical solution. The simulation used centered-in-time and centered-in-space weighting for the difference equations. Results of a simulation using upstream-in-space (**SOLUTION_METHOD** data block, **-space_differencing** 0) and backward-in-time (**-time_differencing** 1 in the **SOLUTION_METHOD** data block) also are plotted for comparison. The PHAST results for centered-in-time and centered-in-space weighting are similar to the analytical solution. The discrepancy with the analytical solution is caused predominantly by the operator-splitting error because the spatial and temporal discretization errors should be minimized by using centered-in-time and centered-in-space weighting. The operator-splitting error for linear sorption and centered-in-time weighting

can be estimated by equation D.33, $\alpha_{nos} = \dfrac{D_{nos}}{v} = \dfrac{(0.1)(0.4)}{(2)(2)}[(2)(0.5)(2) - 1] = 0.01$, which indicates

that the numerical dispersivity caused by operator splitting is one tenth of the specified dispersivity (0.1 m [meter]). For this weighting, the numerical dispersivity caused by operator splitting (and time discretization) can be reduced most effectively by decreasing the time step.

The PHAST results for backward-in-time and upstream-in-space weighting show additional numerical dispersion. Here, numerical dispersion is caused by errors due to spatial and temporal discretization in addition to the operator-splitting error. The total numerical dispersion is estimated from equation D.7, equation D.10, and equation D.33,

$$\alpha_{ns} + \alpha_{nt} + \alpha_{nos} = \frac{0.2}{2} + \frac{(0.1)(0.4)}{2} + \frac{(0.1)(0.4)}{(2)(2)}[(2)(1)(2) - 1] = 0.1 + 0.02 + 0.03 = 0.15.$$ Thus, for

backward-in-time and upstream-in-space weighing, the numerical dispersivity is larger than specified dispersivity (0.1 m), which accounts for the large deviation from the analytical solution. For this weighting, the numerical solution can be improved both by refining the spatial discretization and by decreasing the time step.

6.3. Example 2: Chain of Four Kinetically Decaying Reactants

Example 2 simulates a four-species decay chain as a function of time and space, with the first species in the decay series introduced over a rectangular patch at the upstream end of a flow system; the flow is one-dimensional and uniform in the X direction, but the transport simulation is three-dimensional because the chemical species disperse in all directions. A method for deriving an analytical solution for any number of species in a chain of first-order decay reactions is presented by Sun and others (1999). They present an analytical solution for this example (Sun and others, 1999, example 6.3, p. 436) that is based on their method for chain decay and the approximate analytical solution for a single species given by Domenico (1987). We derived an alternative analytical solution for comparison with PHAST simulation results that uses the method of Sun and others (1999) but implements the analytical solution for a single species given by Wexler (1992). The Wexler (1992) analytical series solution is regarded as more accurate than the single-term approximation of Domenico (1987). The parameters for this example are given in table 6.5. Simulation and analytical results are presented for a 400-day simulation.

Table 6.5. Chemical and physical parameters for example 2.

[m, meter; d, day]

Parameter	Value
Simulation region length, m	100.0
Simulation region width, m	41
Simulation region height, m	25
Interstitial velocity, m/d	.2
Porosity, unitless	.1
Longitudinal dispersivity, m	1.5
Horizontal transverse dispersivity, m	.3
Vertical transverse dispersivity, m	.1
Decay rate constant A to B, 1/d	.05
Decay rate constant B to C, 1/d	.02
Decay rate constant C to D, 1/d	.01
Decay rate constant D, 1/d	.005
Source-patch location (Y direction) m	15 to 26
Source-patch location (Z direction), m	10 to 15
Duration of simulation, d	400

Table 6.6. Chemistry data file for example 2.

```
SOLUTION_MASTER_SPECIES
         [A]     [A]    1       1       1
         [B]     [B]    1       1       1
         [C]     [C]    1       1       1
         [D]     [D]    1       1       1
SOLUTION_SPECIES
         [A] = [A]
                 log_k   0
         [B] = [B]
                 log_k   0
         [C] = [C]
                 log_k   0
         [D] = [D]
                 log_k   0
SOLUTION 1
END
SOLUTION 2
         units    mmol/kgw
         pe       12.0    O2(g)    -0.67
         [A]      1.0
END
RATES
         [A]_decay
         -start
10 rate = -TOT("[A]") * .05/(3600*24)
20 moles = rate * TIME
30 SAVE moles
         -end
         [B]_decay
         -start
10 rate = -TOT("[B]") * .02/(3600*24)
20 moles = rate * TIME
30 SAVE moles
         -end
         [C]_decay
         -start
10 rate = -TOT("[C]") * .01/(3600*24)
20 moles = rate * TIME
30 SAVE moles
         -end
         [D]_decay
         -start
10 rate = -TOT("[D]") * .005/(3600*24)
20 moles = rate * TIME
30 SAVE moles
         -end
END
KINETICS 1
         [A]_decay
                 -formula [A] 1 [B] -1.0
                 -m 0
         [B]_decay
                 -formula [B] 1 [C] -1.0
                 -m 0
         [C]_decay
                 -formula [C] 1 [D] -1.0
                 -m 0
         [D]_decay
                 -formula [D] 1
                 -m 0
END
SELECTED_OUTPUT
         -file ex2.dummy.sel
         -reset false
USER_PUNCH
         -heading A B C D
         -start
10 PUNCH TOT("[A]")*1000
20 PUNCH TOT("[B]")*1000
30 PUNCH TOT("[C]")*1000
40 PUNCH TOT("[D]")*1000
         -end
END
```

The chemistry data file is presented in table 6.6. All of the chemical information for the four species of the chain decay series are defined in the chemistry data file. **SOLUTION_MASTER_SPECIES** and **SOLUTION_SPECIES** data blocks define the name and a single aqueous species for each of four elements: [A], [B], [C], and [D]. (Using brackets in the names avoids conflicts with elements carbon (C) and boron (B), which are defined in the thermodynamic database file.) **SOLUTION** *1* is defined as pure water and **SOLUTION** *2* contains 1 mmol/kgw of [A], which will be applied over the rectangular patch at the upstream end of the grid region.

The **RATES** data block defines rate expressions for the decay of each species in the decay series. Each rate expression converts decay constants in units of per day to units of per second and multiplies by the respective species concentration to obtain the rate of decay. In the **KINETICS** *1* data block, the stoichiometry of each decay reaction is defined. The sign of the stoichiometric coefficient in **-formula** in **KINETICS** combined with the sign of the mole transfer calculated in the **RATES** expression (30 **SAVE** *moles*) gives the direction of the transfer relative to the solution. In this example, *moles* as calculated in **RATES** is negative. Therefore, constituents that have a positive coefficient for **-formula** identifier in the **KINETICS** data block will be removed from solution (*moles* times coefficient is negative); constituents that have a negative coefficient in **-formula** will be added to the solution (*moles* times coefficient is positive). Thus, the rate expression "[A]_decay" converts [A] to [B]; "[B]_decay" converts [B] to [C]; "[C]_decay" converts [C] to [D]; and "[D]_decay" removes [D]. The **SELECTED_OUTPUT** and **USER_PUNCH** data blocks define data to be written to output files as described in example 1.

For the definitions of the flow and transport data file (table 6.7), it is useful to take advantage of the symmetry of the problem definition. A horizontal plane of symmetry is located at Z=12.5 m, and a vertical plane of symmetry is located at Y=20.5 m. All concentration, head, and velocity fields on either side of each of these two planes are mirror images. Thus, it is possible to run a simulation on a grid region that is one-quarter of the original problem (Y=0 to 20.5 m, Z=0 to 12.5 m) and reflect the results to the other three quadrants. In this example, all properties, boundary conditions, and initial conditions are defined for the original problem (Y=0 to 41 m, Z = 0 to 25 m). Only the grid, which defines the extent of the model domain, is restricted to the front lower quadrant (Y=0 to 20.5 m, Z=0 to 12.5 m).

The uses of the **TITLE, SOLUTE_TRANSPORT, UNITS, FREE_SURFACE_BC**, and **PRINT_FREQUENCY** data blocks are the same as described in example 1. For demonstration purposes, steady-state flow (**STEADY_FLOW** data block) is not specified, so head and velocity are calculated at each time step of the simulation, even though the flow field is specified by the initial and boundary conditions to be at steady state.

Specified-head boundaries apply to entire cells, so to ensure that the rectangular patch over which species [A] is introduced into the region corresponds exactly to the edges of cell faces, it is necessary to define cell boundaries that coincide with Y=15 m and Z=10 m. Cell boundaries occur halfway between nodes, so for equally spaced nodes 1 m apart, Y nodes need to be placed at Y=14.5 m and Y=15.5 m. Similarly, Z nodes need to be placed at 9.5 m and 10.5 m. The **GRID** data block defines the locations of these nodes and the extent of the grid region by using a combination of uniform grid spacings. For the Y dimension, nodes are placed at Y=0 m and Y=20.5 m with the **-uniform** *Y* identifier. Nodes are placed at 0.5, 1.5, 2.5, …19.5 m, which satisfy the required cell boundary criterion, by overlaying additional equally spaced nodes with the **-overlay_uniform** *Y* identifier. Similarly, **-uniform** *Z* and **-overlay_uniform** *Z* identifiers place nodes in the required locations in the vertical. Nodes in the X direction have 4-meter spacing from 0 to 100 m. (Note that this alignment of cell boundaries and boundary conditions is not necessary for flux and leaky boundaries. These boundaries are applied to areas that may intersect all or part of a cell face. PHAST Version 1 applied leaky and flux boundaries to entire cell faces, but PHAST Version 2 applies these boundaries only to the part of the cell face that intersects the area defined by the boundary condition.)

Table 6.7. Flow and transport data file for example 2.

```
TITLE
.           3D analytic, continuous injection, 4 species chain decay
.           Sun, Peterson, Clement (1999) J. Cont. Hyd., Example 6.3
.           Documentation compares PHAST with chain decay from
.           Sun and others combined with Wexler (1992) solution
.           to patch source
SOLUTE_TRANSPORT                        true
UNITS
            -time                       day
            -horizontal                 m
            -vertical                   m
            -head                       m
            -hydraulic_conductivity     m/d
            -specific_storage           1/m
            -dispersivity               m
GRID
# Symmetry is used, so that grid is only defined on 1/4 of YZ face
            -uniform X  0.0     100     26      # 4.0  m spacing
            -uniform Y  0.0     20.5    2
            -overlay_uniform Y
# nodes are set so that cell face is at 15 m to allow source patch definition
                    0.5     19.5    20      # 1.0  m spacing
            -uniform Z  0.0     12.5    2
            -overlay_uniform Z
# nodes are set so that cell face is at 10 m to allow source patch definition
                    0.5     11.5    12      # 1.0  m spacing
            -print_orientation XY
MEDIA
            -box    0 0 0 100 41 25
# V = (1/e)*K*dh/dx, .2 = 1/.1*K*1/100, K = 2
                    -Kx                 2
                    -Ky                 2
                    -Kz                 2
                    -porosity           0.1
                    -storage            0
                    -longitudinal_dispersivity  1.5
                    -horizontal_dispersivity    0.45
                    -vertical_dispersivity      0.15
FREE_SURFACE_BC false
SPECIFIED_HEAD_BC
# first define x=0 head, inflow to be pure water
            -box    0 0 0 0 41 25
                    -head               0   1
                    -fixed_solution     0   1
# now define patch source of contaminant A
            -box    0 15 10 0 26 15
                    -fixed_solution     0   2
# define x=100 head
            -box    100 0 0 100 41 25
                    -head               0   0
                    -associated_solution 0  1
HEAD_IC
            -box    0 0 0 100 41 25
                    -head   X 1.0 0.0 0.0 100.0
CHEMISTRY_IC
            -box    0 0 0 100 41 25
                    -solution           1
                    -kinetics           1
SOLUTION_METHOD
            -iterative                  true
            -tolerance                  1e-14
            -space                      0.5
            -time                       0.5
            -cross_dispersion           false
PRINT_FREQUENCY
            0
            -HDF_chemistry              400 day
            -vel                        400 day
            -xyz_chemistry              400 day
            -xyz_head                   400 day
            -xyz_velocity               400 day
TIME_CONTROL
            -time_step                  0   10  day
            -time_end                   400 day
END
```

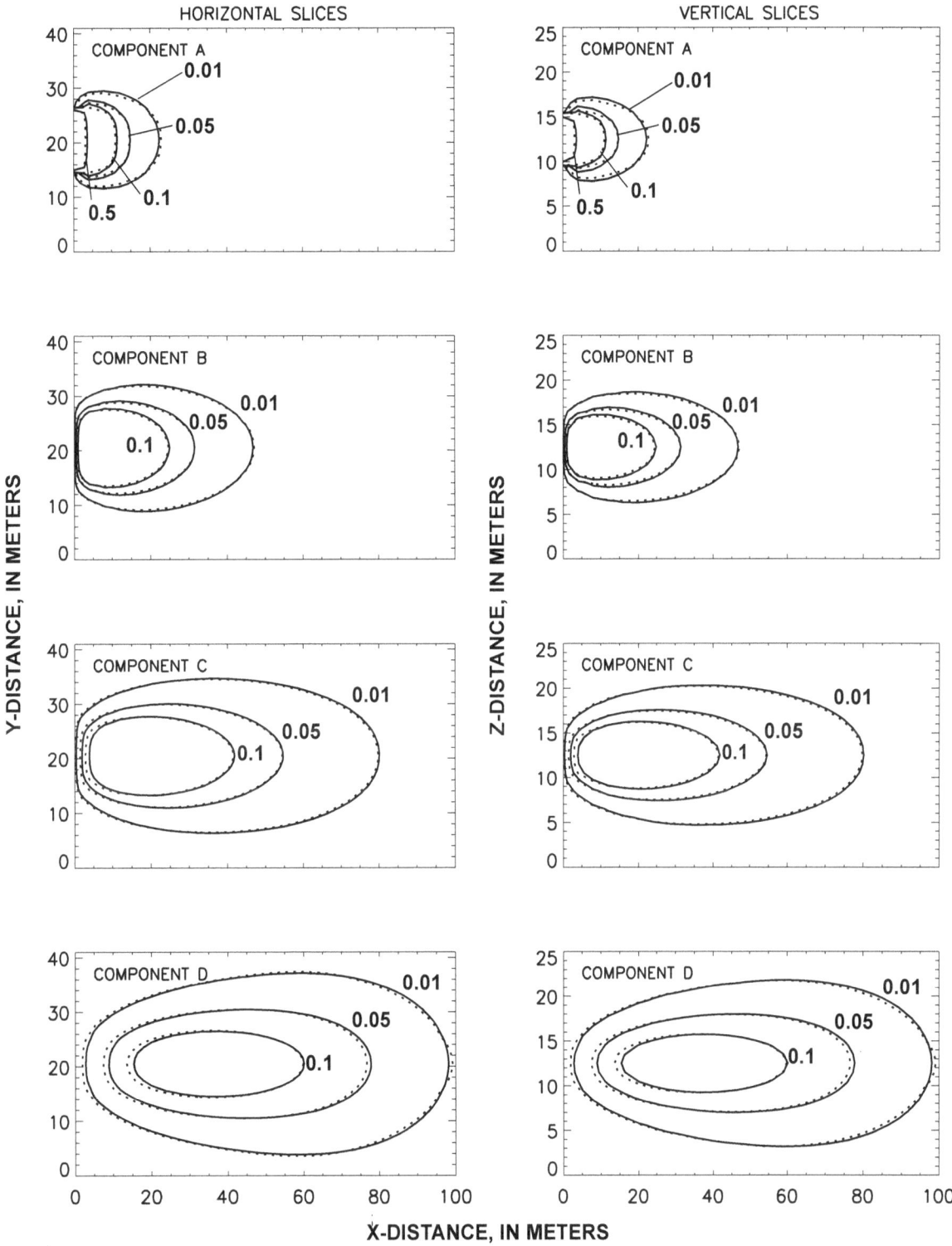

Figure 6.2. Example 2, analytical solution (solid lines) compared to simulation results (dashed lines) for concentrations at 400 days and Z=12.5 meters of a chain of decay products. Component A decays sequentially to components B, C, and D.

An interstitial velocity of 0.2 m/d (meter per day) is used in the analytical solution for this example simulation. From Darcy's law (equation D.3), if the head gradient is defined to be 0.01 (1 m per 100 m) by the boundary conditions and the porosity is 0.1, then a hydraulic conductivity of 2 m/d will produce the desired interstitial velocity of 0.2 m/d. The MEDIA data block defines the hydraulic conductivity, porosity, storage, and the three dispersivities (table 6.5) from Sun and others (1999). The SPECIFIED_HEAD_BC data block defines the head to be 1 m at the upstream end of the grid region and 0 m at the downstream end of the grid region to produce a head gradient of 0.01 for time 0 seconds onward. Also in the SPECIFIED_HEAD_BC data block, the water composition at the boundary (-**fixed_solution**) is defined to be pure water (SOLU-TION *1* in chemistry data file) from time 0 seconds onward, except over a rectangular patch at X=0. For the patch from Y=15 to 26 m and Z=10 15 m, the water composition is specified to be solution 2 (water containing [A], SOLUTION *2* in chemistry data file) from time 0 seconds onward.

The initial head condition is defined to be linear from 1 m at X=0 m to 0 m at X=100 (HEAD_IC data block), which is consistent with the desired steady-state flow system. The initial water composition within the grid region is specified by the CHEMISTRY_IC data block to be pure water (-**solution** *1*), and four kinetic decay reactions are defined by the identifier -**kinetics** *1*, where *1* corresponds to the definition of KINETICS *1* in the chemistry data file.

The iterative linear-equation solver is specified to be used for the finite-difference equations with the SOLUTION_METHOD data block and the identifier -**tolerance** is used to define the relative residual tolerance for convergence of the iterative solver. Centered-in-space (-**space**) and centered-in-time (-**time**) weightings are used to form the finite-difference equations. The TIME_CONTROL data block specifies the time step to be 10 d (days), starting at time 0 seconds. The time at the end of the simulation is 400 days (-**time_end**). The time step is less than the estimate of the time step needed to avoid oscillations due to tem-

poral discretization (see section D.1.3. Temporal Discretization), $\Delta t = 10 < \dfrac{4^2}{1.5(0.2)} = 53$ days. The cell

Peclet number (equation D.5) in the X direction is $\dfrac{4}{1.5} = 2.7$ (see section D.1.2. Spatial Discretization),

which is greater than 2, the criterion to avoid oscillations (equation D.8) due to spatial discretization; however, in this example no oscillations appear in the results. Cross-dispersive terms are not included in the difference equations (-**cross_dispersion** *false*). The cross-dispersion identifier was not necessary because the default is not to include the cross-dispersive terms. Cross-dispersive terms were not used in this example because the analytical solution from Wexler (1992) does not include cross dispersion.

Results of the simulation at 400 days for the plane Z=12.5 are shown in figure 6.2 along with the analytical solution. The similarity of results indicates that PHAST is performing the calculations correctly. The minor discrepancies between the two sets of curves are ascribed to truncation errors and operator-splitting errors in PHAST, which could be decreased by decreasing spatial and temporal discretization.

6.4. Example 3: Aerobic Consumption of a Substrate with Biomass Growth

Example 3 simulates flow in a one-dimensional column that contains a bacterial community that consumes oxygen and degrades a substrate. PHAST is used to model the kinetics of aerobic bacterial growth in a flowing system, while considering cell death and growth limitation by oxygen and substrate availability. The example is taken from Kindred and Celia (1989, section 4.1, p. 1154).

Table 6.8. Chemical and physical parameters for example 3.

[mg, milligrams; L, liter; m, meter; s, second; d, day]

Parameter	Value
$V_{max}(S)$, d^{-1}	1
$K_h(S)$, mg/L	.1
Y_S, L/mg	.25
$V_{max}(O)$, d^{-1}	1
$K_h(O)$, mg/L	.1
Y_O, L/mg	.125
k_m, d^{-1}	.01
$k_{biomass}$, mg/L	1
X_0, initial biomass, mg/L	.2
O_0, initial dissolved oxygen, mg/L	3.0
Porosity, unitless	.38
Dispersivity, m	.2
Hydraulic conductivity, m/s	1.e-4
Column length, m	50
Time simulated, d	37
Interstitial velocity, m/d	1.0

The rate equations for oxygen consumption are given by

$$R_{oxygen,\, O} \;=\; V_{max}(O)\frac{X}{I_b}\left(\frac{O}{K_h(O)+O}\right), \tag{6.4}$$

$$R_{oxygen,\, S} \;=\; \frac{Y_S}{Y_O}V_{max}(S)\frac{X}{I_b}\left(\frac{S}{K_h(S)+S}\right), \text{ and} \tag{6.5}$$

$$R_{oxygen} \;=\; Minimum(R_{oxygen,\, O}, R_{oxygen,\, S}), \tag{6.6}$$

where $R_{oxygen,\, O}$ is the limiting rate of oxygen consumption due to oxygen availability; $R_{oxygen,\, S}$ is the limiting rate of oxygen consumption due to substrate availability; R_{oxygen} is the limiting rate of oxygen consumption, which is the minimum of the two rates, $R_{oxygen,\, O}$ and $R_{oxygen,\, S}$; O, S, and X are the concentrations of oxygen, substrate, and aerobic biomass; $V_{max}(O)$ and $K_h(O)$ are the kinetic parameters for oxygen; $V_{max}(S)$ and $K_h(S)$ are the kinetic parameters for the substrate; Y_O and Y_S are yield coefficients for oxygen and organic substrate, and I_b is the biomass inhibition factor. $I_b = 1 + \dfrac{X}{k_{biomass}}$, where

$k_{biomass}$ is the aerobic biomass inhibition constant. The rate of biomass production is

$$R_{biomass} = Y_O R_{oxygen} - k_m X, \qquad (6.7)$$

where k_m is the specific death rate or maintenance coefficient for aerobic biomass. The parameters for equations 6.4–6.7 and the flow and transport parameters for the simulation are given in table 6.8.

The chemistry data file for this example is given in table 6.9. The **TITLE**, **SELECTED_OUTPUT**, and **USER_PUNCH** data blocks are used in the same ways as the previous two examples. The **SOLUTION_MASTER_SPECIES** and **SOLUTION_SPECIES** data blocks define a new chemical component to represent the substrate that is used by the bacteria. **SOLUTION** *1* defines the composition of the water initially present in the column, which contains no substrate; **SOLUTION** *2* defines the water that is introduced into the column, which contains 10 mg/L (milligram per liter) of the substrate. All concentrations are entered as milligrams per liter.

Two rate expressions are defined in the **RATES** data block, which are named "Aerobic" and "Aerobic_biomass". The Basic-language statements for the "Aerobic" rate expression calculate the rate of oxygen consumption as defined by equations equations 6.4–6.6. The Basic-language statements for the "Aerobic_biomass" rate expression calculate the rate of biomass production (or decay) as defined by equation 6.7. The **KINETICS** *1* data block defines the stoichiometry of the reactions and the amount of reactants initially present in the rest of the column ($X > 0$), which includes oxygen and substrate consumption ("Aerobic") and biomass growth ("Aerobic_biomass"). The **KINETICS** *2* data block defines the stoichiometry of the reactions and the amount of reactants that will be present in the cell at $X = 0$; biomass growth ("Aerobic_biomass") is included (Kindred and Celia [1989] show aerobic biomass at $X = 0$), but oxygen and substrate consumption ("Aerobic") are not included to maintain the specified-head boundary condition in this cell for dissolved oxygen and substrate concentrations. Considering the definitions in **KINETICS** and **RATES** for "Aerobic", the sign and magnitude of the transfers need explanation. First, the sign of the mass transfer calculated in the "Aerobic" rate expression (the quantity *mg_O* in the **RATES** data block) is negative, and the coefficients (**-formula**) in the **KINETICS** data blocks are positive. Therefore, the product of mass transfer (*mg_O* from **RATES**) and coefficient (from **KINETICS**, **-formula**) is negative and the "Aerobic" kinetic reaction removes oxygen and substrate from solution. The rate expressions from Kindred and Celia (1989) calculate mole transfer in terms of milligrams, which is assumed to be equal to milligrams per kilogram of water; however, for PHAST, mass-transfers must have units of moles. The coefficients in the identifier **-formula** have been adjusted to provide molar mass-transfer quantities. If *mg_O* is the mass-transfer calculated by the **RATES** expression, the mole transfer is

$$mg\_O \frac{1}{f_O 1,000} = mg\_O \frac{1}{(16)1,000} = 6.25 \times 10^{-5} mg\_O,\ \text{where } f_O \text{ is the gram formula weight of elemen-}$$

tal oxygen g/mol (grams per mole). The number of milligrams of substrate removed is $\dfrac{Y_O}{Y_S} mg\_O$, and the

number of moles of substrate removed is $\dfrac{Y_O}{Y_S} mg\_O \dfrac{1}{f_S 1,000} = \dfrac{0.5}{(1)1,000} mg\_O = 5 \times 10^{-4} mg\_O,\ \text{where } f_S$

Table 6.9. Chemistry data file for example 3.

```
TITLE
        Kindred and Celia, WRR, 1989, v. 25, p. 1154
        Problem 4.1. Aerobic Biodegradation
SOLUTION_MASTER_SPECIES
        Substrate          Substrate       0        1.0       1.0
SOLUTION_SPECIES
        Substrate = Substrate
        log_k    0.0
SOLUTION 1  Initial condition
        units    mg/L
        pH       7.0
        pe       8
        Substrate          0.0
        O(0)               3
SOLUTION 2  Infilling
        units    mg/L
        pH       7.0
        pe       8
        Substrate          10.
        O(0)               3
END
RATES
Aerobic
        -start
200 vmaxO = 1/(24*3600)
210 vmaxS1 = 1/(24*3600)
240 Kh1O = .1
250 Kh1S1 = .1
290 kbio = 1
320 km = .01/(24*3600)

410 O = TOT("O(0)")*16*1000
430 S1 = TOT("Substrate")*1000
450 x1 = KIN("aerobic_biomass")
470 Ib = 1 + x1 / kbio

500    REM  **  Oxygen rate ***
510 rateO = (vmaxO * x1 / Ib) * (O / ( kh1O + O))
520    REM  **  Substrate rate **
530 rateS1 = 2*(vmaxS1 * x1 / Ib) * (S1 / ( kh1S1 + S1))
560 rate = rateO
570 if rateS1 < rate then rate = rateS1
590 mg_O = -TIME*rate
600 SAVE mg_O
        -end

Aerobic_biomass
        -start
200 vmaxO = 1/(24*3600)
210 vmaxS1 = 1/(24*3600)
240 Kh1O = .1
250 Kh1S1 = .1
290 kbio = 1
320 km = .01/(24*3600)

410 O = TOT("O(0)")*16*1000
430 S1 = TOT("Substrate")*1000
450 x1 = KIN("aerobic_biomass")
470 Ib = 1 + x1 / kbio

520    REM  **  Substrate rate ***
530 rateS1 = .25*(vmaxS1 * x1 / Ib) * (S1 / ( kh1S1 + S1))
540    REM  **  Oxygen rate ***
550 rateO = 0.125*(vmaxO * x1 / Ib) * (O / ( kh1O + O))
560 rate = RateO
570 if rateS1 < rate then rate = rateS1
590 mg_O_biomass = -TIME*(rate - km * x1)
600 SAVE mg_O_biomass
        -end

KINETICS 1
Aerobic
#     mg_O * 1/gfw * 1/1000 = mg_O * 1/(16*1000) = mg_O * 6.25e-5
#     mg_O * (mg_S/mg_O)* 1 / gfw * 1/1000 = mg_O * 0.5 * 1/(1*1000) = mg_O * 5e-4
        -formula  O 6.25e-5 Substrate 5e-4
        -m 0
```

Table 6.9. Chemistry data file for example 3.—Continued

```
Aerobic_biomass
        -formula  Substrate  0
        -m 0.2
KINETICS 2
Aerobic_biomass
        -formula  Substrate  0
        -m 0.2
SELECTED_OUTPUT
        -file ex3.dummy.sel
        -reset false
USER_PUNCH
-headings Substrate O(0) O_biomass
10 PUNCH TOT("Substrate")*1000,  TOT("O(0)")*16*1000
20 PUNCH KIN("Aerobic_biomass")
END
```

is the gram formula weight of substrate (g/mol). Thus, the coefficients in **-formula** for "Aerobic" in the **KINETICS** data blocks are 6.25×10^{-5} for oxygen and 5×10^{-4} for substrate.

The coefficient for Substrate in **-formula** for "Aerobic_biomass" in the **KINETICS** data blocks is zero. Thus, the "Aerobic_biomass" reaction does not affect the solution composition in any way, but the mass transfer (*mg_O_biomass*) does affect the calculation through the cumulative amount of a reactant that is present. Initially, the amount of biomass is set to 0.2 mg (**-m** identifier) per kilogram of water. As biomass increases or decreases according to the *mg_O_biomass* calculated in the rate expression, the variables KIN("Aerobic_biomass") and "M" (only available within the Basic-language program for "Aerobic_biomass") are updated. Note the sign convention for calculating the cumulative moles of reactant: positive *mg_O_biomass* (from **RATES** expressions) decreases the amount of reactant (biomass) and negative *mg_O_biomass* increases the amount of reactant (biomass). (For dissolved Substrate, the gram formula weight was defined to be 1.0, which means that moles equal grams.) At any time in the simulation, the cumulative amount of biomass is known through either of the two variables KIN("Aerobic_biomass") or "M".

The flow and transport file corresponding to the problem description is given in table 6.10. The **TITLE, UNITS, SOLUTE_TRANSPORT, MEDIA,** and **FREE_SURFACE_BC** data blocks have the same functions as previously described in the first two examples. A 50-m column is defined with 101 nodes at a uniform spacing of 0.5 m (**GRID** data block). Because the example is one-dimensional for flow and transport and symmetric in the Y and Z directions, chemistry needs to be calculated for only one line of nodes in the X direction (**-chemistry_dimensions** X). The head gradient is calculated such that the interstitial velocity is 1 m/d. By assuming a hydraulic conductivity of 1×10^{-4} m/s and a porosity of 0.38, Darcy's law (equation D.3) is used to determine that a decrease in head of 2.199074 m over the 50-m grid region will result in an interstitial velocity of 1 m/d. Specified heads are defined at each end of the column (**SPECIFIED_HEAD_BC** data block) to maintain the head gradient and any water entering the column at the upstream end will have the composition of solution 2 (water containing substrate) as defined in the chemistry data file. Both the heads and solution compositions for the specified-head boundary conditions apply from time 0 days to the end of the simulation. The initial head condition is defined to be consistent with the boundary conditions and the steady-state flow at the specified velocity; the initial head is 2.199074 m at X=0, decreasing linearly to 0 m at X=50 m (**HEAD_IC** data block). The **CHEMISTRY_IC** data block defines the water composition and reactants that initially are present in the column. Solution 1 fills the column and a set of kinetic reactions (**KINETICS** *1*, in the chemistry data file) are present throughout the column, except at the inlet (X=0). At the inlet, biomass is allowed to accumulate, but the "Aerobic" reaction, which consumes oxygen, is not included because the initial concentration of oxygen is specified by the problem definition to be fixed at 3.0 mg/L at the boundary (table 6.8).

Table 6.10. Flow and transport data file for example 3.

```
TITLE
        Kindred and Celia, WRR, 1989, v. 25, p. 1154
        Problem 4.1. Aerobic Biodegradation
UNITS
        -time                           days
        -horizontal_grid                meters
        -vertical_grid                  meters
        -head                           meters
        -hydraulic_conductivity         m/s
        -specific_storage               1/m
        -dispersivity                   m
SOLUTE_TRANSPORT                        true
GRID
        -uniform x      0.0     50.     101
        -uniform y      0.0     1.0     2
        -uniform z      0.0     1.0     2
        -chemistry_dimensions           X
        -print_orientation              XZ
MEDIA
        -box    0. 0. 0. 50.  1. 1.
                -porosity               0.38
                -long_dispersivity      0.2
                -horizontal_dispersivity 0.2
                -vertical_dispersivity  0.2
                -Kx                     1e-4
                -Ky                     1e-4
                -Kz                     1e-4
                -specific_storage       0
HEAD_IC
#velocity 1 m/day    v*por/K*L = delta H
# 1/(24*3600)*.38*50./1e-4 = 2.199074
        -box    0. 0. 0. 100. 1. 1.
                -head   X   2.199074 0.  0. 50.
SPECIFIED_HEAD_BC
        -box    0. 0. 0. 0.   1.  1.
                -head                   0       2.199074
                -associated_solution    0       2
        -box    50. 0. 0. 50. 1.  1.
                -head                   0       0.
                -associated_solution    0       1
FREE_SURFACE_BC                         false
CHEMISTRY_IC
        -box    0. 0. 0. 50. 1. 1.
                -solution               1
                -kinetics               1
        -box    0. 0. 0. 0. 1. 1.
                -kinetics               2
SOLUTION_METHOD
        -direct_solver                  true
        -space_differencing             0.5
        -time_differencing              0.5
PRINT_INITIAL
        -components                     true
TIME_CONTROL
        -delta_time                     0       0.2 day
        -end_time                       37  day
PRINT_FREQUENCY
        0
        -xyz_chemistry                  10  day
        -HDF_chemistry                  1   day
        -HDF_velocity                   100 day
END
```

The **SOLUTION_METHOD** data block specifies use of the direct linear-equation solver and centered-in-space, centered-in-time weighting for the finite-difference equations. Oscillations are possible with this weighting because the cell Peclet number (2.5, equation D.5) is greater than 2.0 (see section D.1.2. Spatial Discretization) for this node spacing (0.5 m) and dispersivity (0.2). The time step (0.2 day) exceeds the time criterion for possible oscillations (0.125 day, equation D.8) (see section D.1.3. Temporal Discretization). The **PRINT_INITIAL** data block specifies that initial concentrations for each component will be written to the *ex3.comps.txt* file. The print frequencies for the *ex3.chem.xyz.tsv* and *ex3.h5* files are specified in the **PRINT_FREQUENCY** data block. The duration of the simulation and the size of the time step are defined in the **TIME_CONTROL** data block.

Results of the simulation (fig. 6.3) appear to be identical to results presented in figure 1 of Kindred and Celia (1989) (not reproduced here). No oscillations are observed in the results, which justifies the use of the centered-in-time, centered-in-space weighting for the difference equations with the spatial and temporal discretization chosen.

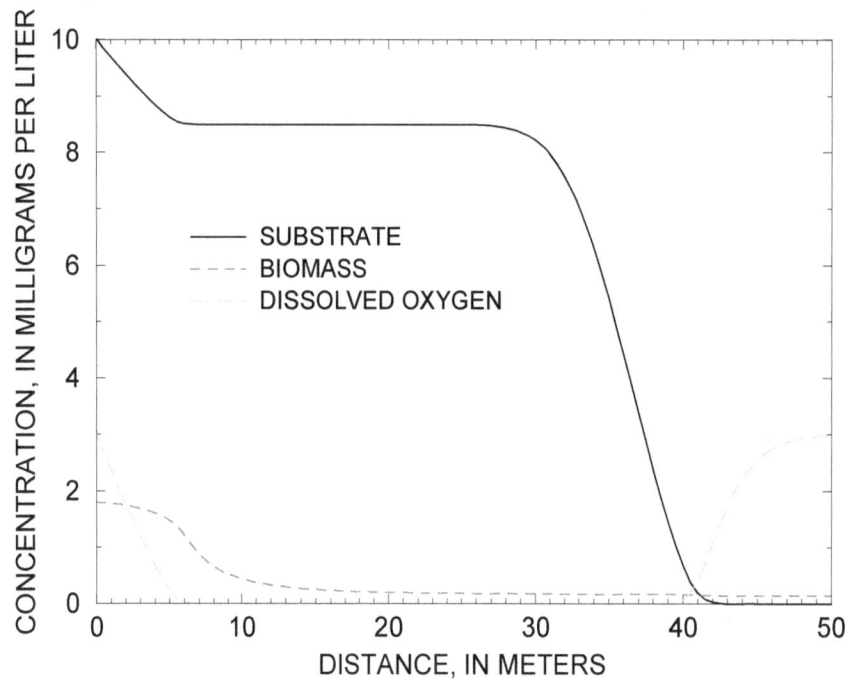

Figure 6.3. Example 3, concentrations of substrate, biomass, and dissolved oxygen at 37 days.

6.5. Example 4: Regional-Scale Transport and Reactions in the Central Oklahoma Aquifer

Example 4 presents a reactive-transport model that simulates the evolution over geologic time of water compositions that are found in the Central Oklahoma aquifer. The model simulates flow, transport, and reactions at a regional scale (90 km [kilometers] by 48 km) for an aquifer with both confined and unconfined regions and a complex three-dimensional flow pattern. Assuming the aquifer initially contained brines similar to those found at depth in the area, the chemical evolution of the aquifer over geologic time is hypothesized to result from a constant influx of freshwater from precipitation and chemical reactions with the aquifer material. The simulated chemical evolution provides an explanation for differing water compositions between confined and unconfined parts of the aquifer and the large, naturally occurring arsenic concentrations in the confined part of the aquifer. Most of the flow and transport capabilities and the equilibrium-phase, ion-exchange, and surface-complexation reaction capabilities of PHAST are used in this example. Because of the variety of boundary conditions and chemical capabilities used, the input files are useful templates for setting up other field-scale reactive-transport simulations.

The geochemistry of the aquifer has been described in Parkhurst and others (1996). Two predominant water types occur in the aquifer, a calcium magnesium bicarbonate water with pH in the range of 7.0 to 7.5

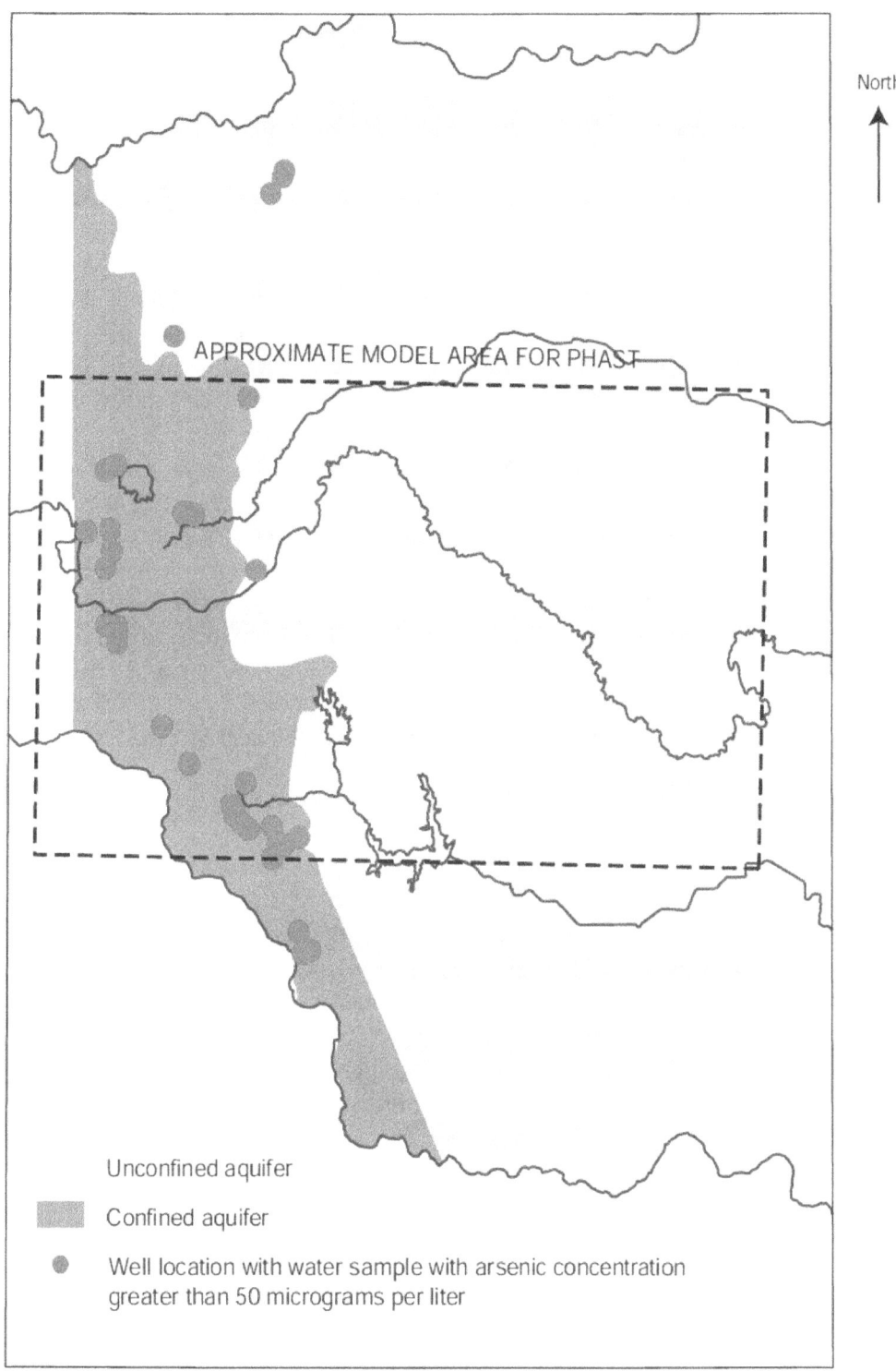

North

Unconfined aquifer

Confined aquifer

Well location with water sample with arsenic concentration greater than 50 micrograms per liter

Figure 6.4. Locations in the Central Oklahoma aquifer with arsenic concentrations greater than 50 micrograms per liter and PHAST model area.

in the unconfined part of the aquifer and a sodium bicarbonate water with pH in the range of 8.5 to 9.2 in the confined part of the aquifer. In addition, marine-derived sodium chloride brines exist below the aquifer and presumably in fluid inclusions within the aquifer. Large concentrations of arsenic, selenium, chromium, and uranium occur naturally within the aquifer, but only arsenic is considered in this simulation. Figure 6.4 shows that locations of large arsenic concentrations are almost exclusively within the confined part of the aquifer, which also is the location of high-pH, sodium bicarbonate water.

The conceptual model for the calculation assumes that brines initially filled the aquifer. The aquifer contains calcite, dolomite, clays with cation exchange capacity, and hydrous ferric oxide surfaces. The initial compositions of the cation exchanger and surfaces are in equilibrium with the brine, which contains arsenic. Arsenic is initially sorbed on the hydrous ferric oxide surfaces. The aquifer is assumed to be recharged with rainwater that is concentrated by evaporation and equilibrated with calcite and dolomite in the unsaturated zone. This water then enters the saturated zone and reacts with calcite and dolomite in the presence of the cation exchanger and hydrous ferric oxide surfaces. A period of 1,000,000 years of flushing the brine-filled aquifer with freshwater is simulated.

In general, the ion-association model of PHAST is not appropriate for brine calculations. However, a brine is a logical initial condition for the chemical composition of the aquifer water in this example. The Pitzer activity model was not used in these calculations because of the need for parameterizations of the aqueous species of iron and arsenic, which are not available in the *pitzer.dat* database of PHREEQC. The use of the ion-association model with the brine initial condition is justified on the basis that concentrations quickly decrease to levels appropriate for the ion-association model and that the brine has a sodium chloride composition. Chloride salts were used to fit the activity coefficient parameters of the major cations in the ion-association aqueous model of PHAST; therefore, thermodynamic calculations should be accurate to higher ionic strengths in sodium chloride dominated systems than in systems dominated by other ions.

6.5.1. Initial Conditions

Parkhurst and others (1996) provide data from which it is possible to estimate the number of moles of calcite, dolomite, and cation exchange sites in the aquifer per liter of water. Point counts from thin sections from aquifer core material indicate the weight percent for calcite ranges from 0 to 2 percent and dolomite from 0 to 7 percent, with dolomite more abundant in all samples. Again from point counts of thin sections, porosity is estimated to be 0.22. Measurements of cation exchange capacity for the clay ranged from 10 to 50 meq/100 g (milliequivalents per 100 grams), with average clay content of 30 percent. For the simulation in this example, calcite was assumed to be present at 0.1 weight percent and dolomite at 3 weight percent, which, assuming a rock density of 2.7 kg/L (kilogram per liter), corresponds to 0.027 mol/L rock (mole per liter of rock) for calcite and 0.44 mol/L rock for dolomite. The number of cation exchange sites was estimated to be 0.14 eq/L rock (equivalents per liter of rock).

The amount of arsenic on the surface of hydrous ferric oxides was estimated from sequential extraction data on core samples (Mosier and others, 1991). Arsenic concentrations in the solid phases generally ranged from 10 to 20 ppm, which correspond to 0.36 to 0.72 mmol/L rock arsenic. The number of surface sites was estimated from the amount of extractable iron in sediments, which ranged from 1.6 to 4.4 percent (Mosier and others, 1991). A content of 2 percent iron for the sediments corresponds to 1 mol/L rock of iron. However, most of the iron is contained in goethite and hematite, which have fewer surface sites than hydrous ferric oxides. The fraction of iron in hydrous ferric oxides, as opposed to other ferric oxides, was arbitrarily assumed to be 0.2. Thus, a total of 0.2 mol of iron per liter of rock was assumed to be in hydrous ferric oxides, and using a value of 0.2 for the number of sites per mole of iron, a total of 0.04 mol of sites per liter of rock was used in the calculations. A gram formula weight of 89 was used to estimate that the mass of hydrous ferric oxides was 17 g/L rock (gram per liter of rock). The specific surface area for hydrous ferric oxides was assumed to be 300 m^2/g (square meter per gram).

6.5.2. Chemistry Data File

The chemistry data file for this example simulation is presented in table 6.11. The default thermodynamic database file, *phast.dat* (which is equivalent to *phreeqc.dat* in the PHREEQC distribution), was augmented with additional thermodynamic data for arsenic and surface complexation in the chemistry data file. The **SURFACE_MASTER_SPECIES** data block defines a new surface named "Surf", which is not present in the thermodynamic database. The **SURFACE_SPECIES** data block defines all of the surface-complexation reactions for the surface *Surf*. **SOLUTION_MASTER_SPECIES** and **SOLUTION_SPECIES** data blocks explicitly define the aqueous reactions of arsenic; these species are not present in *phast.dat*, but the definitions would supersede any definitions of these species had they been present in the thermodynamic database file. To maintain consistency between the aqueous model and the surface complexation model, both the aqueous model for arsenic and the equilibrium constants for surface species are taken from Dzombak and Morel (1990).

The water entering the saturated zone of the aquifer was assumed to be the composition of rainwater concentrated twentyfold (**SOLUTION** *1*) to account for evapotranspiration and then equilibrated with calcite and dolomite at an unsaturated-zone partial pressure of carbon dioxide (P_{CO_2}) of $10^{-1.5}$ atm (atmosphere).

The batch-reaction calculation (**USE** and **EQUILIBRIUM_PHASES** *1* data blocks) is defined, and the resulting solution is stored as solution 1 (**SAVE solution** *1*).

The composition of the brine that represents the water that initially filled the aquifer was selected from Parkhurst and others (1996) and is defined in the **SOLUTION** *2* data block. An equilibrium-phase assemblage containing calcite and dolomite is defined with the **EQUILIBRIUM_PHASES** *2* data block. The brine and equilibrium-phase assemblage are allowed to react to equilibrium, and the result of the batch-reaction equilibration is stored as solution 2 (**SAVE solution** *2*).

The number of cation exchange sites is defined with the **EXCHANGE** *2* data block, and the number of surface sites, specific surface area, and mass of surface material are defined with the **SURFACE** *2* data block. Both the initial exchange and the initial surface compositions are specified to be in equilibrium with the brine (**-equil solution** *2*). The concentration of arsenic in the brine is set to 0.01 µmol/kgw (micromole per kilogram of water), which results in a total of approximately 2.3 mmol arsenic on the surface (from the *ex4*.**chem.txt** file). This concentration on the surface is the same order of magnitude, but greater than the upper end of the range of the sequential extraction data.

The concentrations of solid reactants (**EQUILIBRIUM_PHASES**, **EXCHANGE**, and **SURFACE**) are entered in moles per liter of rock in the chemistry data file. At the beginning of the PHAST simulation, a conversion to moles per liter of water is performed due to the specifications for **-equilibrium_phases**, **-exchange**, and **-surface** in the UNITS data block in the flow and transport data file.

The **SELECTED_OUTPUT** and **USER_PUNCH** data blocks are used to specify the data that will be written to the *ex4*.**h5** and *ex4*.**chem.xyz.tsv** files. Writing of data selected by default is suppressed (**-reset false**) and only pH is selected for writing (**-pH**) in the **SELECTED_OUTPUT** data block. Concentrations of calcium, magnesium, sodium, chloride, carbon, and sulfur are specified to be printed in units of milligrams per kilogram of water and arsenic is specified to be printed in units of micrograms per kilogram of water in the Basic-language statements of the **USER_PUNCH** data block.

Table 6.11. Chemistry data file for example 4.

```
SURFACE_MASTER_SPECIES
        Surf    SurfOH
SURFACE_SPECIES
        SurfOH = SurfOH
                log_k   0.0
        SurfOH  + H+ = SurfOH2+
                log_k   7.29
        SurfOH = SurfO- + H+
                log_k   -8.93
        SurfOH + AsO4-3 + 3H+ = SurfH2AsO4 + H2O
                log_k   29.31
        SurfOH + AsO4-3 + 2H+ = SurfHAsO4- + H2O
                log_k   23.51
        SurfOH + AsO4-3 = SurfOHAsO4-3
                log_k   10.58
SOLUTION_MASTER_SPECIES
        As      H3AsO4          -1.0    74.9216         74.9216
SOLUTION_SPECIES
#
#   Dzombak and Morel
#
#H3AsO4 primary master species
        H3AsO4 = H3AsO4
        log_k           0.0
#H2AsO4-            482
        H3AsO4 = AsO4-3 + 3H+
        log_k   -20.7
#HAsO4-2            483
        H+ + AsO4-3 = HAsO4-2
        log_k   11.50
#AsO4-3             484
        2H+ + AsO4-3 = H2AsO4-
        log_k           18.46

SOLUTION 1 20 x precipitation
        pH      4.6
        pe      4.0     O2(g)   -0.7
        temp    25.
        units   mmol/kgw
        Ca      .191625
        Mg      .035797
        Na      .122668
        Cl      .133704
        C       .01096
        S       .235153         charge
END
USE solution 1
EQUILIBRIUM_PHASES 1
        Dolomite        0.0     1.0
        Calcite         0.0     1.0
        CO2(g)          -1.5    1.0
SAVE solution 1
END
SOLUTION 2 Brine
        pH      5.713
        pe      4.0     O2(g)   -0.7
        temp    25.
        units   mol/kgw
        Ca      .4655
        Mg      .1609
        Na      5.402
        Cl      6.642           charge
        C       .00396
        S       .004725
        As      .01 umol/kgw
END
USE solution 2
EQUILIBRIUM_PHASES 2
        Dolomite        0.0     0.44    # mol/Lrock
        Calcite         0.0     0.027   # mol/Lrock
SAVE solution 2
END
EXCHANGE 2
        -equil with solution 2
        X       0.14                    # eq/L rock
```

Table 6.11. Chemistry data file for example 4.—Continued

```
SURFACE 2
        -equil solution 2
        SurfOH          0.04     300.    17.
END
SELECTED_OUTPUT
        -file ex4.dummy.sel
        -reset false
        -pH
USER_PUNCH
# Prints concentrations in mg/kgw to ex4.xyz.chem and ex4.h5
-heading        Ca       Mg       Na       Cl       C(4)     SO4      As
10 PUNCH TOT("Ca")*1e3*40.08
20 PUNCH TOT("Mg")*1e3*24.312
30 PUNCH TOT("Na")*1e3*23.
40 PUNCH TOT("Cl")*1e3*35.45
50 PUNCH TOT("C(4)")*1e3*61.    # as HCO3-
60 PUNCH TOT("S(6)")*1e3*96.    # as SO4
70 PUNCH TOT("As")*1e6*74.296   # ug/L
END
```

6.5.3. Flow and Transport Data File

The flow and transport file (table 6.12) specifies the input units, grid, the porous-media properties, boundary conditions, initial conditions, print control, time step, and duration of the simulation through a series of keyword data blocks. A two-line title is defined with the **TITLE** data block. Units for all of the *input* data are defined in the **UNITS** data block; note that the definition of ROCK for **-equilibrium_phases**, **-exchange**, and **-surface** causes the input data for these data blocks in the chemistry data file to be converted from moles per liter of rock to moles per liter of water for the PHAST simulation. All *output* data are in SI units, with the exception of time, which will be years as specified by the **-time** identifier. A grid (**GRID** data block) containing 31 nodes in the X direction (3,000-m node spacing), 17 nodes in the Y direction (3,000-m node spacing), and 9 nodes in the Z direction (50-m node spacing) is used to represent the part of the Central Oklahoma aquifer selected for modeling. The northern and southern boundaries of the model are near rivers that provide satisfactory boundary conditions. The eastern boundary of the model coincides with the eastern extent of the geologic units of the aquifer. The extent of freshwater in the aquifer is used to set the western boundary of the model.

Both flow and reactive-transport are simulated, which is specified with **SOLUTE_TRANSPORT** *true*. The **STEADY_FLOW** data block specifies that the program perform an initial flow calculation to obtain the steady-state head and velocity fields, which will be used for all transport calculations during the simulation. Criteria for attainment of steady flow are defined by the identifiers in the **STEADY_FLOW** data block. The properties of the porous media are defined in the **MEDIA** data block. Hydraulic conductivity was taken from Parkhurst and others (1996), but the horizontal hydraulic conductivity was decreased to attain a maximum head in the aquifer that was consistent with the measured water table (Parkhurst and others, 1996). The longitudinal dispersivity (2,000 m) and horizontal and vertical transverse dispersivities (50 m) are set arbitrarily to be less than or equal to the node spacing. The dispersivity specifies the amount of physical dispersion for the simulation; however, numerical dispersion also is present in the numerical solution, as discussed subsequently in this section. The **MEDIA** data block excludes a wedge of nodes in the western part of the grid region (**-active** *0*) to represent a completely impermeable confining layer.

Boundary conditions are defined with the **RIVER**, **FLUX_BC**, **SPECIFIED_HEAD_BC**, **LEAKY_BC**, **FREE_SURFACE_BC**, and **WELL** data blocks. This example demonstrates the use of all of these data blocks; the nodes affected by these boundary conditions are shown in figure 6.5. Three rivers are defined in the **RIVER** data blocks. The location of each river is defined with X, Y locations. Head, width, depth, riverbed thickness, riverbed hydraulic conductivity, and associated solution are defined for the

Table 6.12. Flow and transport data file for example 4.

```
TITLE
 Central Oklahoma aquifer,
 demonstration of PHAST
UNITS
        -time                                   years
        -horizontal_grid                        m
        -vertical_grid                          m
        -map_horizontal                         m
        -map_vertical                           m
        -head                                   m
        -hydraulic_conductivity                 meters/day
        -specific_storage                       1/m
        -dispersivity                           m
        -flux                                   meters/year
        -leaky_hydraulic_conductivity           meters/day
        -leaky_thickness                        m
        -well_diameter                          inches
        -well_flow_rate                         liter/day
        -well_depth                             m
        -river_bed_hydraulic_conductivity meters/day
        -river_bed_thickness                    m
        -river_width                            m
        -river_depth                            m
        -equilibrium_phases                     rock
        -exchange                               rock
        -surface                                rock
GRID
        -uniform X 0 90000 31
        -uniform Y 0 48000 17
        -uniform Z 0 400    9
        -print_orientation XY
SOLUTE_TRANSPORT                                true
        -diffusivity                            1e-009
STEADY_FLOW                                     true
        -head_tolerance                         1e-6
        -flow_balance                           1e-6
MEDIA
        -domain
                -Kx                             0.5
                -Ky                             0.5
                -Kz                             0.00014
                -porosity                       0.22
                -specific_storage               0
                -long_dispersivity              2000
                -horizontal_dispersivity        50
                -vertical_dispersivity          50
        -wedge 0 0 150 27000 48000 400 Y4 GRID
                -active                         0
RIVER 1 Little River
        -coordinate_system                      GRID
        -point 44000 15000
                -head                           0          335
                -solution                       0          1
                -width                          200
                -bed_hydraulic_conductivity     1
                -bed_thickness                  1
                -depth                          1
        -point 44000 0
        -point 90000 0
                -head                           0          275
                -solution                       0          1
                -width                          200
                -bed_hydraulic_conductivity     1
                -bed_thickness                  1
                -depth                          1
RIVER 2 North Fork River
        -coordinate_system GRID
        -point 30000 36000
                -head                           0          335
                -solution                       0          1
                -width                          200
                -bed_hydraulic_conductivity     1
                -bed_thickness                  1
                -depth                          1
        -point 30000 48000
```

Table 6.12. Flow and transport data file for example 4.—Continued

```
            -point 90000 48000
                  -head                         0        280
                  -solution                     0        1
                  -width                        200
                  -bed_hydraulic_conductivity   1
                  -bed_thickness                1
                  -depth                        1
RIVER 3 North Canadian River
      -coordinate_system GRID
      -point 60000 30000
                  -head                         0        350
                  -solution                     0        1
                  -width                        200
                  -bed_hydraulic_conductivity   1
                  -bed_thickness                1
                  -depth                        1
      -point 90000 20000
                  -head                         0        305

                  -solution                     0        1
                  -width                        200
                  -bed_hydraulic_conductivity   1
                  -bed_thickness                1
                  -depth                        1
FLUX_BC
      -box 27000 0 400 90000 48000 400  GRID
                  -description Recharge on unconfined area
                  -face                         Z
                  -associated_solution          0        1
                  -flux                         0 years  -0.0071
SPECIFIED_HEAD_BC
      -box 29000 14000 300 31000 16000 400 GRID
                  -description Simulated lake
                  -head                         0 years  348
                  -associated_solution          0        1
LEAKY_BC
      -box 0 48000 0 27000 48000 400 GRID
                  -description North boundary of confined aquifer
                  -face                         Y
                  -associated_solution          0 years  1
                  -head                         0 years  305
                  -hydraulic_conductivity       0.5
                  -thickness                    30000
LEAKY_BC
      -box 0 0 0 39000 0 400 GRID
                  -description South boundary of confined aquifer
                  -face                         Y
                  -associated_solution          0        2
                  -head                         0 years  320
                  -hydraulic_conductivity       0.5
                  -thickness                    20000
FREE_SURFACE_BC true
WELL 1 Observation well 1 in arsenic zone
      -xy_coordinate_system           GRID
      -z_coordinate_system            GRID
      -location                       12000    36000
      -diameter                       2
      -elevation                      110      90
      -allocate_by_head_and_mobility  False
      -pumping_rate                   0        1
HEAD_IC
      -domain
                  -head                         380
CHEMISTRY_IC
      -domain
                  -solution                     2
                  -equilibrium_phases           2
                  -exchange                     2
                  -surface                      2
```

Table 6.12. Flow and transport data file for example 4.—Continued

```
SOLUTION_METHOD
     -iterative_solver          true
     -tolerance                 1e-012
     -save_directions           20
     -maximum_iterations        500
     -space_differencing        0
     -time_differencing         1
     -cross_dispersion          false
     -rebalance_fraction        0.5
     -rebalance_by_cell         false
TIME_CONTROL
     -time_step                 0           20000 years
     -time_change               1000000     years
PRINT_INITIAL
     -heads                     true
     -velocities                true
     -xyz_head                  true
     -xyz_ss_velocities         true
PRINT_FREQUENCY
     -save_final_heads true
     0
          -HDF_chemistry        1           step
          -XYZ_chemistry        500000      y
          -XYZ_well             1           step
END
```

beginning and ending points for each river; in this example, values for intermediate points are not specified but are calculated by interpolation along the line segments that connect the river points. All rivers are expected to be gaining streams for this simulation, but if any water recharges from the river to the aquifer, its composition will be the composition of solution 1 (**-solution** *1* is used for all river end points). The hydraulic conductivity of the riverbed is large (**-bed_hydraulic_conductivity**), which will cause the head in the aquifer to be nearly equal to the head in the river for each cell that contains a river. All heads and solution compositions for the rivers are defined to apply from time 0 yr (year) through the end of the simulation.

A flux boundary condition (**FLUX_BC** data block) represents recharge from precipitation of 0.0071 m/yr (meter per year) (Parkhurst and others, 1996) over the eastern two-thirds of the aquifer with water composition defined by solution 1. Note the sign of the flux is negative, which indicates that the flux is in the negative Z direction. The **SPECIFIED_HEAD_BC** data block is used for demonstration purposes to represent a lake. Leaky boundary conditions (**LEAKY_BC** data blocks) are defined on the north and south boundaries of the western part of the grid region. These leaky boundaries allow flow out of the grid region in response to the difference between the head at the boundary and a specified head at a given distance from the boundary. If flow into the active grid region should occur through these boundaries, the composition of the water will be given by solution 1. All heads, solution compositions, and fluxes are defined to apply from time 0 years through the end of the simulation.

Unconfined flow with a water-table boundary condition is simulated (**FREE_SURFACE_BC** *true*). One well is included (**WELL** *1* data block) in a zone that has large arsenic concentration; the well is pumped at a small rate, which is constant from 0 years through the end of the simulation. The well is included to demonstrate how a well could be used to monitor concentrations at a given point in the aquifer. Component concentrations for wells are written to the file *ex4*.**wel.xyz.tsv**. More detailed information for a node can be extracted from the *ex4*.**h5** and *ex4*.**chem.xyz.tsv** files, but the *ex4*.**wel.xyz.tsv** file contains fewer data items and thus can be written at a higher frequency without generating large files.

Initial conditions for head are defined in the **HEAD_IC** data block. Although this initial-condition head distribution is uniform (380 m), the initial steady-state flow calculation will calculate the head distribution to achieve steady-state flow for the specified boundary conditions. Initial conditions for chemistry (**CHEMISTRY_IC** data block) include not only the composition of the water in the active grid region, but also the initial amounts of reactive minerals, the amount and composition of the cation exchanger, and the

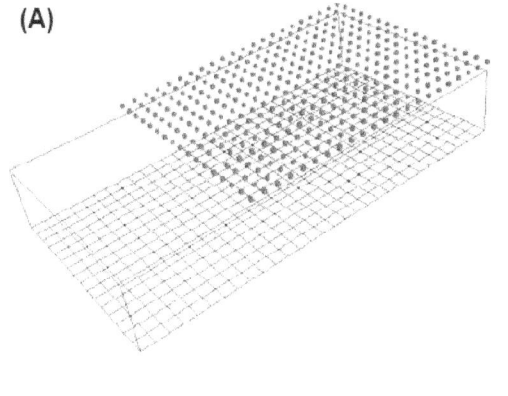

(A)

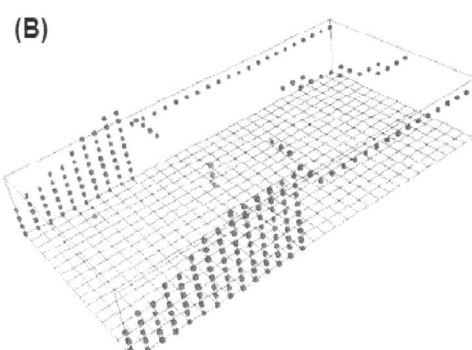

(B)

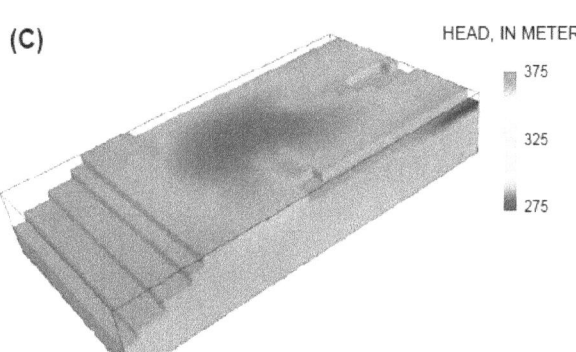

(C)

HEAD, IN METERS

375
325
275

Figure 6.5. Example 4, locations of boundary conditions and steady-state head distribution: (A) flux boundary-condition nodes are shown in orange; (B) river boundary-condition nodes are shown in blue, leaky boundary-condition nodes are shown in red, three specified-head boundary-condition nodes are shown in brown, and one node containing the well screen is shown in pink; and (C) color rendition of steady-state head distribution in the active grid region. View is from the southwest looking to the northeast.

directly in the flow and transport file, but by index numbers that refer to definitions in the chemistry data file. In this simulation, initial chemical conditions are uniform throughout the active grid region, but heterogeneous distributions may be defined by using multiple zone definitions in the **CHEMISTRY_IC** data block.

The **SOLUTION_METHOD** data block is used to specify that the iterative linear-equation solver is used in this example. The difference equations are formulated with upstream-in-space and backward-in-time weighting (as opposed to centered-in-time and centered-in-space weighting for examples 1, 2, and 3), which avoids the possibility of oscillations due to spatial and temporal discretization. However, this differencing scheme introduces numerical dispersion into the solution to the flow and transport equations. For this example, the maximum calculated velocity is about 15 m/yr; but typically, velocities are about 1 m/yr or less, the time step is 20,000 years, and the horizontal discretization is 3,000 m. The numerical dispersivity (equation D.7 and D.10) is on the order of $\frac{3,000}{2} + \frac{1(20,000)}{2} = 11,500$ m (see sections D.1.2. Spatial Discretization and D.1.3. Temporal Discretization), which is much larger than the physical dispersivity specified in the **MEDIA** data block. Thus, the simulations have an effective dispersivity that is approximately six times the specified dispersivity. To reduce the numerical dispersion, the time step and grid could be refined.

The length of simulation (1,000,000 years) and uniform time step (20,000 years) for the transient transport and reaction simulation are defined with **TIME_CONTROL** data block. The length of time simulated was determined by the time needed for the simulated water compositions to resemble the observed distribution of water compositions in the aquifer and to nearly remove all of the brine. Finally, **PRINT_INITIAL** specifies the data for the active grid region that are to be printed after initialization, and **PRINT_FREQUENCY** specifies the transient data to be printed at specified time intervals.

6.5.4. Simulation Results

Selected results at the end of the 1,000,000-year simulation are plotted in figure 6.6. The three-dimensional view of the active grid region is from the southwest, which shows a part of the grid region is missing—the inactive grid region—at the western end of the grid region. Also, the top layer of cells is not shown because the cells are dry. Parts of the second layer of cells near the rivers are also not shown because they are dry.

Figure 6.6A shows the concentration of chloride in the active grid region. After 1,000,000 years, the large-concentration brine has been flushed from the eastern two-thirds of the active grid region, and only traces of chloride remain at the western edge of the active grid region.

Initially, the exchange sites in each active cell were in equilibrium with the brine, which caused sodium to be the dominant cation on the exchanger. The yellow areas in the eastern two-thirds of the active grid region in figure 6.6B represent areas where sodium has been removed from the exchanger, and calcium and magnesium are the dominant cations in solution and on the exchanger. In the western one-third of the active grid region, brines have been mostly removed, but sodium persists as the dominant cation in solution and on the exchanger. The blue areas of figure 6.6B correspond to high pH zones in figure 6.6C. The pH is high in these zones because of dissolution of calcite and dolomite, which is enhanced because of the exchange of calcium and magnesium for sodium on the exchanger.

The zones of high pH correspond closely with large arsenic concentrations in figure 6.6D. The carbonate and exchange reactions drive the pH to higher than 8.5 and arsenic concentrations increase to as much as

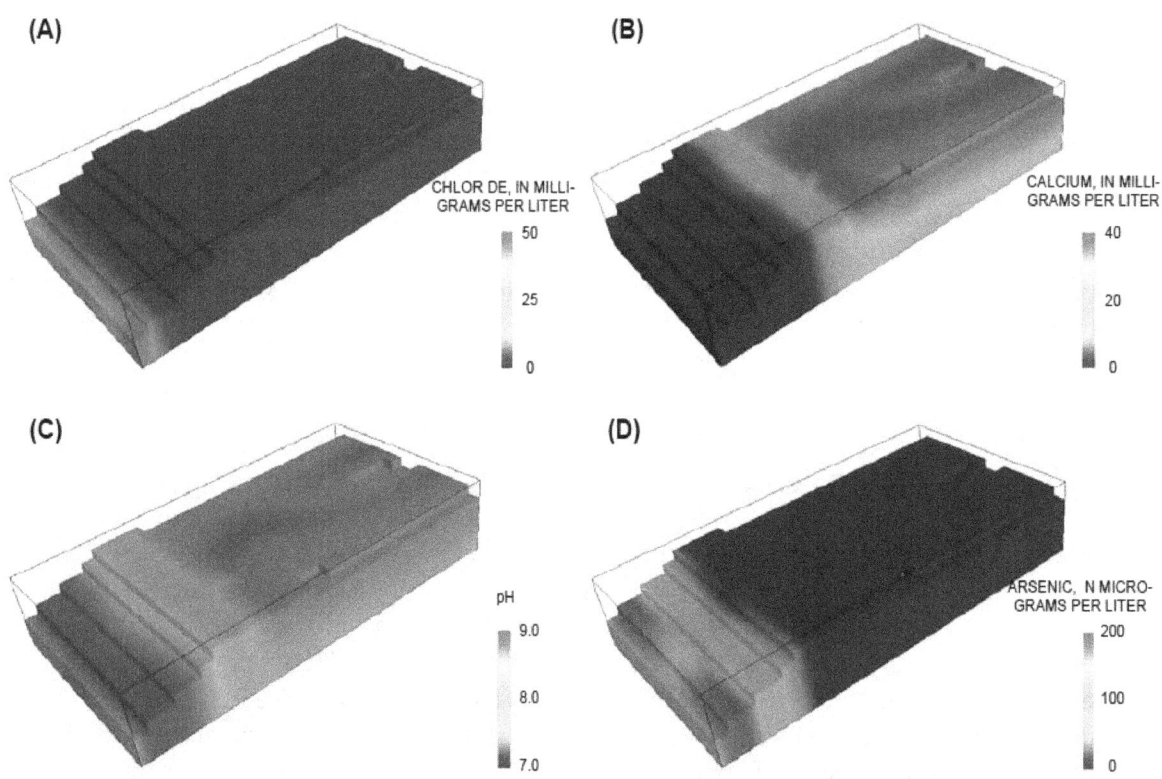

Figure 6.6. Example 4, distribution of (A) chloride concentrations, (B) calcium concentrations, (C) pH, and (D) arsenic concentrations after 1,000,000-year simulation of the evolution of water in the Central Oklahoma aquifer. View is from the southwest looking to the northeast.

200 µg/L in the red zones, which is consistent with the locations of large arsenic concentrations shown in figure 6.4. In the blue zones of figure 6.6D, the concentrations of arsenic are small. In the east it is because the exchanger has been flushed of sodium and the pH has stabilized at about 7.0 to 7.5; in the west it is because lower pH values associated with the brine still persist.

The transport calculations show three types of water in the aquifer, the remnants of the initial brine, sodium bicarbonate water, and calcium and magnesium bicarbonate water, all of which are similar to the observed water types in the aquifer. Arsenic concentrations are consistent with values observed in the aquifer. The time of evolution to calcium-magnesium-dominated exchange sites is dependent on the number of exchange sites. A larger number of exchange sites will produce lower maximum pH values and lower maximum arsenic concentrations at the end of a 1,000,000-year simulation. The calculation is sensitive to the stability constant for the surface-complexation reactions for arsenic (**SURFACE_SPECIES** data block in the chemistry data file), the number of surface complexation sites (**SURFACE** data block in the chemistry data file), the specific area of the surface complexer, and the concentration of arsenic in the brine; a decrease in the log K for the predominant arsenic complexation reaction, a decrease in the number of sites, an increase in the surface area, or a decrease in the arsenic concentration in the brine tends to decrease the maximum arsenic concentrations. The large arsenic concentrations are dependent on the flow field; changes in the flow model cause the position of the peak arsenic concentrations to occur at different times and in different locations. Finally, the peaks in arsenic concentrations are affected by the effective dispersivity (specified plus numerical) in the calculation; smaller effective dispersivity tends to increase the maximum arsenic concentrations.

The model results, which are based largely on measured values and literature thermodynamic data, provide a satisfactory explanation of the variation in major ion chemistry, pH, and arsenic concentrations within the Central Oklahoma aquifer. However, the time scale for the simulation (1,000,000 years) is shorter than the length of time over which the aquifer probably evolved (millions of years) and the initial amount of sorbed arsenic in the model is greater than the selective extractions from modern sediments. It is the nature of modeling that even satisfactory results give rise to more detailed questions and suggest new hypotheses about the groundwater system. One of the keys to a successful modeling study is to know when to stop.

6.6. Example 5: Simulation of Groundwater Flow for a Sewage Wastewater Plume at Cape Cod, Massachusetts

This example uses many of the new features in PHAST Version 2, including importing spatial data, defining arbitrary spatial volumes for model properties, using the drain boundary condition, and writing transient heads for subsequent simulations on a restricted domain. Example 5 computes a flow field for a regional-scale model for a domain that includes the wastewater plume from a sewage treatment plant; example 6 uses the heads from the regional model to establish boundary conditions for a simulation of reactive transport of nitrogen for the part of the aquifer that contains the wastewater plume.

The wastewater plume has developed in a permeable sand and gravel glacial outwash aquifer. A sewage treatment plant released treated wastewater to infiltration beds, through which the wastewater percolated into the aquifer (fig. 6.7). The plant released water at varying rates, depending on water use, from 1936 through 1995. Horizontal interstitial water velocities are as large as 200 m/yr, and, over the 60 years, conservative constituents produced a plume several kilometers long. Reactive constituents were retarded, consumed, or sequestered by a variety of biogeochemical reactions (LeBlanc, 1984a).

Only a flow and transport data file is developed for example 5 (table 6.13). Two simulations are performed: first, a steady-flow simulation is used to calculate initial-condition heads for subsequent simulations (both flow and reactive transport); second, a transient simulation, with a 60-year time series of loading of

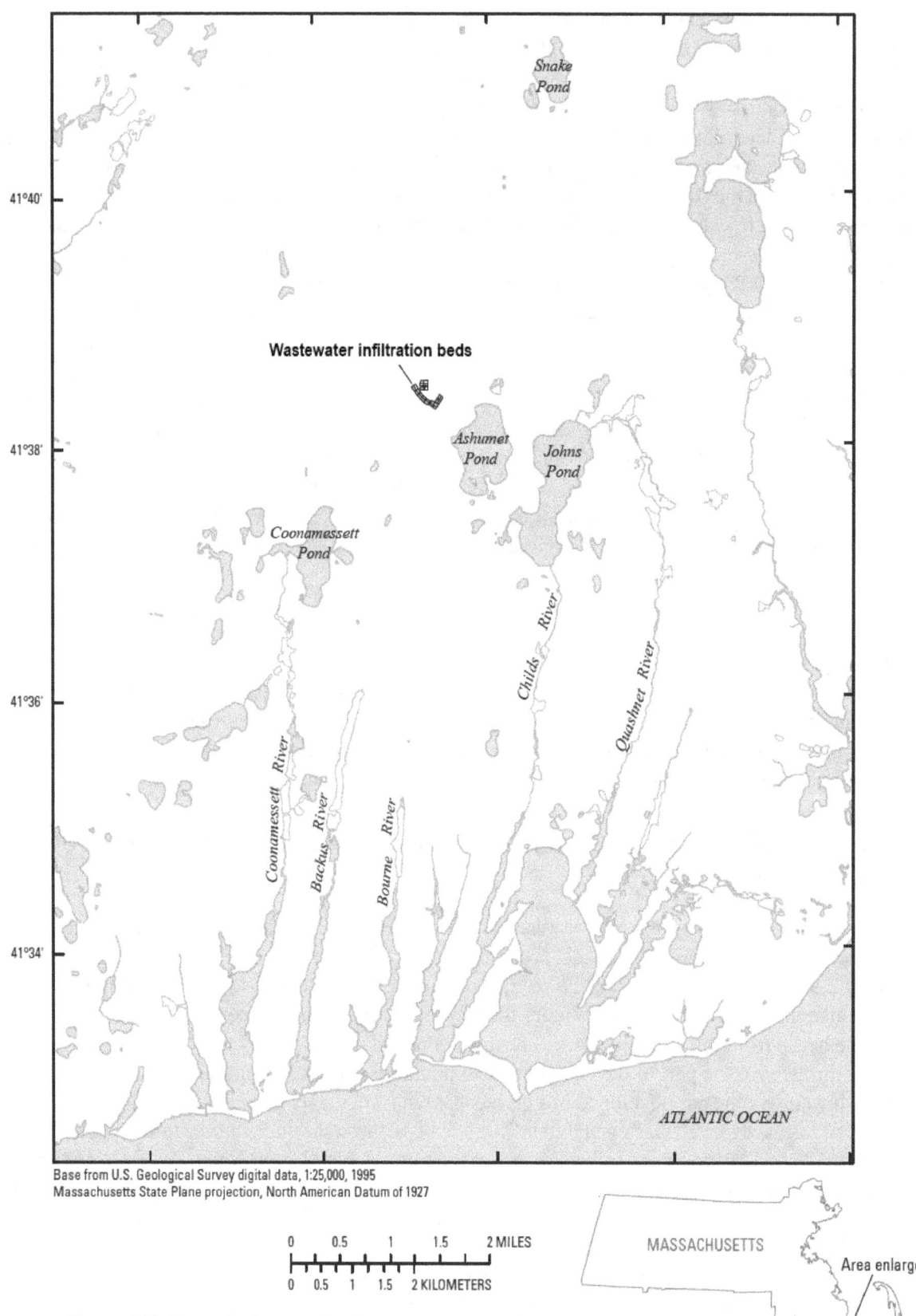

Figure 6.7. Example 5, map showing wastewater infiltration beds and selected surface-water features of Cape Cod, Massachusetts.

wastewater infiltration beds, is used to produce a time-series of heads that define head boundary conditions for example 6. The input files for the two simulations of example 5 differ only in two lines, one that defines whether the simulation is steady flow or transient, and one that defines the initial head condition.

Three shapefiles were developed in ArcInfo to help define active regions of the model grid and model boundary conditions. One shapefile, named *ModelExtent.shp*, defines a model outline that includes all active areas. The edges of this area were selected to be two rivers (the Coonamessett and Quashnet Rivers) that should constrain the flow to the southwest and southeast of the active region. North of the rivers, the outline was defined to be perpendicular to the contours of a head map of the aquifer (Timothy McCobb, U.S. Geological Survey, written commun., 2008). To the south, the edge of the active area was extended a short distance beyond the coastline. The other two shapefiles subdivide the model outline (*ModelExtent.shp*) into that area that is inside the coastline (*OnShore.shp*) and that area that is outside the coastline (*OffShore.shp*).

6.6.1. Flow and Transport Data File

In table 6.13, a title is defined with the TITLE data block, the simulation is specified to be flow-only with the SOLUTE_TRANSPORT data block, and the iterative solver is chosen in the SOLUTION_METHOD data block. Units for all of the *input* data are defined in the UNITS data block; all *output* data are in SI units, with the exception of time, which will be years as specified by the **-time** identifier in the UNITS data block. The STEADY_FLOW data block is used to specify a steady-flow calculation for the initial simulation. An unconfined flow system is defined with the FREE_SURFACE_BC data block.

All horizontal locations are defined in State plane coordinates (in meters) for Massachusetts. A grid with 200-m horizontal node spacing is defined with the GRID data block. Vertically, the grid extends from −120 m to 20 m relative to sea level, with 10-m node spacing. The grid is situated to contain plausible hydrologic boundaries for the part of the flow system that includes the wastewater plume.

The active grid region and media properties within the grid are defined with the MEDIA data block. The first zone within the MEDIA uses **-domain** to define media properties for all elements in the grid (fig. 6.8A). The horizontal and vertical hydraulic conductivities (Kx, Ky, and Kz) are interpolated from XYZ files that contain hydraulic conductivity at a series of X–Y–Z coordinates (Donald Walter, U.S. Geological Survey, written commun., 2009), which were derived from a MODFLOW (Harbaugh and others, 2000) model. Data for the elements of the grid are interpolated from the data in the files by closest point interpolation. The first zone within the MEDIA data block also defines constant values for porosity and specific storage, and defines the entire grid volume to be inactive. Note that the next zone definitions will define parts of the grid to be active, in which case, the media properties of the first zone definition will be used unless property values are redefined in the subsequent zone definitions.

Table 6.13. Flow and transport data file for example 5.

```
TITLE Flow model for Cape Cod Massachusetts
SOLUTE_TRANSPORT                    false
SOLUTION_METHOD
        -iterative_solver           true
        -tolerance                  1e-012
        -time_differencing          1
UNITS
        -time                       yr
        -horizontal_grid            m
        -vertical_grid              m
        -map_horizontal             m
        -map_vertical               m
        -head                       m
        -hydraulic_conductivity     m/d
        -specific_storage           1/m
        -flux                       m/d
        -leaky_hydraulic_conductivity m/d
        -leaky_thickness            m
        -drain_hydraulic_conductivity m/d
        -drain_thickness            m
        -drain_width                m
#----- Use the following line for EXAMPLE 5A, steady state simulation
STEADY_FLOW true
#----- Use the following line for EXAMPLE 5B, transient simulation
#STEADY_FLOW false
FREE_SURFACE_BC true
GRID
        -uniform X 275000 285000  51        # 200 m grid
        -uniform Y 810000 830000 101        # 200 m grid
        -uniform Z -120    20      15       # 10  m grid
        -snap    X                  0.001
        -snap    Y                  0.001
        -snap    Z                  0.001
        -print_orientation XY
        -grid_origin       0  0  0
        -grid_angle        0
MEDIA
        # Default values
        -domain
                -active             0
                -Kx        XYZ      MAP    ./Parameters/kh_meters
                -Ky        XYZ      MAP    ./Parameters/kh_meters
                -Kz        XYZ      MAP    ./Parameters/kz_meters
                -porosity           0.39
                -specific_storage   0.39
        # Activate inside model outline, above bedrock to sea level
        -prism
                -perimeter SHAPE    GRID   ./ArcData/ModelExtent.shp
                -bottom    ARCRASTER GRID  ./ArcData/bedrock.txt
                -top       CONSTANT  GRID  0
                -active             1
        # Activate inside coast, sea level to top of grid
        -prism
                -bottom    CONSTANT  GRID  0
                -perimeter SHAPE     GRID  ./ArcData/OnShore.shp
                -active             1
        # Add ponds as high K zones
        -prism
                -top       CONSTANT  GRID  20
                -bottom    SHAPE     GRID  ./ArcData/SimpleBath.shp 10
                -perimeter SHAPE     GRID  ./ArcData/SimplePonds.shp
                -Kx                 30000
                -Ky                 30000
                -Kz                 30000
                -porosity           1
                -specific_storage   1
FLUX_BC
        -prism
                -description Rainfall
                -bottom    CONSTANT  GRID  0
                -perimeter SHAPE     GRID  ./ArcData/OnShore.shp
                -face Z
                -flux
                        0          -0.00188
        -box 279725 821500 0 279825 821600 50
                -description Beds 1-4
                -face Z
```

Table 6.13. Flow and transport data file for example 5.—Continued

```
                  -flux
                          0               0
                          1936            -0.037
                          1941            -0.114
                          1946            -0.023
                          1956            -0.126
                          1971            -0.05
                          1978            -0.114
                          1984            0
        -box 279625 821425 0 279725 821525 50
                  -description Beds 5-8
                  -face Z
                  -flux
                          0               0
                          1941            -0.114
                          1946            -0.023
                          1956            -0.126
                          1971            -0.05
                          1978            0
                          1984            -0.038
                          1996            0
        -box 279525 821475 0 279625 821575 50
                  -description Beds 9-12
                  -face Z
                  -flux
                          0               0
                          1941            -0.114
                          1946            -0.023
                          1956            -0.126
                          1971            -0.05
                          1978            0
                          1984            -0.038
                          1996            0
        -box 279400 821575 0 279500 821675 50
                  -description Beds 13-16
                  -face Z
                  -flux
                          0               0
                          1941            -0.114
                          1946            -0.023
                          1956            0
        -box 279500 821700 0 279600 821800 50
                  -description Beds 19-25
                  -face Z
                  -flux
                          0               0
                          1941            -0.114
                          1946            -0.023
                          1956            0
SPECIFIED_HEAD_BC
        -prism
                  -description Constant head ocean
                  -top         CONSTANT   GRID   0
                  -bottom      CONSTANT   GRID   -10
                  -perimeter SHAPE        GRID   ./ArcData/OffShore.shp
                  -exterior_cells_only  Z
                  -head
                          0               0.0
DRAIN 1 East Boundary Quashnet R
        -point  282107.6    820940.0
                  -width             20
                  -bed_hydraulic     10.00
                  -bed_thickness     1
                  -z                 11.83
        -point  282487.8    821288.7
        -point  282757.6    821288.7
        -point  282919.5    820702.8
        -point  283205.1    819653.4
        -point  283300.6    818796.2
        -point  283204.1    818082.7
        -point  283164.3    817427.9
        -point  282915.4    816494.7
        -point  282209.6    814714.3
                  -width             20
                  -bed_hydraulic     10.00
                  -bed_thickness     1
                  -z                 0
```

Table 6.13. Flow and transport data file for example 5.—Continued

```
DRAIN 2 Childs River, Johns Pond drain
        -point 281619.9    818867.5
                -width                 20
                -bed_hydraulic         10.00
                -bed_thickness         1
                -z                     10.56
        -point  281589.3    818251.7
        -point  281488.0    817897.2
        -point  281342.9    817343.9
        -point  281272.3    816543.5
        -point  281319.4    816213.9
        -point  281211.6    815490.0
        -point  280897.2    814875.9
                -width                 20
                -bed_hydraulic         10.00
                -bed_thickness         1
                -z                     0
DRAIN 3 Bourne River
        -point  279187.9    815453.3
                -width                 20
                -bed_hydraulic         1
                -bed_thickness         1
                -z                     2.97
        -point  279141.9    814825.4
        -point  279108.7    814446.3
                -width                 20
                -bed_hydraulic         1
                -bed_thickness         1
                -z                     0
DRAIN 4 Backus River (Cranberry bogs)
        -point  278572.0    817160.5
                -width                 20
                -bed_hydraulic         10.00
                -bed_thickness         1
                -z                     7.16
        -point  278404.6    816798.4
        -point  278260.8    816171.1
        -point  278104.0    815347.7
        -point  278146.1    814774.8
        -point  278077.8    814393.5
                -width                 20
                -bed_hydraulic         10.00
                -bed_thickness         1
                -z                     0
DRAIN 5 West Boundary Coonamessett River
        -point 277365.6    819130.2
                -width                 20
                -bed_hydraulic         10.00
                -bed_thickness         1
                -z                     9.92
        -point  277224.7    818453.8
        -point  277381.6    817842.1
        -point  277365.6    817312.8
        -point  277461.8    816574.9
        -point  277349.5    815885.2
        -point  277301.4    814730.3
        -point  277260.0    814502.1
                -width                 20
                -bed_hydraulic         10.00
                -bed_thickness         1
                -z                     0
ZONE_FLOW 1
        -box 276400 811000 -200 281300 822300 20
                -description Plume submodel
                -write_heads_xyzt      plume.heads.xyzt
HEAD_IC
        -domain
                #----- Use the following line for EXAMPLE 5A, steady state
simulation
                -head              10
              #----- Use the following line for EXAMPLE 5B, transient simulation
                #-head     XYZ      GRID   ex5.head.200.dat
```

Table 6.13. Flow and transport data file for example 5.—Continued

```
TIME_CONTROL
        -time_step
                0                       0.5
        -time_change                   1998
        -start_time                    1935
PRINT_INITIAL
        -HDF_media_properties          true
        -heads                         false
PRINT_FREQUENCY
        -save_final_heads true
        0 years
                -HDF_heads             5        years
                -progress_statistics   1        step
                -zone_flow_heads       1        step
END
```

The second zone within the **MEDIA** data block begins the definition of the active grid region. It uses a prism definition to specify as active the elements of the grid that are contained within the model outline (*ModelExtent.shp*), above the bedrock surface (*bedrock.txt*), and below or equal to 0-m elevation (fig. 6.8B). The prism bottom is defined by an ArcInfo ASCII raster file, which contains a row and column array representing the bedrock elevation at a series of equidistant points. The bedrock surface is interpolated to horizontal target points (X–Y element centroids) by natural neighbor interpolation. The prism top is, by default, the top of the grid (20 m). Grid-element centroids that are inside the perimeter, above the bottom, and below the top are specified to be active.

The third zone within the **MEDIA** data block defines as active elements that are inside the coastline (*OnShore.shp*) and above or equal to 0-m elevation (fig. 6.8C). Elements that are above 0-m elevation and outside the coast remain inactive because of the first zone definition of the **MEDIA** data block.

The fourth and final zone within the **MEDIA** data block defines four ponds—Snake, Ashumet, Johns, and Coonamessett Ponds—to be high hydraulic conductivity zones. The ponds (fig. 6.8D) are defined as a prism by ArcInfo coverages (Timothy McCobb, U.S. Geological Survey, written commun., 2008). A shapefile (*SimplePonds.shp*) defines the prism perimeter, which contains the X–Y locations of the edges of the four ponds. Note that multiple disconnected ponds are defined with a single shapefile. Another shapefile (*SimpleBath.shp*) is used to define the bathymetry of the ponds. The elevations of the bottoms of the ponds (in meters) at the X–Y locations defined in the shapefile (*SimpleBath.shp*) are stored in the 10th attribute in the *SimpleBath.dbf* file that is associated with the *SimplePonds.shp* file. The elevations at the X–Y locations are used as a set of scattered points that define a surface, which is the bottom of the ponds. The top of the prism is the top of the grid by default. The elements that are within the perimeter and above the bottom (that is, within the ponds) are assigned hydraulic conductivities of 30,000 m/d, porosity of 1.0 (unitless), and storage coefficient of 1.0 (unitless). All zones that define media properties are shown simultaneously in figure 6.8E.

Boundary-condition definitions follow the media properties (table 6.13). Figure 6.9 shows the zones that are used to define the boundary conditions for the model. The **FLUX_BC** data block defines flux boundaries including rainfall infiltration and wastewater infiltration bed loading (fig. 6.9). The first zone in the **FLUX_BC** data block defines a constant rainfall infiltration in the negative Z direction over the entire domain of 0.00188 m/d. The remaining zone definitions (table 6.13, fig. 6.9) define wastewater infiltration at five infiltration beds. The locations of the beds are defined with boxes, and a time series of fluxes (D.R. LeBlanc, U.S. Geological Survey, written commun., 2001; Parkhurst and others, 2003) is associated with each bed.

In the **SPECIFIED_HEAD_BC** data block (table 6.13), a prism is defined in which the perimeter includes the bays and inlets at the southern coast (fig. 6.9). A specified-head boundary condition is applied to all cells that have an exterior Z face within the limits of −10 to 0 elevation. Because the active region was

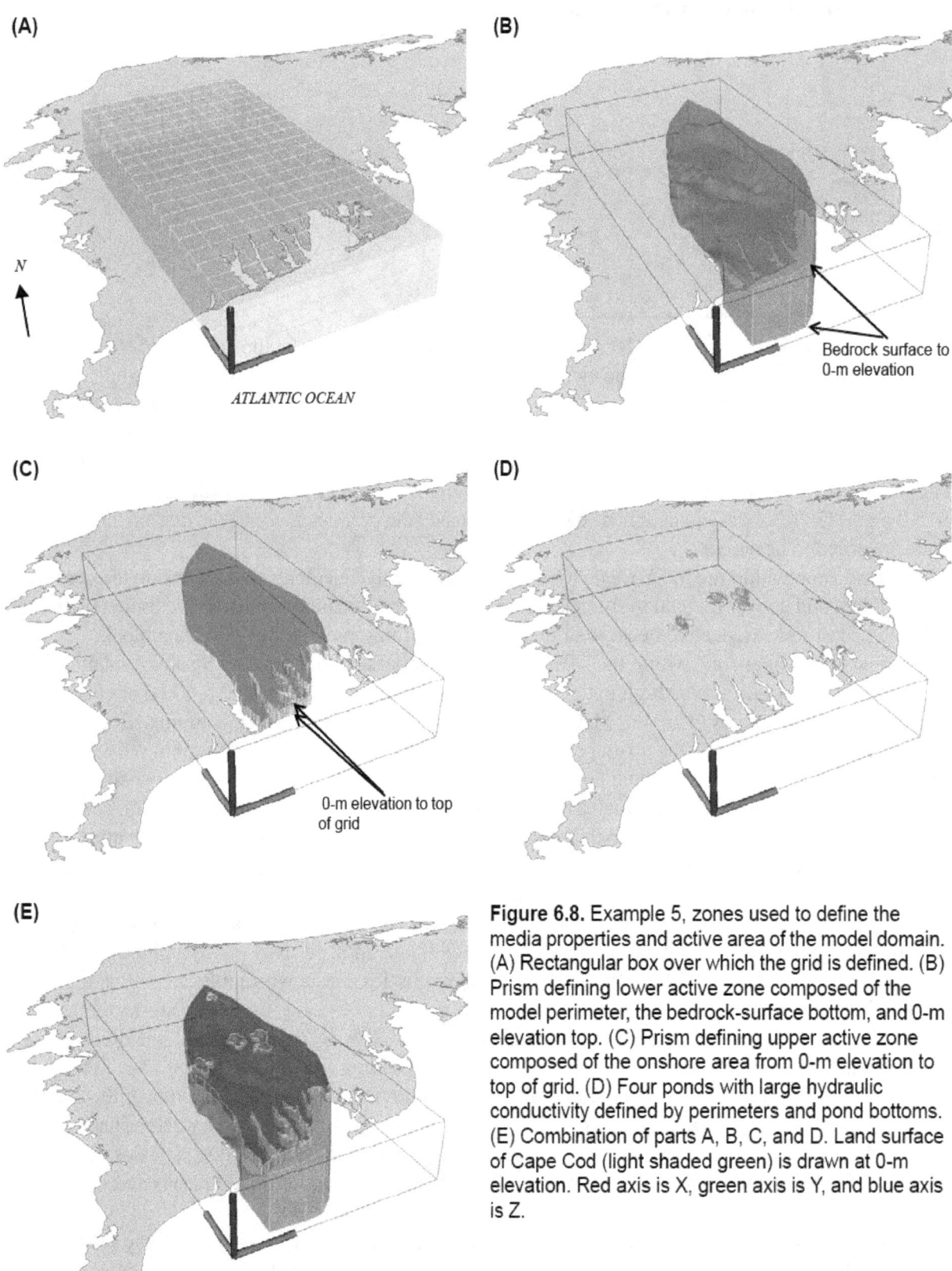

(A)

N

ATLANTIC OCEAN

(B)

Bedrock surface to
0-m elevation

(C)

0-m elevation to top
of grid

(D)

(E)

Figure 6.8. Example 5, zones used to define the
media properties and active area of the model domain.
(A) Rectangular box over which the grid is defined. (B)
Prism defining lower active zone composed of the
model perimeter, the bedrock-surface bottom, and 0-m
elevation top. (C) Prism defining upper active zone
composed of the onshore area from 0-m elevation to
top of grid. (D) Four ponds with large hydraulic
conductivity defined by perimeters and pond bottoms.
(E) Combination of parts A, B, C, and D. Land surface
of Cape Cod (light shaded green) is drawn at 0-m
elevation. Red axis is X, green axis is Y, and blue axis
is Z.

defined to include the zone from 0 m down to bedrock in the area off the coast, this boundary condition applies a 0-m head to all cells at the top of the active region offshore.

A series of drains are defined with **DRAIN** data blocks to represent five rivers in the southern part of the model domain—the Coonamessett, Backus, Bourne, Childs, and Quashnet Rivers (table 6.13). The drains are defined by a sequential series of points that run from the headwaters to the ocean (fig. 6.9). The elevation, width, boundary-layer thickness, and boundary-layer hydraulic conductivity are defined at the first and last river points for each river. Intermediate points define the course of the river. Data values for elevation and the other parameters for these intermediate points are linearly interpolated by river length from the end points. Drain boundary conditions remove water from the aquifer if the head in the aquifer is greater than the elevation of the drain; drain boundary conditions do not contribute water to the aquifer. Drain boundary conditions were used instead of river boundary conditions to simplify the calculation of budgets. Rivers allow water to flow into and out of the aquifer and could allow water to be added to the domain in addition to the rainfall and wastewater infiltration. Because drain boundaries allow no inflow, the inflow part of the water budget is exclusively the rainfall and wastewater infiltration; no other sources of water are defined.

The grid defined for the example and the boundary-condition cells resulting from intersecting the grid with the boundary-condition zone definitions are shown in figure 6.10, excluding the cells to which rainfall infiltration are applied (all onshore areas). Three planes of cells (intersections of lines) are shown in the figure to indicate the grid density for this example.

The **ZONE_FLOW** data block (table 6.13) defines a zone for which boundary-condition flow information will be written. In addition, the **-write_heads_xyzt** identifier requests that the head at each node in the zone be written to the file *plume.heads.xyzt* at a frequency specified by **-zone_flow_heads** in the **PRINT_FREQUENCY** data block. The output from the **ZONE_FLOW** data block is not used in the initial steady-flow calculation. In the second, transient calculation of this example (example 5), the time series of heads that is produced will be used to define heads for specified-head boundary conditions in the transport and reaction simulation (example 6).

For a steady-flow calculation, the initial conditions are not critical, but reasonable initial values are needed to avoid numerical instability. The **HEAD_IC** data block defines a uniform head of 10 m for the

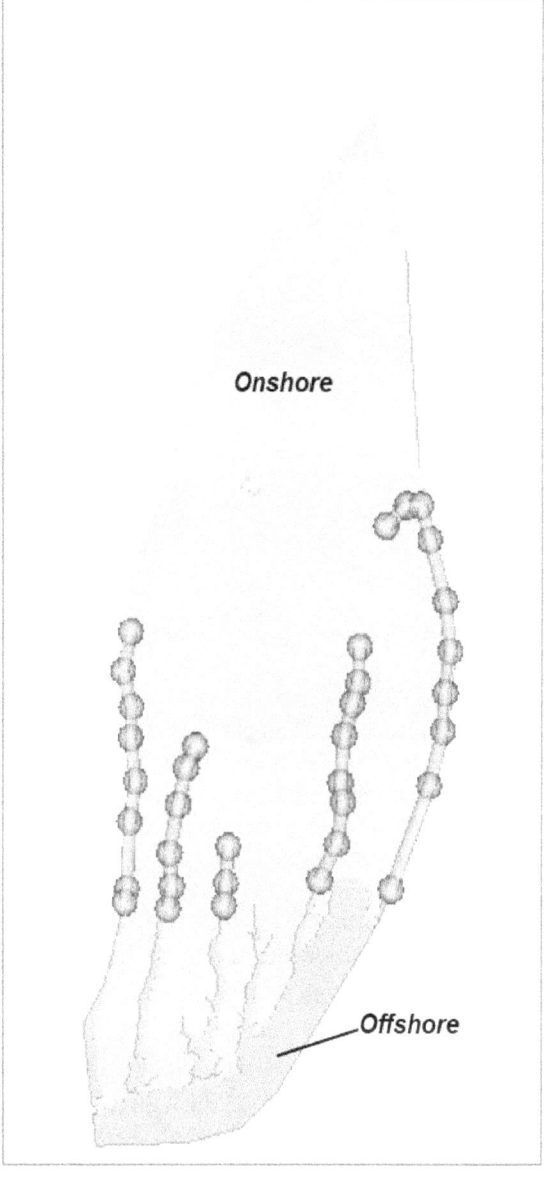

Figure 6.9. Example 5, boundary condition definitions. Rainfall flux is applied over the onshore region. Small orange zones represent wastewater infiltration beds. Specified head of 0 meters is applied over the offshore region. Green points define drain boundary conditions representing rivers. Red rectangle indicates the extent of the grid.

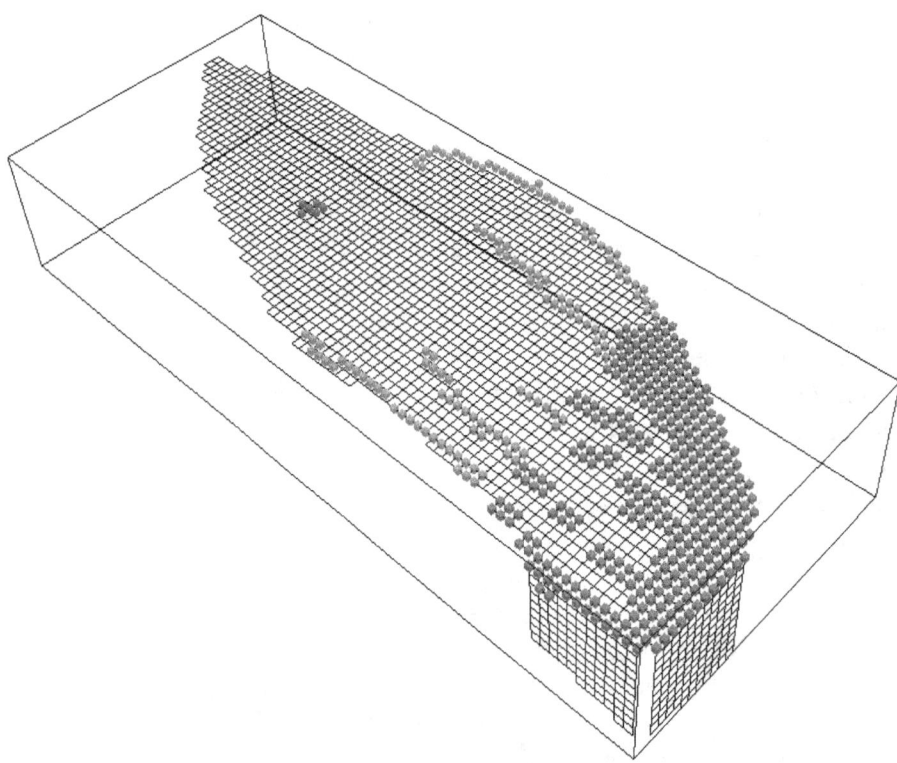

Figure 6.10. Example 5, grid and boundary conditions for the transient simulation viewed from the southwest. Tan nodes are specified-head boundary conditions offshore; green nodes are drain boundaries representing rivers; and orange nodes are flux boundaries representing the wastewater infiltration beds. Not shown are the flux boundaries for rainfall that cover the entire upper surface onshore.

entire model domain. The **TIME_CONTROL** data block specifies the beginning and ending times for the calculations and the time step to be used. The starting and ending times are not used in the steady-flow calculations. The time step is used as a guide for minimum and maximum time steps in the automatic time-stepping algorithm for the steady-flow calculation (unless explicit minimum and maximum time steps are defined in the **STEADY_FLOW** data block). During the second (transient) simulation of this example, the simulation will run from 1935 to 1998 with a fixed time step of 0.5 year. The simulation is run 3 years past the cessation of wastewater disposal to allow the system to attain a steady-state flow condition. The heads are saved (**ZONE_FLOW** data block) every 0.5 years (**-zone_flow_heads** in the **PRINT_FREQUENCY** data block) and will be applied as the time-varying specified-head boundary-condition values in the example 6 simulation; specified-head boundary-condition values will be constant after 1998 in the example 6 simulation.

The **PRINT_INITIAL** data block specifies that media properties (hydraulic conductivities, porosities, and others) will be written to the HDF file (**-HDF_media_properties**) and the initial heads (defined by **HEAD_IC**) will not be written to the *ex5.head.txt* file (**-heads**). Defaults will be used for all other values for **PRINT_INITIAL**.

The **PRINT_FREQUENCY** data block specifies (**-save_final_heads**) that heads at the end of the simulation will be saved to a file (*ex5.head.dat*). The file saved by the steady-flow calculation will be used as the initial head condition for the transient simulation and for the initial head condition in example 6. The remaining identifiers in the **PRINT_FREQUENCY** data block are defined to apply from the beginning of the simulation (time 0) and are not changed at any subsequent time. For the steady-flow simulation, the

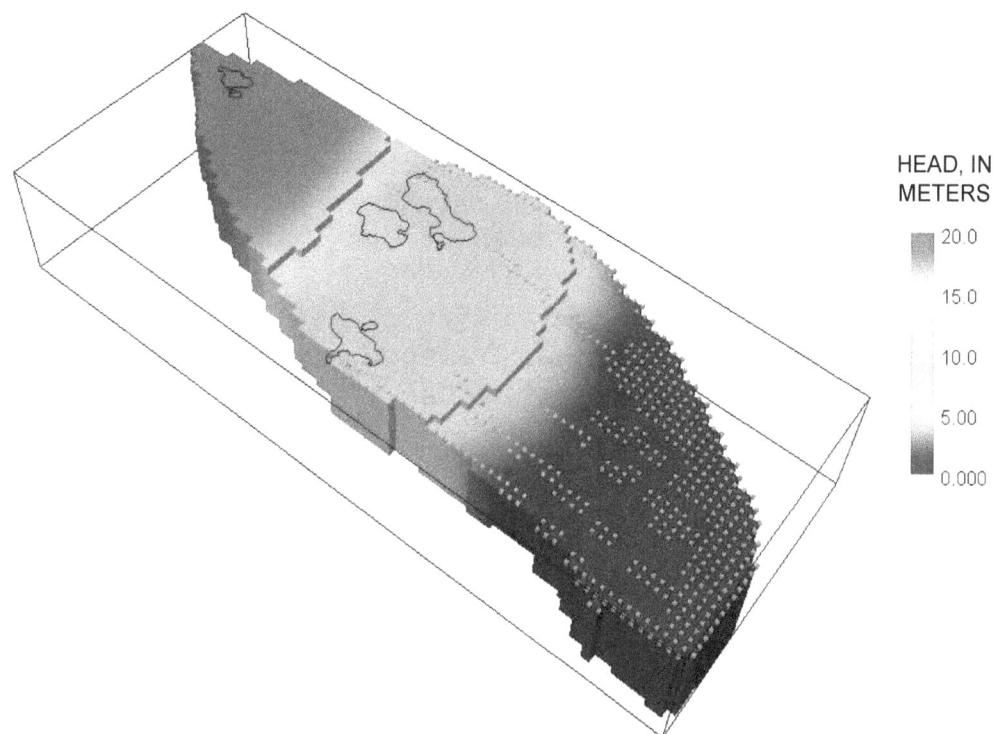

Figure 6.11. Example 5, steady-flow head distribution. Green nodes are drain boundary conditions representing rivers and tan nodes are specified-head boundary conditions offshore. Ponds are outlined in black.

-HDF_heads identifier specifies that the steady-flow heads will be written to the HDF file; for the transient simulation, heads will be written to the HDF file after each 5 years of simulation time. Progress statistics will be written to the screen and the *ex5*.**log.txt** file after each time step of the simulation. The heads for zones defined with **ZONE_FLOW** data blocks, for which **-write_heads_xyzt** is defined, will be written after each time step (**-zone_flow_heads**). Defaults will be used for all other values for **PRINT_FREQUENCY**.

6.6.2. Steady-Flow Simulation Results

The steady-flow simulation uses the *ex5.trans.dat* data file as defined in table 6.13. The simulation calculates steady-flow head and velocity fields based on the specified boundary conditions. The only source of water in the simulation is rainfall recharge, which is specified with a flux boundary condition. All other flux boundaries are inactive because the initial fluxes at time 0, which apply to the steady-flow calculations, are all 0.0 m/d. The sinks for water are the drain and specified-head boundaries. The head distribution at steady state is shown in figure 6.11. After the steady-flow simulation, the file *ex5.head.dat* contains the head distribution in a format that can be used as the initial head condition for subsequent runs. This file was renamed to *ex5.200.head.dat* and is used in the following transient simulation.

6.6.3. Transient Simulation Results

Two modifications are made to the flow and transport data file (table 6.13) for the transient simulation. First, the setting in the **STEADY_FLOW** data block is changed to false. Second, the initial head condition (**HEAD_IC**) is changed from a constant 10 m to the head distribution from the steady-flow simulation (renamed from *ex5.head.dat* to *ex5.200.head.dat*). With these modifications, a transient simulation from 1935 to 1998 is performed. The heads and velocities within the model domain vary with time, depending on the variations in the flux of wastewater through the infiltration beds.

The purpose of the transient simulation is to save a time-varying set of heads that can be used to define boundary conditions for a reactive transport model on a smaller region (example 6). The zone defined in **ZONE_FLOW** 1 encloses the observed wastewater plume in the Ashumet Valley (LeBlanc, 1984b), and a time series of head distributions is saved to the file *plume.heads.xyzt*. Heads for all cells within the zone are saved every 0.5 year.

6.7. Example 6: Simulation of Ammonium Transport and Reactions for a Sewage Wastewater Plume at Cape Cod, Massachusetts

This example builds on the flow simulations of example 5 to model transport processes combined with a simple set of chemical reactions affecting ammonium in the wastewater plume at Cape Cod. The example uses time-varying head distributions from example 5 to specify boundary conditions for a model area that is restricted to the part of the aquifer containing the wastewater plume. The reduced area allows a finer model grid to be used, which increases numerical accuracy and decreases the effects of numerical dispersion and numerical oscillations in the transport solution.

In this example, oxygenated water fills the aquifer and oxygenated water enters the aquifer from rainfall. Water containing ammonium is introduced at the wastewater infiltration beds in a time-varying sequence that approximates the time history of water releases to the individual infiltration beds (D.R. LeBlanc, U.S. Geological Survey, written commun., 2001; Parkhurst and others, 2003). The ammonium-containing wastewater is introduced over the period from 1935 through 1995, at which point all recharge to the aquifer is through rainfall infiltration. As the oxygen and ammonium come into contact through dispersion and advection, ammonium is oxidized to nitrate. Ammonium is assumed to move with a retardation factor of 2, while nitrate is assumed to move unretarded. The simulation continues for an additional 104 years after the cessation of wastewater disposal to evaluate the movement of the ammonium and nitrate plumes. The purpose of the simulation is to calculate the time-dependent loading of nitrogen at the ends of flow paths, which in this simulation are the ponds, rivers, and ocean.

6.7.1. Chemistry Data File

The chemical information for the example is defined in the chemistry data file listed in table 6.14. A title is defined with the **TITLE** data block. The next three data blocks define changes to the *phast.dat* thermodynamic database (same as *phreeqc.dat* in PHREEQC distribution). Dissolved nitrogen and nitrite are effectively eliminated by using small log Ks for the aqueous reactions in the **SOLUTION_SPECIES** data block. These definitions allow ammonium to oxidize to nitrate (nitrification), but do not allow nitrite or dissolved nitrogen to form from nitrate or ammonium. Thus, denitrification (reduction of nitrate to dissolved nitrogen) and intermediate oxidation of ammonium to dissolved nitrogen and nitrite are eliminated as possible reactions. The **SURFACE_MASTER_SPECIES** and **SURFACE_SPECIES** data blocks define reactions that sorb ammonium on the solids to produce a distribution coefficient of 1 and a retardation factor of 2 (see example 1 for further explanation). The **SURFACE** data block completes the definition of the surface reactions that are used to define the retardation of ammonium.

Next, three solutions are defined that are used for the initial composition of (1) water residing in the aquifer, (2) wastewater inflow, and (3) rainwater infiltration. The initial groundwater and rain compositions are the same for this example and have oxygen concentrations in equilibrium with atmospheric oxygen [0.2 atm;

log10(0.2) = -0.7]. The wastewater concentration of ammonium is 180 μmol/L and chloride is 990 μmol/L (Parkhurst and others, 2003); chloride is included as a conservative tracer.

The last keyword data block in the chemistry data file is **SELECTED_OUPUT**. This data block specifies that the total concentrations of dissolved oxygen, nitrate, ammonium, and chloride will be written to the HDF file. The HDF file will contain these concentrations for each cell at time planes defined by the **PRINT_FREQUENCY** data block in the flow and transport data file. Results in the HDF file can be visualized and animated with the program Model Viewer.

Table 6.14. Chemistry data file for example 6.

```
TITLE Input file for example 6, ammonium/nitrate plume
SOLUTION_SPECIES
        NO3- + 2 H+ + 2 e- = NO2- + H2O
                log_k           -50
        2 NO3- + 12 H+ + 10 e- = N2 + 6 H2O
                log_k            0
SURFACE_MASTER_SPECIES
        Surf Surf
SURFACE_SPECIES
        Surf = Surf
                log_k 0.0
        Surf + NH4+ = SurfNH4+
                log_k -100
        Surf + NH3 = SurfNH3
                log_k -100
END
SURFACE 1 linear ammonia adsoprtion
        Surf      1e100    1.0      1e100
END
SOLUTION 1 Uncontaminated groundwater, O2 only
        pH         5.6
        pe         7.0
        temp       14.0
        units      umol/L
        O(0)       250    O2(g)    -0.7
SOLUTION 2 Sewage effluent without oxygen
        pH         6.00
        pe         7.0
        temp       14.0
        units      umol/L
        Cl         990
        N(-3)      180
SOLUTION 3 Rain, O2 only
        pH         5.6
        pe         7.0
        temp       14.0
        units      umol/L
        O(0)       250    O2(g)    -0.7
END
SELECTED_OUTPUT
        -reset     false
        -totals    O(0)   N(5)     N(-3) Cl
END
```

6.7.2. Flow and Transport Data File

The flow, transport, and initial conditions for the example are defined in the flow and transport data file listed in table 6.15. The definitions are similar to the flow and transport data file for example 5 (table 6.13), but are modified to apply a finer grid to a subdomain of the example 5 domain. The main differences are that (1) the model domain is limited to the southeastern part of the example 5 domain, (2) additional specified-head boundaries are used at the north and east edges of the subdomain to impose the flow produced in example 5, (3) the ponds are converted to specified-head boundaries, which causes all solutes that enter a pond to be removed from the system, (4) only drains in the subdomain are included in the definition, and (5) a 100-m grid (X, Y) is used instead of a 200-m grid.

The first four data blocks of the flow and transport data file (table 6.15) define a title, a transient solute transport simulation, and a free-surface boundary condition. The time and space weighting for the difference equations is upstream in space and backward in time, which results in no numerical oscillations but causes numerical dispersion in the solution to the transport equation. The units of the input data—meters with time units of days or years—are defined in the **UNITS** data block.

The **GRID** data block defines 100-m node spacing in the horizontal directions and 10-m node spacing in the vertical direction. The extent of the grid is restricted to a southwestern section of the domain used in example 5. The drain and specified-head boundary conditions for the rivers and coastal areas will be retained from example 5, but new specified-head boundaries will be applied at the new (relative to example 5) northern and eastern boundaries.

The active model region and media properties defined with the **MEDIA** data block are the same as example 5, with the exception that the last zone definition makes part of the example 5 domain inactive. The fourth and last zone of the media definitions is a prism that encloses the northern and eastern areas of example 5, and defines it to be inactive. The prism is defined by a perimeter that is a series of X, Y points and a bottom and top that default to the bottom and top of the grid.

The **FLUX_BC** data block contains the same definitions as example 5. One zone defines rainfall recharge over the onshore region with composition defined by solution 3 (chemistry data file). The other zones define the location of the wastewater infiltration beds and the schedule of wastewater fluxes for each bed. The flux through all infiltration beds is defined to have the composition of solution 2.

The first zone in the **SPECIFIED_HEAD_BC** data block is the same as example 5 and defines a constant head of 0 m on the top of the domain (elevation 0) for all offshore regions. The second and third zones define specified-head boundaries for the north and part of the east side of the model domain. The heads for the cells are defined by a time series of heads produced in the transient example 5 simulation that are stored in the file *plume.heads.xyzt*. The file contains quintuplets of X, Y, Z, time, and head. These heads are then interpolated to the current time and grid by closest point spatial interpolation for the closest time in the XYZT data file that is either equal to or precedes the current time of the simulation. The final zone in the **SPECIFIED_HEAD_BC** data block defines the cells contained in the ponds to be specified-head boundary conditions with heads defined by the file *plume.heads.xyzt*. Any solute transported into the ponds is removed from the model system. Any water entering from the specified-head boundaries has the composition of solution 1, the background water defined in the chemistry data file.

The **DRAIN** data blocks define the three drains that are within the model domain. Definitions are identical to example 5. Initial head conditions (**HEAD_IC**) are the same as for the transient simulation of example 5. Uniform chemistry is defined initially throughout the model domain (**CHEMISTRY_IC**). All cells have a water composition defined by solution 1 and a reactive surface defined by surface 1. The surface initially has no ammonium sorbed, but will sorb ammonium as it is transported from the wastewater infiltration beds. In addition, aqueous equilibrium reactions will cause ammonium to oxidize to nitrate whenever oxygen contacts ammonium in the aquifer.

Initial heads will not be printed to a text file according to the **-heads** identifier in the **PRINT_INITIAL** data block. All other initial data will be printed according to default values. The **PRINT_FREQUENCY** data block sets print intervals to nondefault values for several output files. Aqueous concentrations (as defined in the **SELECTED_OUTPUT** data block of the chemistry data file) will be written to the HDF file every 5 years, heads will be written to the HDF file every 10 years, and velocities will not be written to the HDF file. Heads will not be written to the *ex6*.**head.txt** file. Progress statistics will be written to the screen following each time step. Zone flow information will be written to the *ex6*.**zf.txt** and *ex6*.**zf.tsv** files after each year of simulation. By default, several sets of data are printed at the end of each simulation period (see the description of the **PRINT_FREQUENCY** data block for default settings); however, the **-end_of_period** identifier eliminates default printing at the end of simulation periods and only explicitly defined print frequencies will be used in the simulation.

The **ZONE_FLOW** data blocks are used to account for flow and transport into and out of specified zones, which includes flow into and out of boundary-condition cells within the zone. Six accounting zones are defined. The first and second zone definitions are used to determine flow of solute into Ashumet and Coonamessett Ponds. These ponds are defined with specified-head boundary conditions, so the flow into specified boundary conditions in these zones, which contain the specified-head cells in the ponds, will be equal to the loss of solute from the aquifer to the ponds (file *ex6*.**zf.tsv**). The third and fourth zones contain the Coonamessett and Backus Rivers and their respective estuaries. The fifth definition combines the areas of the third and fourth definitions to provide a summary of all solute lost to the rivers and estuaries. In this example, no use was made of the flows to the individual rivers and estuaries, and only the combined results are plotted. The last definition of the data block accumulates all the sources and sinks of solutes in the grid region. Results for this zone include all of the flow of solutes into the region through the infiltration beds (total flux-boundary inflows). Results of water and solute flow (kilograms per year) are written to the file *ex6*.**zf.tsv** every year (table 6.15, **PRINT_FREQUENCY**, **-zone_flow_tsv** identifier). This file can be imported into a spreadsheet, and flows to various boundary conditions within the specified zones can be plotted.

The **TIME_CONTROL** data block specifies that the simulation will run from 1935 until 2200 for a total duration of 265 years. The time step will be 0.5 year for the entire simulation.

6.7.3. Simulation

Several simulations were performed by varying the grid resolution, time step, and weighting factors for the spatial and temporal discretization. Results are presented for the simulation described by the flow and transport data file defined in table 6.15, with mention of results of other simulations.

6.7.3.1. Accuracy of the Transport Solution

The degree to which errors are present in the numerical solution to the transport equations in this example can be estimated by investigating the behavior of the conservative solute chloride. The file *ex6*.**bal.txt** contains summary information for solute flow rates and cumulative solute mass entering and leaving the domain. Included in the information is the change in mass of solutes due to chemical reactions. For a conservative solute the cumulative change in mass due to reactions should be zero. For upstream-in-space (**SOLUTION_METHOD**, **-space_weighting** 0.0) and backward-in-time weightings (**SOLUTION_METHOD**, **-time_weighting** 1.0), as expected, the cumulative change in chloride from reactions at the end of the run is small (0.08 kg, *ex6*.**bal.txt**).

If other weightings are used (**-space_weighting** > 0.0 or **-time_weighting** <1.0) and numerical oscillations in concentration occur (see Appendix D for a discussion of criteria necessary to avoid oscillations), the oscillations may have minor or catastrophic effects on the simulation. If oscillations are large, then the solver may fail to find a solution to the transport calculation. It is also possible that oscillations will cause large concentrations that result in a failed chemistry calculation. Either of these conditions will cause the simulation to abort. If the simulation completes successfully, it is still possible that numerical oscillations will cause unreliable results. In the transport solution total mass is conserved, but it is possible to generate negative concentrations that are balanced by excess positive concentrations. When the concentrations for each cell are passed to the reaction calculations, negative concentrations are set to zero, but the excess positive concentrations remain. Thus, the result of censoring the negative concentrations is a net increase in mass in the system, which is manifested in an increase due to reactions in *ex6*.**bal.txt**. The extent of increase of a

Table 6.15. Flow and transport data file for example 6.

```
TITLE Modeling nitrogen reactions in sewage plume
SOLUTE_TRANSPORT                    true
STEADY_FLOW                         false
FREE_SURFACE_BC                     true
SOLUTION_METHOD
        -iterative_solver           true
        -tolerance                  1e-12
        -space_differencing         0
        -time_differencing          1
        -cross_dispersion           false
        -rebalance_fraction         0.0
UNITS
        -time                       yr
        -horizontal_grid            m
        -vertical_grid              m
        -map_horizontal             m
        -map_vertical               m
        -head                       m
        -hydraulic_conductivity     m/d
        -specific_storage           1/m
        -dispersivity               m
        -flux                       m/d
        -leaky_hydraulic_cond       m/d
        -leaky_thickness            m
        -drain_hydraulic_cond       m/d
        -drain_thickness            m
        -drain_width                m
GRID
        # 100 m grid
        -uniform X 276400 281300  50  # 100 m grid
        -uniform Y 811000 822300 114  # 100 m grid
        -uniform Z -120    20     15  # 10  m grid
        -snap    X                  0.001
        -snap    Y                  0.001
        -snap    Z                  0.001
        -chemistry_dimensions       XYZ
        -print_orientation          XY
        -grid_origin                0  0  0
        -grid_angle                 0
MEDIA
        # Default is inactive
        -domain
                -active             0
                -porosity           0.39
                -specific_storage   0.39
                -Kx      XYZ        MAP   ../ex5/Parameters/kh_meters
                -Ky      XYZ        MAP   ../ex5/Parameters/kh_meters
                -Kz      XYZ        MAP   ../ex5/Parameters/kz_meters
                -long_dispersivity  0
                -horizontal_disp    0
                -vertical_disp      0
        # Activate inside model outline, above bedrock to sea level
        -prism
                -perimeter SHAPE       GRID   ../ex5/ArcData/ModelExtent.shp
                -bottom    ARCRASTER   GRID   ../ex5/ArcData/bedrock.txt
                -top       CONSTANT    GRID   0
                -active                1
        # 0 m to top, inside coast
        -prism
                -perimeter SHAPE       GRID   ../ex5/ArcData/OnShore.shp
                -bottom    CONSTANT    GRID   0
                -active                1
        -prism
                -description Plume submodel outline
                -perimeter POINTS GRID
                        281100.   822300.    -200
                        281100.   821100.    -200
                        280500.   819000.    -200
                        280000    817400.    -200
                        279900.   816300.    -200
                        279700.   814100.    -200
                        279500.   811400.    -200
                        279500.   809400.    -200
                        285000.   809400.    -200
                        285000.   822300.    -200
                        end_points
                -active                    0
```

Table 6.15. Flow and transport data file for example 6.—Continued

```
FLUX_BC
    -prism
        -description Rainfall
        -bottom    CONSTANT    GRID    0
        -perimeter SHAPE       GRID    ../ex5/ArcData/OnShore.shp
        -face Z
        -flux
            0                  -0.00188
        -associated_solution
            0                  3
    -box 279725 821500 0 279825 821600 50
        -description Beds 1-4
        -face Z
        -flux
            0                  0
            1936               -0.037
            1941               -0.114
            1946               -0.023
            1956               -0.126
            1971               -0.05
            1978               -0.114
            1984               0
        -associated_solution
            0                  2
    -box 279625 821425 0 279725 821525 50
        -description Beds 5-8
        -face Z
        -flux
            0                  0
            1941               -0.114
            1946               -0.023
            1956               -0.126
            1971               -0.05
            1978               0
            1984               -0.038
            1996               0
        -associated_solution
            0                  2
    -box 279525 821475 0 279625 821575 50
        -description Beds 9-12
        -face Z
        -flux
            0                  0
            1941               -0.114
            1946               -0.023
            1956               -0.126
            1971               -0.05
            1978               0
            1984               -0.038
            1996               0
        -associated_solution
            0                  2
    -box 279400 821575 0 279500 821675 50
        -description Beds 13-16
        -face Z
        -flux
            0                  0
            1941               -0.114
            1946               -0.023
            1956               0
        -associated_solution
            0                  2
    -box 279500 821700 0 279600 821800 50
        -description Beds 19-25
        -face Z
        -flux
            0                  0
            1941               -0.114
            1946               -0.023
            1956               0
        -associated_solution
            0                  2
```

Table 6.15. Flow and transport data file for example 6.—Continued

```
SPECIFIED_HEAD_BC
        -prism
                -description Constant head ocean
                -top          CONSTANT   GRID    0
                -bottom       CONSTANT   GRID    -10
                -perimeter SHAPE        GRID    ../ex5/ArcData/OffShore.shp
                -exterior_cells_only   Z
                -head
                        0            0.0
                -associated_solution
                        0            1
        -box 275000 822300 -200 285000 822300 20 MAP
                -description Constant head at north end of sub region
                -head
                        0  XYZT        GRID    ../ex5/plume.heads.xyzt
                -associated_solution
                        0            1
        -box 279500 815300 -50 285000 822300 20 MAP
                -description Constant head on part of east edge of sub region
                -exterior_cells_only   X
                -head
                        0  XYZT        GRID    ../ex5/plume.heads.xyzt
                -associated_solution
                        0            1
        -prism
                -description Constant head for ponds
                -bottom       SHAPE      GRID    ../ex5/ArcData/SimpleBath.shp 10
                -perimeter SHAPE        GRID    ../ex5/ArcData/SimplePonds.shp
                -head
                        0  XYZT        GRID    ../ex5/plume.heads.xyzt
                -associated_solution
                        0            1
DRAIN 3 Bourne River
        -point   279187.9      815453.3
                -width                20
                -bed_hydraulic        1
                -bed_thickness        1
                -z                    2.97
        -point   279141.9      814825.4
        -point   279108.7      814446.3
                -width                20
                -bed_hydraulic        1
                -bed_thickness        1
                -z                    0
DRAIN 4 Backus River (Cranberry bogs)
        -point   278572.0      817160.5
                -width                20
                -bed_hydraulic        10.00
                -bed_thickness        1
                -z                    7.16
        -point   278404.6      816798.4
        -point   278260.8      816171.1
        -point   278104.0      815347.7
        -point   278146.1      814774.8
        -point   278077.8      814393.5
                -width                20
                -bed_hydraulic        10.00
                -bed_thickness        1
                -z                    0
DRAIN 5 West Boundary Coonamessett River
        -point 277365.6        819130.2
                -width                20
                -bed_hydraulic        10.00
                -bed_thickness        1
                -z                    9.92
        -point   277224.7      818453.8
        -point   277381.6      817842.1
        -point   277365.6      817312.8
        -point   277461.8      816574.9
        -point   277349.5      815885.2
        -point   277301.4      814730.3
        -point   277260.0      814502.1
                -width                20
                -bed_hydraulic        10.00
                -bed_thickness        1
                -z                    0
```

Table 6.15. Flow and transport data file for example 6.—Continued

```
HEAD_IC
      -domain
              -head        XYZ        GRID   ../ex5/ex5.head.200.dat
CHEMISTRY_IC
      -domain
              -solution          1
              -surface           1
PRINT_INITIAL
      -heads                     false
PRINT_FREQUENCY
      -save_final_heads          true
      0
              -end_of_period     false
              -flow_balance      1  years
              -HDF_chemistry     5  years
              -HDF_heads         10 years
              -HDF_velocities    0  years
              -heads             0  years
              -progress_statistics 1 step
              -zone_flow         1  years
              -zone_flow_tsv     1  years
ZONE_FLOW 1 Ashumet Pond
      -prism
              -bottom    SHAPE      GRID   ../ex5/ArcData/AshumetBath.shp 10
              -perimeter SHAPE      GRID   ../ex5/ArcData/Ashumet.shp
ZONE_FLOW 2 Coonamessett Pond
      -prism
              -bottom    SHAPE      GRID   ../ex5/ArcData/CoonaBath.shp 10
          -perimeter SHAPE      GRID   ../ex5/ArcData/Coona.shp
ZONE_FLOW 3 Coonamessett River
      -box 276400 811000  -200 277499  818000 20
ZONE_FLOW 4 Backus River and southern specified bc
      -box 277500 811000  -200 281300  818000 20
ZONE_FLOW 5 Coonamessett R, Backus R, and ocean
      -combination 3 4
ZONE_FLOW 6 Entire domain
      -domain
TIME_CONTROL
      -time_step
              0                      0.5   # num disp, v dt / 2 ~ 50
      -time_change               2200
      -start_time                1935
END
```

conservative solute is one estimate of the extent of numerical errors due to the oscillations in the transport solution.

For the aquifer at Cape Cod, estimates of the longitudinal dispersivity are on the order of 1 m for a scale of several hundred meters (Garabedian and others, 1991). Scaling linearly to a region on the order of several kilometers, the estimated dispersivity is on the order of 10 m. The grid resolution necessary to attain this dispersivity for the model domain in this example is about 20 m, which corresponds to a grid with several million nodes. Although the calculation is feasible for a fast computer with approximately 20 GB of memory (allowing for many days of computing time), the current (2009) limitation for a desktop computer with 8 GB of memory is about 1 million nodes (allowing for several days of computing time). Given the limitations, a coarser grid and a larger dispersivity must be accepted. For this example, it was reasoned that the best nonoscillatory solution could be obtained by using upstream-in-space, backward-in-time weightings and allowing the numerical dispersivity to account for dispersive processes. Estimates of the dispersivity for different grid and time step combinations were determined from the formulas in Appendix D. In this example, the grid is relatively coarse to allow shorter computing time, and the estimated numerical dispersivity is on the order of 100 m, half from the spatial discretization and half from the temporal discretization. Smaller dispersivities are possible at the cost of a finer grid and longer computing time; however, many important features of the transport and reaction processes are evident in even the coarse-grid results.

(A)

(B)

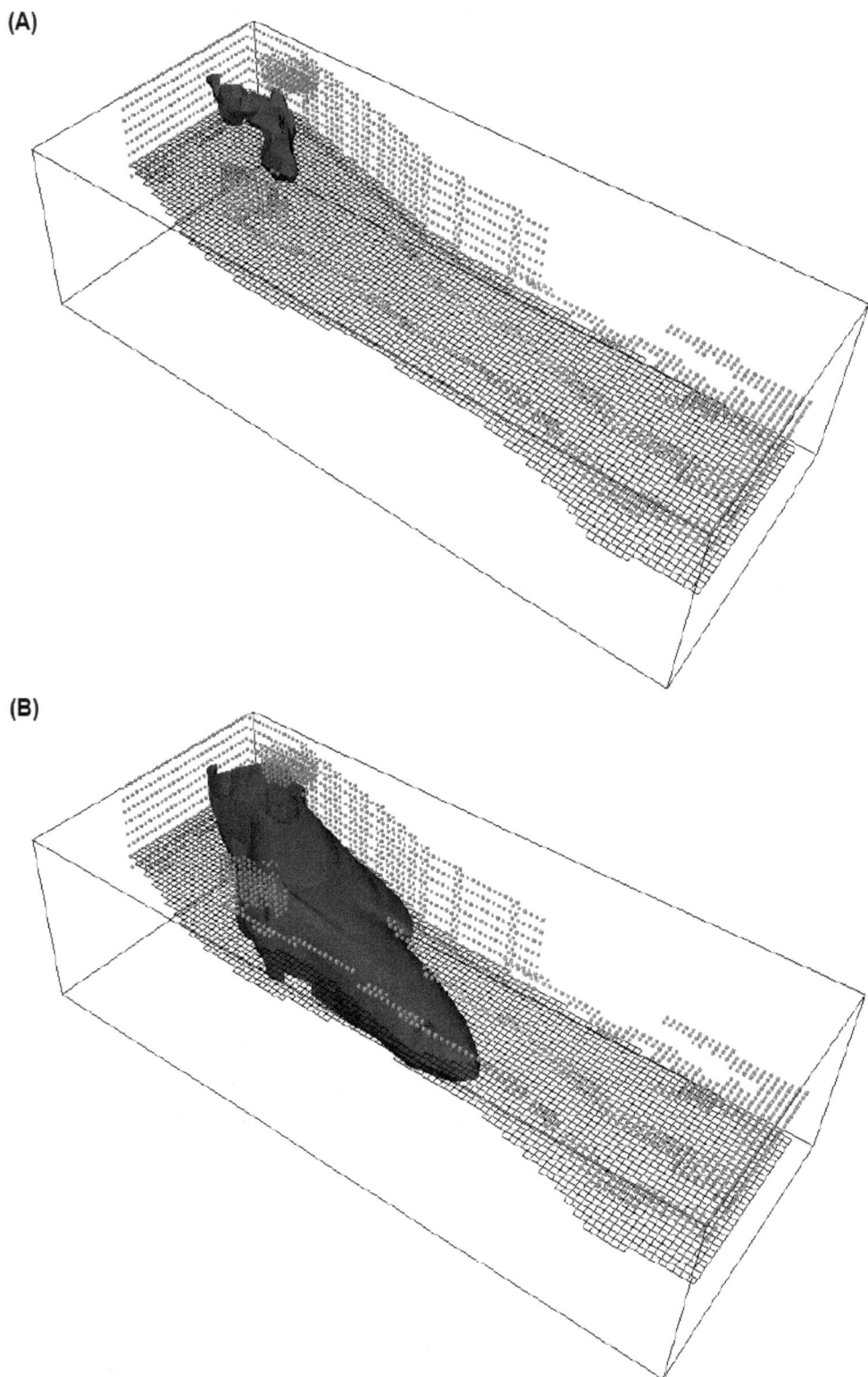

Figure 6.12. Example 6, distribution of (A) ammonium and (B) nitrate after year 1995. Blue surface is 1 micromolar concentration, approximately one-hundredth of the maximum concentration in the simulated wastewater plume. Green nodes are drain boundaries and tan nodes are specified-head boundaries. Black mesh shows horizontal grid resolution.

6.7.3.2. Simulation Results

This example simulates the introduction of ammonium at the wastewater infiltration beds, the oxidation of ammonium to nitrate by oxygen, and the transport of both ammonium and nitrate in the glacial outwash aquifer. The distribution of ammonium and nitrate are shown in figure 6.12 at the end of 1995, the time disposal of wastewater ceased. The ammonium plume is nearer the infiltration beds both because its transport is retarded by sorption and because it is oxidized at the edges. The nitrate has all been produced by oxidation of ammonium, but once it is produced, it moves faster than ammonium because it does not sorb on the sediments.

The oxidation of ammonium depends on the dispersion and advection of oxygen into the ammonium plume, and dispersion for this simulation is relatively large. If dispersion were smaller, then the ammonium plume might oxidize more slowly, advance farther, and persist longer than in the current simulation. However, with the time-varying flux of water at the infiltration beds, some oxygenated water is transported into the plume by advection, and once wastewater disposal ceases, oxygenated water will be advected into the trailing edges of all of the ammonium-containing zones. Additional simulations indicate that even with smaller numerical dispersion (~25 m), the ammonium plume does not advance much farther than is shown in figure 6.12 and the advection and dispersive mixing of oxygen into the ammonium plume still lead to its disappearance over the course of several decades. In the simulations, nitrate produced by the oxidation of ammonium continues to move conservatively through the aquifer and ultimately reaches the rivers or ocean over the next two centuries.

Figure 6.13 shows the source and sinks for nitrogen (ammonium and nitrate undifferentiated) over the period 1935 through 2200 as derived from the flow zones tabulated in *ex6*.**zf.tsv**. The inflow from the infiltration beds is the inflow from the flux boundaries of flow zone 6; the flow to Ashumet Pond is the outflow to specified-head boundaries in flow zone 1; and the flow to the rivers and ocean is the sum of the outflow to drains and specified-head boundaries of flow zone 5. In this simulation, most of the nitrogen (~70 percent) enters Ashumet Pond (fig. 6.7). The fate of nitrogen in the pond-bed sediments and the pond is not simulated. Ammonium could be sequestered in the pond bottom, could reenter the aquifer as ammonium or nitrate, or could be lost to nitrification and denitrification in the pond. The amount of nitrogen reaching the pond is dependent on the recharge and hydraulic properties of the aquifer, which determine flow directions, and on the loading of the infiltration beds, which is poorly constrained. If the western beds were loaded preferentially relative to the current simulation, more nitrogen would skirt the pond and ultimately reach the rivers and ocean.

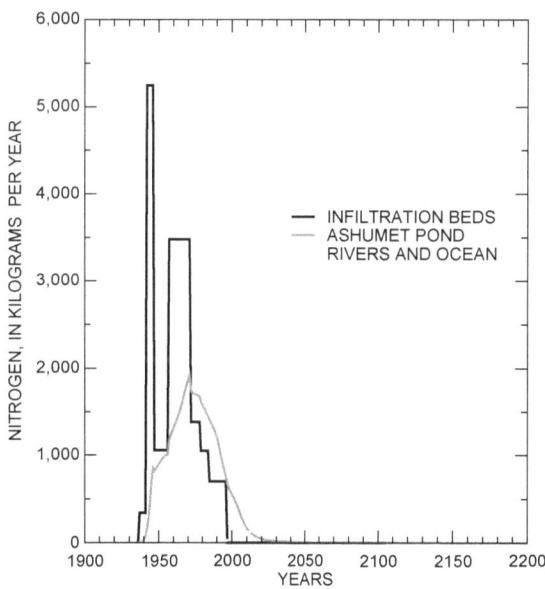

Figure 6.13. Example 6, input of nitrogen through the infiltration beds and simulated discharge of nitrogen to Ashumet Pond and rivers and ocean.

The fraction of nitrogen that is not intercepted by the pond eventually discharges to the drains (representing rivers) and specified-head cells (representing the ocean) of the simulation. Figure 6.13 indicates that the time until the initial discharge to the rivers and ocean is on the order of 70 years and that the nitrogen from the infiltration beds will continue to discharge for two centuries. The amount of dispersion in the simulation will affect the shape and duration of the nitrogen discharge. Smaller dispersivity will tend to lower the rate of nitrate production and decrease the nitrate concentration in the water that discharges to the streams and ocean, but the smaller dispersivity will also decrease the dilution of nitrate concentrations by mixing.

It must be noted that many chemical processes are ignored in this simulation; only ammonium is introduced in the wastewater, and dissolved oxygen in the aquifer can only react with ammonium. In reality, nitrate, oxygen, dissolved organic carbon, organic solids, and a host of other constituents were introduced through the infiltration beds. This mixture of constituents would produce a complex oxidation and reduction environment and a large set of chemical reactions that affect the transport of nitrogen species. However, treating the ammonium-nitrate system in isolation provides useful information on the magnitude of fluxes and the timing of breakthrough fronts.

Although the details of the simulation are subject to uncertainties, the larger issues are clear. A large fraction of the nitrogen derived from ammonium in wastewater enters Ashumet Pond and the fate of this nitrogen is tied to pond and pond-bottom processes. A smaller fraction of nitrogen derived from ammonium in wastewater flows to the rivers and ocean over a period of many decades (~ 200 years). Although there are uncertainties in the loading of the infiltration beds, the chemical reactions, and the flow system in the current simulation, the results indicate that the loss of nitrogen to Ashumet Pond and the relatively long duration of nitrogen discharge produce a nitrogen discharge to the rivers and ocean of about 100 to 400 kilograms per year.

Chapter 7. Notation

This section presents the definitions of the notation used in this report. Units used internally by PHAST and for the output files are given where applicable.

7.1. Roman Characters

A matrix of coefficients of the difference equations (appropriate units)

A^{-1} inverse of the matrix of coefficients of the difference equations (appropriate units)

a activity

b right-hand-side vector of the difference equations (appropriate units)

b_L thickness of a leaky-boundary semiconfining layer (m)

b_{Ls} thickness of the semiconfining layer for segment s (m)

b_R thickness of the riverbed (m)

b_{Rs} thickness of the riverbed for segment s (m)

C_f flow storage factor (m^2)

Cr Courant number (unitless)

C_s component storage factor (m^3)

c_j total aqueous concentration of component j (mol/kgw)

$\hat{c}_j$ total aqueous concentration of component j in the source fluid (mol/kgw)

$c_j^{(*)}$ intermediate total aqueous concentration of component j (mol/kgw)

$\bar{c}_e$ concentration of solid reactant e (mol/kgw)

D dispersion coefficient tensor (m^2/s)

D_M effective molecular diffusion coefficient (m^2/s)

D_{nos} numerical dispersion coefficient from operator splitting (m^2/s)

D_{ns} numerical dispersion coefficient from spatial discretization (m^2/s)

D_{nt} numerical dispersion coefficient from temporal discretization (m^2/s)

D_{ij} dispersion coefficient tensor element (m^2/s)

$e^{(v)}$ error in the solution vector at iteration v (appropriate units)

f_j gram formula weight of component j (g/mol)

f_p fraction of the total set of chemistry calculations to be performed by process p (unitless)

g gravitational acceleration (m/s^2)

gfw gram formula weight (g/mol)

h potentiometric head (m)

h_B potentiometric head at the boundary of the model domain (m)

h_R potentiometric head of the river (m)

h_{eL} potentiometric head at the external boundary of a leaky-boundary semiconfining layer (m)

h_{eLs} potentiometric head at the external boundary of a leaky-boundary segment s (m)

h_m potentiometric head at the boundary of the active grid region for cell m (m)

h_{wt} potentiometric head at the water table (m)

I_b biomass inhibition factor (unitless)

$\boldsymbol{K}$ hydraulic conductivity tensor (m/s)

K equilibrium constant (unitless)

K_d linear sorption distribution constant [(mol/kgw)/(mol/kgw)];

K_e linear equilibrium sorption coefficient (m^3/kgw)

$K_h(O)$ half-saturation constant for oxygen (mg/L)

$K_h(S)$ half-saturation constant for substrate (mg/L)

K_L hydraulic conductivity of a leaky-boundary semiconfining layer (m/s)

K_{Ls} hydraulic conductivity of the semiconfining layer for segment s (m/s)

K_R hydraulic conductivity of the riverbed (m/s)

K_{Rs} hydraulic conductivity of the riverbed for segment s (m/s)

$k_{biomass}$ aerobic biomass inhibition constant (mg/L)

kgw, $kgwater$ kilogram of water (kg)

k_m specific death rate or maintenance coefficient for aerobic biomass (d^{-1})

M_j total mass of component j in the aqueous phase (kg)

m molality (mol/kgw)

N total number of cells

N_c number of chemical components

N_E number of heterogeneous equilibrium reactions

N_K number of kinetic reactions

N_p cell number limit for process p; process p will run cells numbered $N_{p-1} + 1$ to N_p

N_s number of chemical species

(n) index of the discrete time value (superscript)

n_j total number of moles of component j in the aqueous phase (mol)

n_p number of cells assigned to process p

O concentration of dissolved oxygen (mg/L)

P number of processes

P_{CO_2} partial pressure of carbon dioxide (atm)

Pe Peclet number (unitless)

p pressure (Pa)

Q_{Ls} volumetric flow rate for a leaky-boundary segment s (m³/s)

Q_{Rs} volumetric flow rate for a river segment s (m³/s)

q source flow rate intensity ($m^3 s^{-1} m^{-3}$)

q_L volumetric flux at a leaky boundary ($m^3\ m^{-2}\ s^{-1}$)

q_R volumetric flux at river boundary ($m^3\ m^{-2}\ s^{-1}$)

R retardation factor for linear sorption (unitless)

$R_{biomass}$ rate of biomass production ($mg\ L^{-1}\ d^{-1}$)

R_i rate of formation of component i by all kinetic reactions (mol/s)

R_k rate of kinetic reaction k ($mol\ kgw^{-1}\ s^{-1}$)

R_{oxygen} limiting rate of oxygen consumption ($mg\ L^{-1}\ d^{-1}$)

$R_{oxygen,\ O}$ limiting rate of oxygen consumption due to oxygen availability ($mg\ L^{-1}\ d^{-1}$)

$R_{oxygen,\ S}$ limiting rate of oxygen consumption due to substrate availability ($mg\ L^{-1}\ d^{-1}$)

$\boldsymbol{r}^{(v)}$ residual vector at iteration v (appropriate units)

r_p time relative to the fastest process to perform a standard task (unitless)

S concentration of substrate (mg/L)

S_{Ls} area of the leaky-boundary segment s (m²)

S_{Rs} area of the river segment s (m²)

S_s storage coefficient (m^{-1})

$Surf$ surface complexation site

$SurfA$ element A sorbed on a surface complexation site

T time for all processes to complete the reaction task (s)

T_i total component concentration in aqueous and solid phases (mol/L) or mass fraction (unitless)

$\hat{T}_p$ time per cell for the reaction task for process p (s/cell)

t time (s, hr, d, yr)

t_i time required to perform the chemistry calculation for cell i (s)

$\hat{t}_i$ time required by the fastest process to perform the chemistry calculation for cell i (s)

t_{min} minimum time among processes for a standard task (s)

t_p time required by process p to perform a standard task (s)

$V_{max}(O)$ maximum rate parameter for oxygen consumption (d^{-1})

$V_{max}(S)$ maximum rate parameter for substrate consumption (d^{-1})

v interstitial velocity vector (m/s)

v_i interstitial velocity component in the i coordinate direction (m/s)

w_j mass fraction of component j (unitless)

X concentration of aerobic biomass (mg/L)

$x^{(v)}$ solution vector at iteration v (appropriate units)

Y_O yield coefficient oxygen (L/mg)

Y_S yield coefficient substrate (L/mg)

z elevation coordinate (m)

7.2. Greek Characters

α dispersivity (m)

α_L longitudinal dispersivity (m)

α_{ns} numerical dispersivity from spatial discretization (m)

α_{nt} numerical dispersivity from temporal discretization (m)

α_T transverse dispersivity (m)

α_{T_H} horizontal transverse dispersivity (m)

α_{T_V} vertical transverse dispersivity (m)

γ activity coefficient (kgw/mol)

δ_{ij} Kronecker delta function (unitless)

δh_m potentiometric head change for the time step at the boundary of the active grid region for cell m (m)

ε saturated porosity (unitless)

θ weighting factor for time differencing (unitless)

(v) iteration counter (superscript)

ρ water density (kg/m^3)

ρ_b dry bulk density of porous medium (kg/m^3)

τ convergence tolerance (unitless)

υ_p stoichiometric coefficient of a product in a chemical reaction (unitless)

υ_r stoichiometric coefficient of a reactant in a chemical reaction (unitless)

$\upsilon_{j,e}^{E}$ stoichiometric coefficient of component j in heterogeneous equilibrium reaction e (unitless)

$\upsilon_{j,k}^{K}$ stoichiometric coefficient of component j in kinetic reaction k (unitless)

ϕ porosity (unitless)

7.3. Mathematical Operators and Special Functions

Δx cell size (m)

Δt time step (s)

L_f spatial discretization of $\nabla \bullet \boldsymbol{K} \nabla$ (m^{-1} s^{-1})

L_s spatial discretization of $\rho \nabla \bullet \boldsymbol{D} \nabla - \rho \nabla \bullet v$ (kg m^{-3} s^{-1})

∇ gradient operator (m^{-1}) scalars and divergence operator for vectors

$|\ |$ magnitude of a vector

$\|\ \|$ Euclidean norm for a vector or Frobenius norm for a matrix

References Cited

Allison, J.D., Brown, D.S., and Novo-Gradac, K.J., 1990, MINTEQA2/PRODEFA2—A geochemical assessment model for environmental systems—version 3.0 user's manual: Athens, Georgia, Environmental Research Laboratory, Office of Research and Development, U.S. Environmental Protection Agency, 106 p.

Aziz, Khalid and Settari, Antonin, 1979, Petroleum reservoir simulation: Barking, Essex, England, Applied Science, 476 p.

Ball, J.W., and Nordstrom, D.K., 1991, WATEQ4F—User's manual with revised thermodynamic data base and test cases for calculating speciation of major, trace and redox elements in natural waters: U.S. Geological Survey Open-File Report 90–129, 185 p.

Barrett, Richard, Berry, M., Chan, T.F., Demmel, J., Donato, J., Dongarra, J., Eijkhout, V., Pozo, R., Romine, C., and van der Vorst, H., 1994, Templates for the solution of linear systems—Building blocks for iterative methods: Philadelphia, Pa., Society for Industrial and Applied Mathematics, 112 p.

Bear, Jacob, 1972, Dynamics of fluids in porous media: New York, American Elsevier, 764 p.

Behie, G.A., and Forsyth, P.A., 1984, Incomplete factorization methods for fully implicit simulation of enhanced oil recovery: SIAM Journal of Scientific and Statistical Computation, v. 5, p. 543–560.

Bethke, C.M., 1996, Geochemical reaction modeling, concepts and applications: New York, Oxford University Press, 397 p.

Burnett, R.D., and Frind, E.O., 1987, Simulation of contaminant transport in three dimensions, 2. Dimensionality effects: Water Resources Research, v. 23, no. 4, p. 695–705.

Burns, Greg, Daoud, Raja, and Vaigl, James, 1994, LAM—An open cluster environment for MPI, *in* J.W. Ross, ed., Proceedings of Supercomputing Symposium '94: University of Toronto, p. 379–386.

Charlton, S.R., and Parkhurst, D.L., 2002, PHREEQCI—A graphical user interface to the geochemical model PHREEQC: U.S. Geological Survey Fact Sheet FS–031–02, 2 p.

Cheng, A.H.-D., and Morohunfola, O.K., 1993, Multilayered leaky aquifer systems, 1. Pumping well solutions: Water Resources Research, v. 29, no. 8, p. 2787–2800.

Cohen, S.D., and Hindmarsh, A.C., 1996, CVODE, A stiff/nonstiff ODE solver in C: Computers in Physics, v. 10, no. 2, p. 138.

de Marsily, Ghislain, 1986, Quantitative hydrogeology—Groundwater hydrology for engineers: New York, Academic Press, 440 p.

Domenico, P.A., 1987, An analytical model for multidimensional transport of a decaying contaminant species: Journal of Hydrology, v. 91, p. 49–58.

Dongarra, J.J., Duff, I.S., Sorensen, D.C., and van der Vorst, H.A., 1991, Solving linear systems on vector and shared memory computers: Philadelphia, Society for Industrial and Applied Mathematics, 256 p.

Dzombak, D.A., and Morel, F.M.M., 1990, Surface complexation modeling—Hydrous ferric oxide: New York, John Wiley, 393 p.

Eagleson, P.S., 1970, Dynamic hydrology: New York, McGraw-Hill, 462 p.

Elman, H.C., 1982, Iterative methods for large sparse nonsymmetric systems of linear equations: New Haven, Conn., Yale University, Ph.D. dissertation, 190 p.

ESRI, 1998, ESRI shapefile technical description, Environmental Systems Research Institute, Inc., accessed December 22, 2009, at *http://www.esri.com/library/whitepapers/pdfs/shapefile.pdf.*

ESRI, 2009, ESRI ASCII raster format, Environmental Systems Research Institute, Inc., accessed December 22, 2009, at *http://resources.esri.com/help/9.3/arcgisengine/java/GP_ToolRef/spatial_analyst_tools/esri_ascii_raster_format.htm.*

Fletcher, C.A.J., 1991, Computational techniques for fluid dynamics, volume 1, 2d ed.: New York, Springer-Verlag, 401 p.

Foster, I.T., 1995, Designing and building parallel programs: Reading, Mass., Addison-Wesley, 381 p.

Garabedian, S.P., LeBlanc, D.R., Gelhar, L.W., and Celia, M.A., 1991, Large-scale natural gradient tracer test in sand and gravel, Cape Cod, Massachusetts, 2. Analysis of spatial moments for a nonreactive tracer: Water Resources Research, v. 27, no. 5, p. 911–924.

Gelhar, L.W., Welty, C., and Rehfeldt, K.R., 1992, A critical review of data on field-scale dispersion in aquifers: Water Resources Research, v. 28, no. 7, p. 1955–1974.

Gropp, W.D., and Lusk, Ewing, 1996, User's guide for MPICH, a portable implementation of MPI: Argonne, Ill., Mathematics and Computer Science Division, Argonne National Laboratory, ANL–96/6, 70 p.

Gropp, W., Lusk, E., and Skjellum, A., 1995, Using MPI—Portable parallel programming with the message passing interface: Cambridge, Mass., MIT Press, 307 p.

Harbaugh, A.W., Banta, E.R., Hill, M.C., and McDonald, M.G., 2000, MODFLOW-2000, the U.S. Geological Survey modular ground-water model—User guide to modularization concepts and the Ground-Water Flow Process: U.S. Geological Survey Open-File Report 2000–92, 121 p.

Hsieh, P.A., and Winston, R.B., 2002, User's guide to Model Viewer, a program for three-dimensional visualization of ground-water model results: U.S. Geological Survey Open-File Report 2002–106, 18 p.

Johnson, J.W., Oelkers, E.H., and Helgeson, H.C., 1991, SUPCRT92—A software package for calculating the standard molal thermodynamic properties of minerals, gases, aqueous species, and reactions from 1 to 5000 bars and 0° to 1000° C: Livermore, Calif., Earth Sciences Department, Lawrence Livermore Laboratory, 168 p.

Kennel, M.B., 2004, KDTREE 2—Fortran 95 and C++ software to efficiently search for near neighbors in a multi-dimensional Euclidean space: accessed September 24, 2008, at *http://arxiv.org/abs/physics/0408067*.

Kindred, J.S., and Celia, M.A., 1989, Contaminant transport and biodegradation 2. Conceptual model and test simulations: Water Resources Research, v. 25, no. 6, p. 1149–159.

Kipp, K.L., 1987, HST3D—A computer code for simulation of heat and solute transport in three-dimensional ground-water flow systems: U.S. Geological Survey Water-Resources Investigations Report 86–4095, 517 p.

Kipp, K.L., 1997, Guide to the revised heat and solute transport simulator HST3D—Version 2: U.S. Geological Survey Water-Resources Investigations Report 97–4157, 149 p.

Kipp, K.L., Russell, T.F., and Otto, J.S., 1992, D4Z—A new renumbering for iterative solution of ground-water flow and solute-transport equations, *in* Russell, T.F., Ewing, R.E., Brebbia, C.A., Gray, W.G., and Pinder, G.F., eds., Computational methods in water resources IX, v. 1, Numerical methods in water resources: Boston, Computational Mechanics Publications and London, Elsevier Applied Science, p. 495–502.

Kipp, K.L., Russell, T.F., and Otto, J.S., 1994, Three-dimensional D4Z renumbering for iterative solution of ground-water flow and transport equations, *in* Peters, A., Wittum, G., Herrling, B., Meissner, U., Brebbia, C.A., Gray, W.G., and Pinder, G.F., eds., Computational methods in water resources X, v. 2: Dordrecht, The Netherlands, Kluwer Academic Publishers, p. 1417–1424.

Kirkner, D.J., and Reeves, H., 1988, Multicomponent mass transport with homogeneous and heterogeneous chemical reactions— Effect of the chemistry on the choice of numerical algorithm 1. Theory: Water Resources Research, v. 24, no. 10, p. 1719–1729.

Konikow, L.F., Goode, D.J., and Hornberger, G.Z., 1996, A three-dimensional method-of-characteristics solute-transport model (MOC3D): U.S. Geological Survey Water-Resources Investigations Report 96–4267, 87 p.

Lapidus, L., and Amundson, N.R., 1952, Mathematics of adsorption in beds, 6. The effect of longitudinal diffusion and ion exchange in chromatographic columns: Journal of Physical Chemistry, v. 56, p. 984–988.

LeBlanc, D.R., 1984a, Description of the hazardous waste research site, *in* LeBlanc, D.R. ed., Movement and fate of solutes in a plume of sewage contaminated ground water, Cape Cod, Massachusetts: U.S. Geological Survey Open-File Report 84–475, 175 p.

LeBlanc, D.R., 1984b, Digital modeling of solute transport in a plume of sewage-contaminated ground water, *in* LeBlanc, D.R. ed., Movement and fate of solutes in a plume of sewage contaminated ground water, Cape Cod, Massachusetts: U.S. Geological Survey Open-File Report 84–475, 175 p.

Maptools.org, 2003, Shapefile C library V1.2, accessed February 26, 2009 at *http://shapelib.maptools.org.*

Meijerink, J.A., and van der Vorst, H.A., 1977, An iterative solution method for linear systems of which the coefficient matrix is a symmetric M-matrix: Mathematics of Computation, v. 31, no. 137, p. 148–162.

Mosier, E.L., Papp, C.S.E., Motooka, J.M., Kennedy, K.R., and Riddle, G.O., 1991, Sequential extraction analyses of drill core samples, Central Oklahoma aquifer: U.S. Geological Survey Open-File Report 91–347, 42 p.

Parkhurst, D.L., 1995, User's guide to PHREEQC—A computer program for speciation, reaction-path, advective-transport, and inverse geochemical calculations: U.S. Geological Survey Water-Resources Investigations Report 95–4227, 143 p.

Parkhurst, D.L., and Appelo, C.A.J., 1999, User's guide to PHREEQC (Version 2)—A computer program for speciation, batch-reaction, one-dimensional transport, and inverse geochemical calculations: U.S. Geological Survey Water-Resources Investigations Report 99–4259, 312 p.

Parkhurst, D.L., Christenson, Scott, and Breit, G.N., 1996, Ground-water-quality assessment of the Central Oklahoma aquifer—Geochemical and geohydrologic investigations: U.S. Geological Survey Water-Supply Paper 2357–C, 101 p.

Parkhurst, D.L., and Kipp, K.L., 2002, Parallel processing for PHAST—A three-dimensional reactive-transport simulator, *in* Hassanizadeh, S.M., Schlotting, R.J., Gray, W.H., and Pinder, G.F., eds., Computational methods in water resources—Developments in water science: Amsterdam, Elsevier, no. 47, v. 1, p. 711–718.

Parkhurst, D.L., Kipp, K.L., Engesgaard, Peter, and Charlton, S.R., 2004, PHAST—A program for simulating ground-water flow, solute transport, and multicomponent geochemical reactions: U.S. Geological Survey Techniques and Methods 6–A8, 154 p.

Parkhurst, D.L., Stollenwerk, K.G., and Colman, J.A., 2003, Reactive-transport simulation of phosphorus in the sewage plume at the Massachusetts Military Reservation, Cape Cod, Massachusetts: U.S. Geological Survey Water-Resources Investigations Report 2003–4017, 33 p.

Parkhurst, D.L., Thorstenson, D.C., and Plummer, L.N., 1980, PHREEQE—A computer program for geochemical calculations: U.S. Geological Survey Water-Resources Investigations Report 80–96, 195 p. (Revised and reprinted August, 1990.)

Patankar, S.V., 1980, Numerical heat transfer and fluid flow: New York, Hemisphere Publishing Corporation, 197 p.

Plummer, L.N., Parkhurst, D.L., Flemming, G.W., and Dunkle, S.A., 1988, A computer program incorporating Pitzer's equations for calculation of geochemical reactions in brines: U.S. Geological Survey Water-Resources Investigations Report 88–4153, 310 p.

Press, W.H., Flannery, B.P., Teukolsky, S.A., and Vetterling, W.T., 1989, Numerical recipes—The art of scientific computing: Cambridge, U.K., Cambridge University Press, 702 p.

Price, H.S., and Coats, K.H., 1974, Direct methods in reservoir simulation: Transactions of the Society of Petroleum Engineers of the American Institute of Mining, Metallurgical, and Petroleum Engineers, v. 257, p. 295–308.

Reed, M.H., 1982, Calculation of multicomponent chemical equilibria and reaction processes in systems involving minerals, gases, and an aqueous phase: Geochimica et Cosmochimica Acta, v. 46, p. 513–528.

Robson, S.G., 1974, Feasibility of digital water-quality modeling illustrated by application at Barstow, California: U.S. Geological Survey Water-Resources Investigations Report 73–46, 66 p.

Robson, S.G., 1978, Application of digital profile modeling techniques to ground-water solute-transport at Barstow, California: U.S. Geological Survey Water-Supply Paper 2050, 28 p.

Sakov, Pavel, 2000, Natural neighbour interpolation C library: Australian Commonwealth Scientific and Research Organization (CSIRO), Marine Research, Hobart, Tasmania, accessed April 15, 2008 at *http://www.marine.csiro.au/~sakov/*.

Scheidegger, A.E., 1961, General theory of dispersion in porous media: Journal of Geophysical Research, v. 66, no. 10, p. 3273–278.

Shewchuk, J.R., 1996, Triangle—Engineering a 2D quality mesh generator and Delaunay triangulator, *in* Ming, C.L., and Dinesh, Manocha, eds., Applied computational geometry—Towards geometric engineering, Lecture Notes in Computer Science, v. 1148: Springer-Verlag, p. 203–222.

Shewchuk, J.R., 2003, A two-dimensional quality mesh generator and Delaunay triangulator: accessed September 24, 2008 at *http://www.cs.cmu.edu/~quake/triangle.html*.

Skiena, Steven, 2008, The Stony Brook Algorithm Repository: Accessed April 18, 2008, at *http://www.cs.sunysb.edu/~algorith/implement/KDTREE/implement.shtml*.

Steefel, C.I. and MacQuarrie, K.T.B., 1996, Approaches to modeling of reactive transport in porous media, *in* Lichtner, P.C., Steefel, C.I., and Oelkers, E.H., eds., Reactive transport in porous media, Reviews in mineralogy v. 34: Washington, D.C., Mineralogical Society of America, p. 83–130.

Stoer, Josef, and Bulirsch, Roland, 1993, Introduction to numerical analysis, 2d edition: New York, Springer-Verlag, 660 p.

Sun, Y., Petersen, J.N., and Clement, T.P., 1999, Analytical solutions for multiple species reactive transport in multiple dimensions: Journal of Contaminant Hydrology, v. 35, no. 4, p. 429–440.

Wang, H.F. and Anderson, M.P., 1982, Introduction to groundwater modeling—Finite difference and finite element methods: New York, W.H. Freeman and Co., 237 p.

Wexler, E.J., 1992, Analytical solutions for one-, two-, and three-dimensional solute transport in ground-water systems with uniform flow: Techniques of Water-Resources Investigations of the U.S. Geological Survey, book 3, chap. B7, 190 p.

Winston, R.B., 2006, GoPhast—A graphical user interface for PHAST: U.S. Geological Survey Techniques and Methods 6–A20, 98 p.

Wolery, T.J., 1992a, EQ3/EQ6, a software package for geochemical modeling of aqueous systems, package overview and installation guide (version 7.0): Lawrence Livermore National Laboratory Report UCRL–MA–110662, part 1, 66 p.

Wolery, T.J., 1992b, EQ3NR, a computer program for geochemical aqueous speciation-solubility calculations—Theoretical manual, user's guide, and related documentation (version 7.0): Lawrence Livermore National Laboratory Report UCRL–MA–110662, part 3, 246 p.

Yeh, G.T., and Tripathi, V.S., 1989, A critical evaluation of recent developments in hydrogeochemical transport models of reactive multichemical components: Water Resources Research, v. 25, no. 1, p. 93–108.

Zhu, Chen, and Anderson, Greg, 2002, Environmental applications of geochemical modeling: Cambridge, UK, Cambridge University Press, 284 p.

Appendix A. Three-Dimensional Visualization of PHAST Simulation Results

The boundary-condition nodes, media properties, and flow and chemical results from a PHAST simulation may be visualized by using a modified version of the Model Viewer software (Hsieh and Winston, 2002), which runs only under Windows operating systems. Model Viewer has capabilities to rotate, translate, and slice a three-dimensional visualization. The Model Viewer software and documentation (PDF file) of Model Viewer are included in the distribution package for PHAST.

Model Viewer displays data that are read from the HDF output file (*prefix*.**h5**). The data written to the HDF file may include heads, velocities, and chemical data as well as boundary-condition features and distributions of media properties. The scalar quantities for chemistry, head, and media properties can be visualized as solid representations (see fig. 6.6 for examples) or isosurfaces (see fig. 6.12 for examples). Interstitial groundwater velocities can be visualized as vectors at each model node. In addition, model features, including specified-head, flux, leaky, river, drain, and well boundary nodes can displayed by Model Viewer.

The active grid region can be dissected parallel to coordinate axes to visualize the interior structure of scalar fields. The color range (color bar) corresponding to the numerical range of scalar values can be adjusted. Options exist to display grids, axes, and boundaries of the grid region, and to change many other aspects of the visualizations. For any variable that can be visualized, an animation may be displayed that shows a scalar (or vector) field for a series of time steps. Visualizations can be exported to graphics files to include in other documents, and animations can be exported to make movies.

Model Viewer can be run from the "Start" menu of Windows by a shortcut written to the desktop or by double clicking on the executable file in a file folder. By default in the PHAST distribution, the Model Viewer executable is located in the directory *C:\Program Files\Usgs\phast-x.x\Model Viewer\bin\modview.exe*, where "x.x" represents the version number of PHAST.

Appendix B. Using PHASTHDF to Extract Data from the HDF Output File

The HDF output file (*prefix*.**h5**) is in compressed, binary format (*http://www.hdfgroup.org/*), which cannot be read by text editors. The contents of the HDF output file can be extracted with the HDF utility *h5dump* and written to an ASCII text file, but the file format produced by the utility is not convenient for use with other programs. PHASTHDF is a graphical, interactive utility program included with the PHAST distribution that is used to extract results from the HDF output file. PHASTHDF is written in the Java language, which can be run on any type of computer that has a Java Runtime Environment, version 1.2 or later. PHASTHDF generates an ASCII text file, referred to as the export file, which is in the same format as the **.xyz.tsv** files. The format of an export file is suitable for importing into spreadsheets, plotting programs, statistics programs, text editors, or for manipulation by scripting languages such as *perl* or *awk*. Export files generated by PHASTHDF contain a set of lines—one line for each active node—for each time step that is selected; time steps for inclusion in the export file are selected from interactive screens of the PHASTHDF program. The lines of the export file include columns for X, Y, and Z coordinates of a node, the time, a cell-saturation identifier that indicates whether the cell for the node is dry (0) or not (1), followed by columns for the scalar and velocity data that are selected from the appropriate screens of the PHASTHDF program.

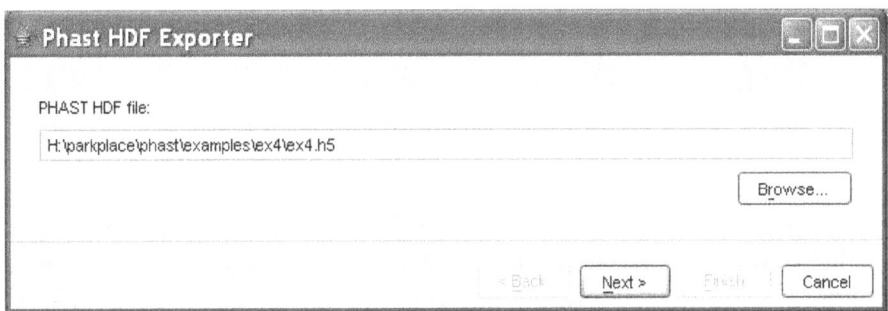

Figure B.1. PHASTHDF input screen for selection of HDF output file from which data are to be extracted.

Running PHASTHDF consists of three steps: (1) selecting the HDF output file from which to extract data (fig. B.1); (2) selecting the X, Y, and Z coordinates of node locations, the time steps, the scalars, and vectors to be extracted (fig. B.2); and (3) specifying the name of the export file (fig. B.3).

Figure B.2 shows the second screen for PHASTHDF, which allows selection of the nodes, time steps, scalars, and vectors that will be extracted from the *prefix*.**h5** file. This screen has five tabs that change the content of the screen. By default, all data contained in the HDF file—all scalars and vectors, all nodes, and all time steps—are selected for extraction. Extracting the complete set of data may generate a large text file, which could exceed the file-size limitation of a 32-bit operating system. Normally, the screens corresponding to the five tabs are used to specify a subset of data that can be exported to a text file of manageable size.

A hyperslab is a set of nodes that is selected by specifying individual X, Y, and Z node locations. The screen on the left of figure B.2 corresponds to the "X-Coor" tab. In the figure, a subset of X coordinates has been selected by using the "Hyperslab selection" check boxes. Beginning with the third coordinate (Start=2 indicates the offset from the first coordinate), every other node (Stride=2) for a total of seven nodes (Count=7) has been unselected. Nodes with these X coordinates will not appear in the export file. Selecting

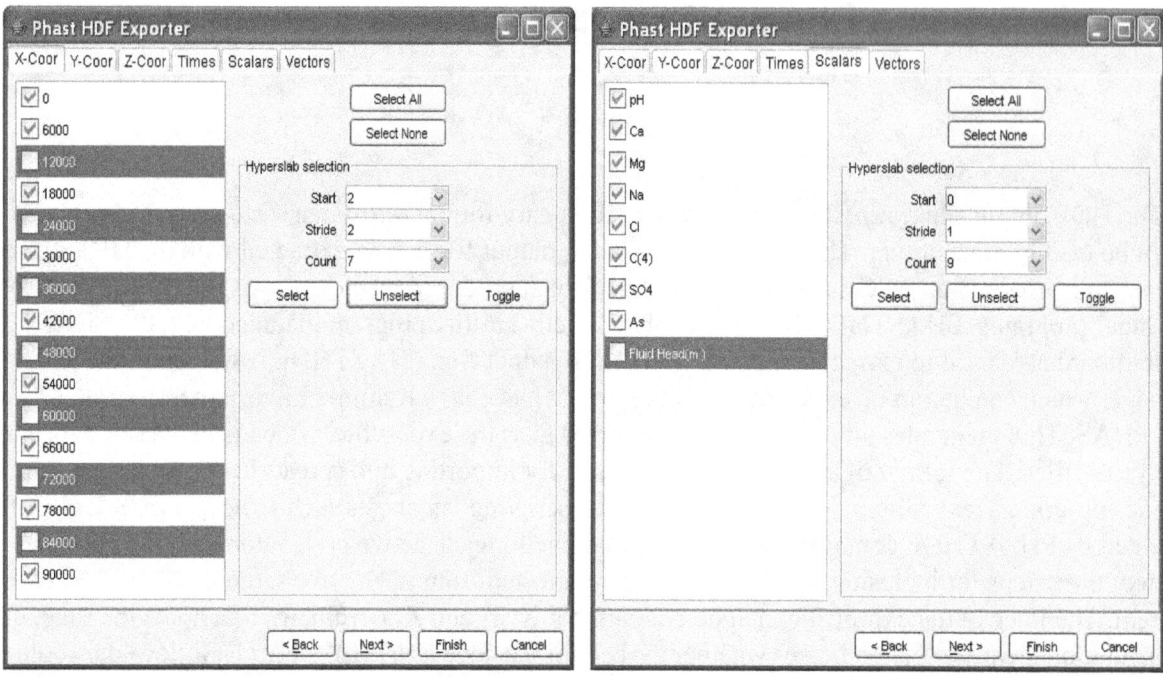

Figure B.2. PHASTHDF input screens associated with the first (X-coor) and fourth (Scalars) of five tabs that allow selection of model nodes and chemical data to be extracted from the HDF output file.

the Y-Coor or Z-Coor tab produces a screen for the respective coordinate that is identical in function to the screen for the X-Coor tab. The HDF output file contains a series of time planes; the "Times" tab allows selection of a subset of time planes to be extracted and written to the export file.

The screen on the right of figure B.2 corresponds to the "Scalars" tab. Selecting this tab produces a list of all of the scalar data that have been stored in the HDF file. It is possible to include or exclude each item of the list individually. For every item that is checked, a column of data will be written to the export file. In this example, all of the data except potentiometric head will be written to the export file. Similarly, the "Vector" tab allows selection of any combination of the X, Y, and Z components of the interstitial velocities at nodes to be written to the export file.

The final screen of PHASTHDF is used to specify the name for the export file (fig. B.3). When the "Finish" button is clicked, the selected data will be written to the export file.

Figure B.3. PHASTHDF input screen used to specify the file name for the export file.

Appendix C. Parallel-Processing Version of PHAST

The numerical simulation of groundwater flow and solute transport with geochemical reactions is computationally intensive and commonly requires the use of large vector or massively parallel computers. Three-dimensional simulation of reactive transport with dozens of components and hundreds of thousands of grid nodes can take days of computing time on desktop workstations or personal computers (PCs). Speeding up multicomponent reactive-transport simulation in a workstation or PC environment makes reactive-transport simulations accessible to a much wider group of groundwater modelers.

One class of parallel computer that is potentially available to groundwater modelers is a cluster of workstations or PCs on a private, high-speed, local area network (LAN). If the Linux operating system is used on the cluster, the cluster is known as a Beowulf cluster. Clusters of computers running Unix, Macintosh, and Windows operating systems also are possible. Another class of parallel computer that is readily available is a single workstation, server, or PC containing multiple (2–8) processors. A parallel version of PHAST has been written that can be run on both of these classes of parallel systems.

C.1. Parallelization of PHAST

Typically, the CPU time spent on the chemistry calculation is at least twice the time spent in the flow and transport calculation, and usually the chemistry calculation requires 90 to 99 percent of the CPU time. A PHAST demonstration problem run on a single processor required 98 percent of the simulation time for the reaction-chemistry calculations; only 2 percent of the time was used for flow and transport calculations (Parkhurst and Kipp, 2002). Thus, parallelization has been applied to the reaction part of the simulator because it could lead to the greatest reduction in CPU time.

The strategy for parallelization of PHAST is based on the parallel machine model of Foster (1995), which is a number of von Neumann computers (nodes) coupled by an interconnection network. Each computer executes its own program (process) on its own local data and communicates with other computers by sending messages over the network or within the parallel workstation. The memory is distributed among the processors. This configuration forms a distributed-memory multiple-instruction, multiple-data (MIMD) computer.

The parallel programming model (Foster, 1995) consists of tasks connected by message queues called channels. One or more tasks can be mapped to a process. The Message Passing Interface (MPI) (Gropp and others, 1995) variant of the task/channel model, as implemented by the Local Area Multicomputer (LAM) (Burns and others, 1994) project and MPICH (Gropp and Lusk, 1996), can be used for communication. Message passing programs create a number of processes at program startup. Each process executes the same program on its own data following a single program multiple data (SPMD) model. A set of tasks is distributed among the available processes.

For the parallel version of PHAST, a fixed number of processes are started at initialization of a simulation to execute the necessary tasks. The tasks for a PHAST simulation include initialization, flow, transport, cell-by-cell reaction, summary calculation and load balancing, and completion. The flow and reactive-transport tasks are logically separate and sequential in constant-density flow. The operator-splitting algorithm in PHAST further separates functions into transport tasks, one transport calculation for each component, and chemical-reaction tasks, one for each cell in the active grid region. The data requirements for the reaction tasks are independent for each cell.

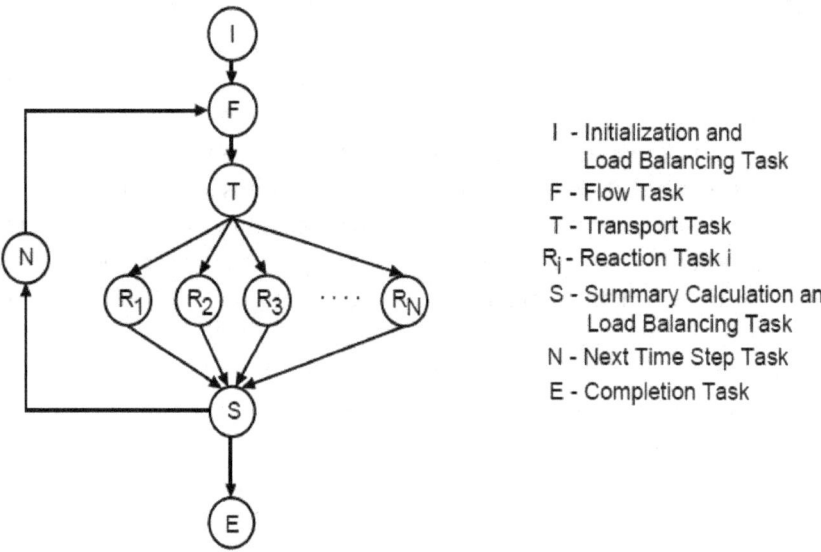

Figure C.1. Flow-control diagram for parallel-processing version of PHAST.

I - Initialization and
    Load Balancing Task
F - Flow Task
T - Transport Task
R_i - Reaction Task i
S - Summary Calculation and
    Load Balancing Task
N - Next Time Step Task
E - Completion Task

Figure C.1 shows the flow-control graph (Dongarra and others, 1991) for the partitioning of tasks for a single time step. The nodes of the graph represent tasks and the arrows represent execution dependencies. Data dependencies do not exist for a task with no incoming arrows. The parallel structure of the reaction tasks is apparent. Additional time steps are indicated by the feedback loop. Data communication is required between the flow, transport, reaction, and summary calculation and load-balancing tasks. The parallel version has a communication pattern that is

- Local with each reaction task communicating only with the transport task;

- Structured with a partitioning that has dynamic mapping;

- Static with reaction and transport tasks as fixed communication partners; and

- Synchronous with communication after the sequential tasks of transport, reaction, and summarization.

Parallel communication includes scattering from the transport task to the reaction tasks and gathering from the reaction tasks to the summary calculation task. Additional communication is needed for the load-balancing task, which is included in the summary calculation task. The flow and transport tasks are combined and are run on a single, root process. Cell-reaction tasks are combined into sets of reaction tasks, which are mapped by the root process—one set to each process (including the root process). Flexibility is maintained by allowing the size of each set of reaction tasks to vary for each time step.

The root process determines the sets of reaction tasks and the mapping of the sets to the processes by a load-balancing algorithm that has a static and a dynamic component. The static component of the algorithm consists of randomly ordering the cells during the initialization task. Much of the computational load may be in a reaction front located in a set of contiguous cells. The randomization is intended to distribute the cells from computationally intensive areas uniformly among all of the processes. To adjust for the heterogeneity in processor capability, a small calculation is run initially on each process to measure its relative speed. The results are used to establish the initial number of cells in each set of reaction tasks.

The dynamic component of the algorithm consists of rebalancing the computational workload at the end of each time step. The sets of reaction tasks are revised by adjusting the number of cells assigned to each reaction-task set. The revision of the number of cells in each set causes cell calculations to be moved from one process to another to equalize the reaction-task computation time for each process. When a cell is moved from one process to another, all of the chemical information for a cell is sent from the old process to the new process by data messages. Two algorithms are implemented to rebalance the load as determined by the identifier **-rebalance_by_cell** in **SOLUTION_METHOD** (section **4.7. Description of Keyword Data Blocks** of this report). The first algorithm is based on measuring for each process the average time per cell for the chemistry calculations. The second algorithm is based on measuring the time spent for the chemistry calculation for each cell for each process. Both algorithms estimate the number of cells to assign to each process by finding a set of n_p integers such that all processes require the same amount of time, T, to complete their set of reaction tasks:

$$n_p \hat{T}_p = T; \text{ for } p = 1, ..., P; \text{ subject to the constraint that } \sum_{p=1}^{P} n_p = N, \tag{C.1}$$

where n_p is the number of cells for process p; $\hat{T}_p$ is the average time per cell for a reaction task for process p (the average time to complete one cell calculation during the last chemical calculation); P is the number of processes; and N is the total number of cells.

In the first algorithm, the following relation is derived from equation C.1:

$$n_p = n_1 \frac{\hat{T}_1}{\hat{T}_p}, \text{ for all } p = 1, ..., P, \tag{C.2}$$

and substituting into $\sum_{p=1}^{P} n_p = N$, the following relation is obtained:

$$n_1 = \frac{N}{\sum_{p=1}^{P} \frac{\hat{T}_1}{\hat{T}_p}}, \tag{C.3}$$

which determines the number of cells to assign to process 1. The number of cells to assign to all other processes can be determined by equation C.2, with some accommodation to account for the fact that each n_p must be an integer. An optimal load balance will satisfy equation C.1 and will reduce the idle time for each of the processes to zero during the simultaneous reaction-task calculations. A damping factor is used to eliminate repeated overcorrection in rebalancing (**SOLUTION_METHOD**, **-rebalance_fraction**); by default, only half of the number of cells to be remapped are actually transferred to a different process at any rebalancing step.

An alternative algorithm for load rebalancing was implemented in PHAST Version 2. Initially, a standard task is performed by each process. If the fastest process requires t_{min} seconds to perform the task, a relative speed $r_p = \frac{t_{min}}{t_p}$ is calculated for each process, where r_p is the relative speed and t_p is the time for process

p to perform the task. The fraction of the chemistry calculation load for process p (f_p) is calculated by the following formula:

$$f_p = \frac{r_p}{\sum_{q=1}^{P} r_q} .$$ (C.4)

During each chemistry calculation, the time required to calculate the chemistry for each cell t_i is measured. These times are normalized to the times $(\hat{t}_i)$ it would take the fastest process to perform the tasks as follows:

$$\hat{t}_i = t_i r_p,$$ (C.5)

where r_p is the relative speed for the process that performed the chemistry calculation for cell i. The total time for the fastest process to do the task is estimated to be $T_{serial} = \sum_{i=1}^{N} \hat{t}_i$. The set of chemistry calculations for process 1 is then determined by the number N_1, such that

$$\sum_{i=1}^{N_1} \frac{\hat{t}_i}{T_{serial}} \sim f_1 .$$ (C.6)

In general, a number N_p is determined such that

$$\sum_{i=N_{p-1}+1}^{N_p} \frac{\hat{t}_i}{T_{serial}} \sim f_p .$$ (C.7)

The number of cells to apportion to each process is $n_p = N_p - N_{p-1}$, where $n_1 = N_1$. Adjustments are made to determine when the relation C.7 is sufficiently close to an equality and to ensure that every process has at least one chemical calculation. This algorithm was implemented to account for the case where the chemistry calculation in one cell may be very slow.

In practice, it is not clear that either rebalancing algorithm is superior to the other. The time to rebalance the load using either algorithm, including communication time to move cell information, is minimal, typically consuming much less than 1 percent of the total computation time.

The parallel algorithm for PHAST is a coarse-grain parallelization in that the ratio of computation time to communication time is large. Overall efficiency, which includes the communication time of distributing data to and gathering data from each process, frequently is in the range of 80 to 95 percent. This efficiency implies that a network of 10 identical single-processor computers can run the same simulation about 8 to 9 times faster than a single computer with one processor.

The results of a parallel simulation should be nearly identical to the results of the same simulation run on a single processor. However, slight differences in the results between the parallel and single-processor versions are caused by rounding and truncation of numbers in the message passing between the parallel processes and by differences in implementation of floating-point arithmetic among the processors of a cluster. The chemistry data file, flow and transport data file, and thermodynamic database file used to run the single-processor version are the same as those used to run the parallel version, and the same output files are produced. The only differences in the content of the output files are some additional data related to the efficiency of the parallel processing that are written to the screen and to the log file.

C.2. Running the Parallel Version

To be able to run the parallel version of PHAST, it is necessary to install a version of MPI on a network of computers (or a single multiprocessor computer). LAM and MPICH are public-domain versions of MPI and other proprietary versions of MPI are available (Burns and others, 1994, *http://www.lam-mpi.org*; *http://www-unix.mcs.anl.gov/mpi/mpich*). A compiled parallel version of PHAST for LAM on Linux computers and MPICH (Gropp and Lusk, 1996) on Windows NT, Windows 2000, Windows XP, Windows Vista, and Windows 7 computers (earlier Windows operating systems cannot run MPICH) is provided in the distribution packages. It also is possible to run MPI on a heterogeneous cluster of Unix computers (SUN, IBM, Linux, and others). This requires building PHAST (compiling and linking to libraries) for each of the operating systems on the network that are used for PHAST simulation.

Installing and running MPI can be difficult. Properly setting up the communication that is needed among computers, including the sets of network files that each computer can access, is not a simple task. It is beyond the scope of this manual to describe all of the ways that MPI can be installed and configured. However, two examples are provided to demonstrate how the parallel version of PHAST can be run under some simplifying assumptions on a Beowulf cluster and a cluster of PCs running Windows.

For the first example, consider a Beowulf cluster of Linux operating system computers on which LAM has been installed and an MPI test problem runs correctly. It is assumed that all of the computers are able to read from and write to a directory named */z/wulf1/home/geek/phast/examples/ex4*. It is further assumed that the executables for PHASTINPUT and the LAM parallel version of PHAST (found in */bin* of the installation directory) are copied into this directory with the names *phastinput* and *phast-lam*. A file named *hosts*, created by the user, is in this directory, which lists the names of a number of computers on the network that will be used for the simulation. Finally, the input files *ex4.chem.dat*, *ex4.trans.dat*, and *phast.dat* also are in the directory. The parallel version of PHAST for LAM can be run with the following sequence of commands:

```
cd /z/wulf1/home/geek/phast/examples/ex4
phastinput ex4
lamboot hosts
mpirun N -wd /z/wulf1/home/geek/phast/examples/ex4 phast-lam
wipe hosts
```

The *phastinput* command runs PHASTINPUT, which reads the data input files and generates an intermediate file named *Phast.tmp*, the same as for single-processor execution. The *lamboot* command initiates the LAM daemon on each computer listed in the *hosts* file. The *mpirun* command initiates the program *phast-lam* on each computer for which a daemon was initiated (*N* option). Each process (program) reads from and writes to the working directory (-*wd* option). Note that the program *phast-lam*, which is run on each computer, also is found in the working directory. The final command (*wipe*) removes the LAM daemon from each computer. Many temporary files are written by each process into the shared working directory but will be deleted if the job completes successfully. The procedure for running PHAST with LAM can be simplified by the use of shell scripts and LAM schema files, but use of these files is not discussed in this documentation.

For the second example, consider a network of computers running the Windows NT, Windows 2000, Windows XP, Windows Vista, or Windows 7 operating system. MPICH has been installed on the computers and an MPI test problem runs correctly. In the process of running a test problem, the computers available to MPICH will be defined by the user when running the MPICH Configuration Tool. It is assumed that all of the computers are able to read from and write to a shared directory named *\\hostname\mpich_share*, which points to the local directory *c:\phast\examples\ex4* residing on the computer named *hostname*. It is further assumed that the executables for PHASTINPUT (named *phastinput.exe*), the MPICH parallel version of PHAST (named *phast-mpich.exe*), *zlib.dll*, and *hdf5dll.dll* (all found in */bin* of the installation directory) are copied into this shared directory. Finally, the input files *ex4.chem.dat*, *ex4.trans.dat*, and *phast.dat* also are in the directory. The parallel version of PHAST for MPICH can be run on *hostname* with the following sequence of commands entered in a DOS command-line window:

```
cd c:\phast\examples\ex4
phastinput ex4
mpirun -np 10 -dir \\hostname\mpich_share \\hostname\mpich_share\phast-mpich
```

The *phastinput* command runs PHASTINPUT, which reads the data input files and generates an intermediate file named *Phast.tmp*, the same as for single-processor execution. The *mpirun* command initiates 10 instances of the program *phast-mpich* distributed among the computers for which MPICH has been configured (*-np* 10). Each process (program) reads from and writes to the shared directory. Many temporary files are written into the shared directory for each process but will be deleted if the job completes successfully. The process of running PHAST with MPICH may be simplified by writing batch scripts, adding directories to the path environmental variable, and by defining additional environmental variables, but use of these features is not discussed in this documentation.

Appendix D. Theory and Numerical Implementation

PHAST is a simulator of multicomponent, reactive transport in three dimensions derived from the flow and solute-transport model HST3D and the geochemical-reaction model PHREEQC. Multicomponent solute transport and chemical reactions are simultaneous processes. However, in PHAST, the solute-transport and chemical-reaction calculations are separated by operator splitting and performed by a sequential-solution approach without iteration.

The governing equations for a reactive solute-transport simulator are a set of partial differential equations describing groundwater flow and solute transport for each aqueous component, a set of nonlinear algebraic equations describing equilibrium chemical reactions, and a set of ordinary differential equations describing rates of kinetic chemical reactions. Multiple aqueous chemical constituents are present in a natural or contaminated groundwater flow system. A subset of the aqueous chemical constituents is chosen to be components. The number of components required is the minimum number of species from the total set necessary to describe the chemical composition of the system (Reed, 1982). Components as defined for this simulator are not the same as thermodynamic components used in connection with phase rule calculations (see Zhu and Anderson, 2002, p. 51–55). A set of aqueous species is chosen as master species (one for each component), such that all other species (aqueous and solid) can be formed by chemical reactions from these master species. The total aqueous concentrations of the chemical components are the dependent variables of the transport equations, and the activities of the master species and the number of moles of solid reactants are the dependent variables for the chemical equations.

This appendix covers the essential equations, features, and methods used in the PHAST simulator, including the assumptions underlying the flow and transport equations and their discretization; the equilibrium and kinetic reaction equations; the property functions and transport coefficients; the well-source conditions; the boundary and initial conditions; the operator-splitting approach and sequential solution of flow, transport, and chemical equations; accuracy of solutions to the coupled equations; the global balance calculation; the nodal velocity calculation; and the zonal flow-rate calculation. The groundwater flow and component-transport differential equations and boundary conditions are summarized, and those that are implemented differently than in HST3D are described in detail. No additions or modifications to the numerical methods of PHREEQC were necessary for the chemical-reaction calculations of PHAST; minor modifications were made to the output functions of PHREEQC. The PHREEQC manual (Parkhurst and Appelo, 1999) covers the numerical implementation details for the chemical-reaction calculations, and only a brief summary will be included here.

D.1. Flow and Transport Equations

The general saturated groundwater flow and component solute-transport equations solved by the PHAST simulator are based on those of the HST3D simulator described in Kipp (1987, 1997). In HST3D, the equations were written for variable fluid density and viscosity, and for a single component solute-transport equation including linear sorption and decay. For PHAST, the simulator has been restricted to constant density and viscosity and the linear sorption and decay terms are eliminated. The following assumptions determine the constitutive equation set for PHAST:

- Groundwater fully saturates the porous medium within the region of groundwater flow, referred to as the simulation region.

- Groundwater flow is described by Darcy's law.

- Groundwater and the porous medium are compressible under confined flow conditions.

- Groundwater and the porous medium are incompressible under unconfined flow conditions.
- Groundwater has constant, uniform density and viscosity.
- Isothermal conditions exist in the simulation region with respect to flow and transport.
- Porosity and permeability are functions of space.
- The coordinate system is orthogonal and aligned with the principal directions of the permeability tensor so that this tensor is diagonal for anisotropic media.
- The coordinate system is right-handed; that is, when the curl of the right hand fingers is from positive X-axis to positive Y-axis, the thumb points in the direction of the positive Z-axis.
- Dispersive-mass fluxes of the bulk fluid from spatial-velocity fluctuations are neglected.
- Contributions to the total fluid mass balance from pure-solute mass sources within the region are neglected.
- Multiple solute components are present.
- Chemistry is defined by a set of chemical reactions, both equilibrium and kinetic.
- Changes in the ratio of solid reactants to mass of water caused by porosity changes in confined systems are neglected.
- Chemical reactions do not induce porosity or permeability changes.
- Dispersivity values are the same for each chemical component.

These assumptions simplify the flow and transport equations relative to those in HST3D but expand the chemical equation set to include multiple components and chemical reactions.

The groundwater flow equation used in PHAST is given in Kipp (1987). The dependent variable internal to the simulator is pressure; however, given the assumptions of PHAST, equivalent equations can be written by using potentiometric head as the dependent variable. The term "pressure" will be used in connection with some of the discussion of the flow equation and boundary conditions, although potentiometric head is used for data input and output. The groundwater flow equation using potentiometric head (Bear, 1972) is as follows:

$$S_s \frac{\partial h}{\partial t} = \nabla \bullet K \nabla h + q, \tag{D.1}$$

$$\text{with } h = \frac{p}{\rho g} + z, \tag{D.2}$$

where S_s is the storage coefficient (per meter, m^{-1}); h is the potentiometric head (m); t is time (s); ∇ is the gradient operator (m^{-1}) for scalars and the divergence operator for vectors; K is the hydraulic conductivity tensor (m/s); q is the source flow-rate intensity ($m^3 s^{-1} m^{-3}$); p is the pressure (Pa); ρ is the water density (kg/m^3); g is the gravitational acceleration (m/s^2); and z is the elevation coordinate (m).

Darcy's law relates the specific discharge field to the gradient of the potentiometric head. Thus, the interstitial velocity is

$$v = -\frac{K}{\varepsilon} \nabla h, \tag{D.3}$$

where v is the interstitial velocity (m/s) and ε is the saturated porosity (unitless).

The equation for conservation of a single solute aqueous component is based on the following assumptions:

- Pressure diffusion is neglected.

- Solute transport by local, interstitial, velocity-field fluctuations and mixing at pore junctions is described by a hydrodynamic dispersion-coefficient tensor.

- Forced diffusion by gravitational, electrical, and other fields is neglected.

- No pure diffusive solute sources occur in the fluid or porous-matrix phases.

The general form of the reactive solute-transport equation for component j in the aqueous phase with both equilibrium and kinetic reactions (modified from Bear, 1972; Kirkner and Reeves, 1988) is as follows:

$$\frac{\partial}{\partial t}(\varepsilon \rho c_j) = \nabla \bullet \varepsilon D \nabla \rho c_j - \nabla \bullet \varepsilon v \rho c_j - \sum_{e=1}^{N_E} \upsilon_{j,e}^E \frac{\partial}{\partial t}(\varepsilon \rho \bar{c}_e) + \sum_{k=1}^{N_K} \upsilon_{j,k}^K \varepsilon \rho R_k + q \varepsilon \rho \hat{c}_j ; \quad j = 1, ..., N_c, \quad \text{(D.4)}$$

where c_j is the total aqueous concentration of component j (mol/kgw); D is the dispersion-coefficient tensor (m^2/s); N_E is the number of heterogeneous equilibrium reactions; $\upsilon_{j,e}^E$ is the stoichiometric coefficient of component j in heterogeneous equilibrium reaction e (unitless); $\bar{c}_e$ is the concentration of solid reactant e (mol/kgw); N_K is the number of kinetic reactions; $\upsilon_{j,k}^K$ is the stoichiometric coefficient of component j in kinetic reaction k (unitless); R_k is the rate of kinetic reaction k (mol kgw^{-1} s^{-1} [moles per kilogram of water per second]); $\hat{c}_j$ is the total aqueous concentration of component j in the source water (mol/kgw); and N_c is the number of chemical components in the system. Equation D.4 is written for each aqueous component. Note that the reaction rate, R_k, may be a function of aqueous species of components other than j, and the R_k and $\bar{c}_e$ may appear in multiple transport equations. Thus, the reaction rate and the rate of solid species formation provide terms that couple the set of reactive-transport differential equations D.4.

The classical method of finite differences (Patankar, 1980) is used to discretize the partial-differential equations and boundary conditions in space and time. Several options are available for the differencing, which are discussed in the following sections. The selection of components for PHAST also is discussed.

D.1.1. Components

For PHAST, the term component is used to refer to an entity that has an aqueous mole-balance equation in a PHREEQC calculation. PHREEQC has an aqueous mole-balance equation for each element that is present in the system, including hydrogen and oxygen, and a mole-balance equation that conserves charge. Therefore hydrogen, oxygen, and charge imbalance are transported, plus any additional elements. It is possible to extend the definition of an "element" to include an organic compound, for example atrazine, or a single redox state of an element, for example ferrous iron, by appropriate definitions in **SOLUTION_MASTER_SPECIES** and **SOLUTION_SPECIES** data blocks. It also is possible to formulate mass balance on individual isotopes through the use of the isotope capabilities of PHREEQC, for example carbon-13. If these definitions are made, then atrazine, ferrous iron, or carbon-13 could be components in a PHAST simulation.

A component may be present in the form of many different aqueous species; however, the transport equation is applied to the total aqueous concentration (sum of moles in all aqueous species) for each component. As in HST3D (Kipp, 1987, 1997) concentration units for the transport equation in PHAST are mass fraction. Mass fraction is calculated for each component as $w_j = \dfrac{M_j}{1 + M_j} = \dfrac{0.001\, n_j f_j}{1 + 0.001\, n_j f_j}$, where w_j is mass fraction of component j (unitless), M_j is the total mass of component j in the aqueous phase (kg), n_j is the total number of moles of component j in the aqueous phase (mol), and f_j is the gram formula weight of component j (g/mol). Note that the masses of other components do not affect the mass fraction of a given component and that the mass of water is assumed to be exactly 1 kg. This formulation is compatible with the single-component transport formulation inherited from the program HST3D (Kipp, 1987, 1997) and is necessary to conserve mass between the transport calculations and the reaction calculations.

The components hydrogen and oxygen are handled in a special way. The two quantities that are transported are the "excess" hydrogen and oxygen relative to 1 kg of water. In other words, the mass of excess hydrogen is calculated by taking the total mass of hydrogen in solution and subtracting the mass of hydrogen in 1 kg of water, similarly for oxygen. PHREEQC allows the mass of water to vary, so the total mass of water in a solution may not be identically 1.0 kg, but variable water mass is not consistent with the constant density assumption for the flow and transport equations. To avoid mass-balance errors, the transport calculations use the mass of each component in solution relative to an assumed 1 kg of water. However, the chemical calculations use the total mass of water, including water that is produced or consumed by reactions. The chemical calculations reconstitute the total hydrogen and oxygen, and implicitly the total mass of water, by adding the excess hydrogen and oxygen from transport to 1.0 kg of water.

D.1.2. Spatial Discretization

Although the spatial discretization is described by Kipp (1987, 1997), a brief summary of grid construction and terminology is given here for reference. The model domain is a bounded volume of three-dimensional space, often irregularly shaped, over which the flow and transport equations are solved. A right-handed, three-dimensional Cartesian coordinate system is applied to the model domain with the Z-axis pointing in the opposite direction of the gravitational force. A cylindrical-coordinate system is not available in PHAST. Spatial discretization is done by constructing a grid of node points that covers the model domain (fig. D.1). This grid is formed by specifying the distribution of node locations in each of the three coordinate directions. Every combination of X, Y, and Z nodal coordinate values gives the location of all the node points in the three-dimensional grid. The grid occupies a rectangular parallelepiped of space, referred to as the grid region. The grid region completely contains the model domain. A set of nodes in the grid is selected to approximate the shape of the model domain, and nodes in the set are referred to as boundary nodes. Boundary nodes and nodes internal to the grid region form the set of active nodes. The remaining nodes of the grid are inactive nodes. Boundaries of the active grid region are planes that are normal to one of the coordinate directions and connect boundary nodes. Simulation results are computed for all active nodes. Cells at edges and corners of the active grid region have two or three boundary faces, respectively (fig. D.1 or fig. 4.2).

The grid region can be partitioned into elements. An element is the volume of a rectangular parallelepiped defined by eight nodes that are adjacent in the grid, with one node at each corner of the element (fig. 4.1). The set of elements fills the entire grid region. An element is referred to as active if all of its nodes are in the

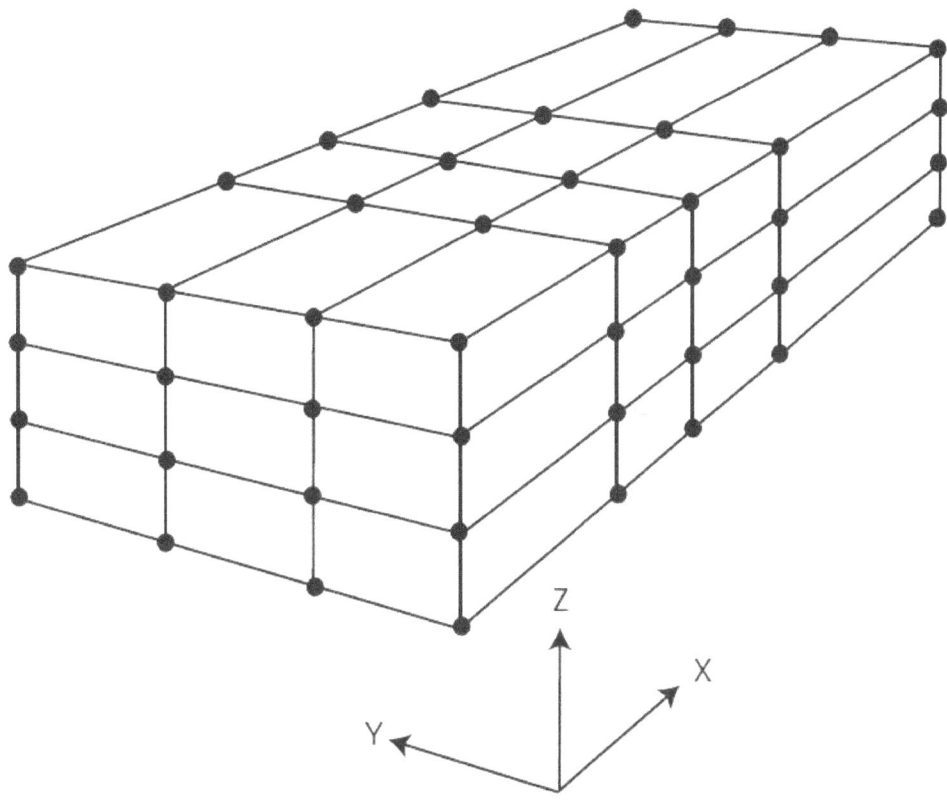

Figure D.1. Grid of node points that covers the model domain.

active grid region, either internal or on the boundary. An element is referred to as inactive if any of its nodes are in the inactive grid region.

The active grid region also can be partitioned into cells (fig. 4.2). A cell is the volume associated with an active node. For interior nodes, a cell is a rectangular parallelepiped with cell boundaries that are planes bisecting the distance between adjacent node points. Interior cells are referred to as full cells and contain portions of eight elements that share the cell node. Cells associated with boundary nodes are referred to as fractional cells. From one to seven active elements may share a boundary node; thus, a fractional boundary cell may consist of various fractions of the full cell (1/8 to 7/8). Parts of one to three boundary faces of a fractional cell do not lie between adjacent node points but, rather, are boundary planes of the active grid region and intersect at the cell node (fig. 4.2). Each of the grid region boundary faces of a given fractional cell is perpendicular to a different coordinate axis. Cells are not defined for inactive nodes.

The type of grid that is described here and is used by PHAST is called a point-distributed grid. A common alternative method for constructing the grid is to specify the locations of the planes that form the cell boundaries. Their intersections form the cells, and the node points are located in the center of each cell. This type of grid is called a cell-centered or block-centered grid.

An element has uniform porous-media properties: porosity, specific storage coefficient, hydraulic conductivity, and dispersivity. However, a cell contains portions of one to eight active elements (fig. 4.6); thus it can contain as many as eight different values for a porous-media property. The finite-difference equations are formed by integrating the partial differential equations over each cell volume, as explained in Kipp (1987). In this integration, the variations in porous-media properties within a cell are arithmetically averaged in proportion to their active element volume in the cell (fig. 4.6). Fluid and solute-dispersive conductances are calculated from porous-media properties by averaging the properties over a given cell face in proportion to the area of the cell face that is within each active element. A cell face is within parts of one to four active elements (fig. 4.6); thus a conductance is calculated from property values of as many as four different elements.

Variable weighting for spatial discretization of the advective term in the transport equations is available, ranging from centered-in-space differencing to upstream-in-space differencing (**SOLUTION_METHOD** data block, **-space_differencing** identifier). Centered-in-space differencing has the potential for causing oscillations in the solutions. This can be particularly problematic for the chemical-reaction calculations if negative concentrations are computed. In PHAST, negative concentrations are set to zero and ignored in the chemical-reaction calculation, but this leads to mass-balance errors because excess positive concentrations that balanced the negative concentrations are preserved. Analysis of one-dimensional flow and transport with constant coefficients (Fletcher, 1991, p. 286–293) shows that numerical oscillation does not occur if

$$\frac{\Delta x}{\alpha} \leq 2 \text{ or } Pe \leq 2, \tag{D.5}$$

where Δx is the cell size (m); α is the dispersivity (m); Pe is the cell Peclet number $(Pe = v_x \frac{\Delta x}{D_{xx}} = \frac{\Delta x}{\alpha})$(unitless); v_x is the component of velocity (m/s) in the X direction; and D_{xx} is the dispersion-coefficient tensor element (m^2/s) appropriate for one-dimensional flow and transport in the X direction.

Upstream-in-space differencing introduces artificial numerical dispersion into the transport solutions. Again, analysis of one-dimensional flow and transport with constant coefficients (Fletcher, 1991, p. 302–303) yields a numerical dispersion coefficient and a numerical dispersivity given by

$$D_{ns} = \frac{v_x \Delta x}{2} \text{ and} \tag{D.6}$$

$$\alpha_{ns} = \frac{\Delta x}{2}, \tag{D.7}$$

where D_{ns} is the numerical dispersion coefficient from spatial discretization (m^2/s) and α_{ns} is the numerical dispersivity from spatial discretization (m). These terms result from the truncation error of the finite-difference approximation.

D.1.3. Temporal Discretization

Variable weighting for temporal discretization is available (**SOLUTION_METHOD** data block, **-time_differencing** identifier). Well-known specific cases include fully implicit (backward-in-time) differencing and Crank-Nicholson (centered-in-time) differencing. Centered-in-time differencing also has the potential for causing oscillations in the solutions. Again, this can be particularly problematic for the chemical-reaction calculations if negative concentrations are computed.

Fully implicit temporal discretization is always used for computing a steady-state solution to the flow equation. This removes the possibility of oscillations in the solution and allows for large time steps to be taken as steady state is approached. For transient-flow simulation, the specified weighting for temporal discretization is used for both the flow equation and the transport equations.

Analysis of one-dimensional flow and transport with constant coefficients shows that numerical oscillation does not occur for centered-in-time differencing (with centered-in-space differencing) if

$$\frac{D_{xx}\Delta t}{\Delta x^2} = \frac{\alpha v_x \Delta t}{\Delta x^2} \leq 1 \ \text{ or } Cr \leq Pe, \tag{D.8}$$

where Δt is the time step (s) and Cr is the cell Courant number ($Cr = \frac{v_x \Delta t}{\Delta x}$) (unitless). This criterion is not as stringent as that for spatial oscillation, and in some cases the method is stable even if the criterion is exceeded.

Backward-in-time differencing also introduces artificial numerical dispersion into the transport solutions. Again, analysis of one-dimensional flow and transport with constant coefficients yields the numerical dispersion coefficient and numerical dispersivity from temporal discretization given by

$$D_{nt} = \frac{v_x^2 \Delta t}{2} \ \text{ and} \tag{D.9}$$

$$\alpha_{nt} = \frac{v_x \Delta t}{2}, \tag{D.10}$$

where D_{nt} is the numerical dispersion coefficient from temporal discretization (m²/s) and α_{nt} is the numerical dispersivity from temporal discretization (m). These terms result from the truncation error of the finite-difference approximation.

For pure transport simulations (no reactions), the total numerical dispersivity is the sum of the spatial (equation D.7) and temporal (equation D.10) numerical dispersivities. For numerical accuracy in transport simulations, the total numerical dispersivity should be much less than the physical dispersivity. Thus, for upstream-in-space and backward-in-time differencing,

$$\frac{\Delta x}{2} + \frac{v_x \Delta t}{2} \ll \alpha \ \text{ or in dimensionless form:} \tag{D.11}$$

$$Pe(1 + Cr) \ll 2. \tag{D.12}$$

If upstream-in-space and centered-in-time differencing are used, then the temporal dispersivity is zero and accuracy depends on the relation:

$$\frac{\Delta x}{2} \ll \alpha. \tag{D.13}$$

If centered-in-space and backward-in-time differencing are used, then the spatial dispersivity is zero and accuracy depends on the relation:

$$\frac{v_x \Delta t}{2} \ll \alpha. \tag{D.14}$$

Although these equations are from a simplified, restricted one-dimensional analysis, they provide guidance for discretization of more complex three-dimensional simulations of transport.

D.1.4. Automatic Time-Step Algorithm for Steady-State Flow Simulation

An option is available that calculates the steady-state flow field by simulating transient flow until steady state is reached. The time steps are automatically adjusted during the steady-flow simulation, which significantly reduces computation time compared to the use of fixed time steps. The groundwater flow simulator in PHAST uses an empirical algorithm for automatic time stepping based on Aziz and Settari (1979, p. 403) as adapted by Kipp (1997, p. 13).

For the automatic time-step option, the user may specify the target change in potentiometric head over a time step that is considered acceptable (STEADY_FLOW data block, **-head_change_target** identifier) as well as the maximum (STEADY_FLOW data block, **-maximum_time_step** identifier) and minimum (STEADY_FLOW data block, **-minimum_time_step** identifier) time steps allowed. At the beginning of each time step, adjustment is made to the size of the time step to try to attain the target change in head. If the maximum change in head for the last time step exceeds the target change in head or is less than 90 percent of the target change in head, the time step is adjusted by the ratio of the target head change to the change in head for the last time step. However, the new time step is constrained by the specified minimum and maximum time-step limits, and further, it may not grow by more than a specified factor (2.0 by default; STEADY_FLOW data block, **-growth_factor** identifier) at each adjustment. Because the actual changes in the potentiometric head during a simulation are not directly proportional to the changes in time-step length, the specified target head change can be considerably exceeded when this linear predictor is used for the adjustment of the time step. Therefore, an additional test is made at the end of each time step. If the maximum change in head is greater than 1.5 times the target head change, the time step is repeated with a shorter time-step length, which is calculated as previously explained.

The minimum allowed time step, optionally set by the user, is maintained for the first two steps of the simulation. If the maximum change in the head during the second time step exceeds the target head change, the simulator stops because it cannot adjust the time step below the minimum allowed value.

D.2. Chemical-Reaction Equations

Both equilibrium and kinetic reactions are simulated in PHAST. Homogeneous equilibrium in the aqueous phase is described with an ion-association aqueous model that includes mass-action equations for ion pairs and complexes, and activity coefficients based on Debye-Hückel or Pitzer specific-ion-interaction theory. Heterogeneous equilibrium reactions are based on equality of activities among phases. For kinetic chemical reactions, the rates of reactions are user-defined functions of solution composition. The equations describing equilibrium reactions and the formulation of kinetic chemical reactions are given in detail in Parkhurst and Appelo (1999).

The chemical equations form a set of nonlinear algebraic equations for equilibrium reactions and ordinary differential equations in time for kinetic reactions. The differential and algebraic equations are coupled through the total aqueous concentration of components. However, the chemical reactions in each cell of the active grid region are solved independently. For a given cell, no information about chemical compositions in other cells is used in the simulation of equilibrium and kinetic reactions.

D.2.1. Equilibrium Reactions

The set of equilibrium reactions includes aqueous-complexation, mineral-phase, ion-exchange, surface-complexation, solid-solution, and gas-phase (rarely useful in PHAST) equilibria. All equilibrium reactions are based on mass-action equations that relate activities of species to an equilibrium constant:

$$K = \frac{\prod\limits_{products} a_p^{\upsilon_p}}{\prod\limits_{reactants} a_r^{\upsilon_r}},$$

(D.15)

where K is the equilibrium constant, a is activity, and υ_r and υ_p are stoichiometric coefficients of the reactants and products in the chemical equation. Equilibrium constants are functions of temperature. Activities of aqueous species are related to molalities through activity coefficients:

$$a = \gamma m,$$

(D.16)

where γ is the activity coefficient and m represents molality. The activity coefficients are functions of species charge and ionic strength and are derived from extensions of the Debye-Hückel or Pitzer specific-ion-interaction activity-coefficient theory. Activities of pure minerals are assumed to be fixed at 1.0. Activities of gas components are equal to their partial pressures. Activities of solid-solution components are equal to their mole fraction or optionally a function of mole fraction. Activities of ion-exchange species are equal to their equivalent fractions, possibly multiplied by an activity coefficient that is a function of solution composition. Activities of surface complexes are equal to mole fraction of the site type (Parkhurst and Appelo, 1999).

D.2.2. Component Mole Balance

For any time step in the flow and transport simulation, the total number of moles of each component in each cell is defined by the sum of moles of component in solution and the moles of component in the solids. At the beginning of the chemistry calculation for a cell, the total aqueous concentration of a component in solution is defined by the concentration resulting from the transport calculation. The moles of component in the solids includes the sum of moles of the component in minerals, exchangers, surfaces, solid solutions, and gas phases in the saturated part of the cell. During the chemistry calculation, the moles of component in solution are augmented by kinetic reactions, after which equilibrium is recalculated among solution and solids to produce a new total aqueous concentration. The total aqueous concentrations of components at the end of the chemistry calculation are then used in the next transport step after adjustment for any change in saturation in the cell (see section D.5.7. Free-Surface Boundary).

D.2.3. Kinetic Reactions

Kinetic reactions (**KINETICS** data block in the chemistry data file) are defined with rate equations that are a function of solution composition. The user defines the rate equations by using Basic-language statements (**RATES** data block in the chemistry data file), which allow general definitions of rate expressions without recompilation of the computer code (Parkhurst and Appelo, 1999).

D.3. Property Functions and Transport Coefficients

Before the flow and component transport equations can be solved, information about the fluid properties, porous-media properties, and transport coefficients needs to be provided. As mentioned previously, fluid density and viscosity are constant and uniform in the PHAST simulator. Porosity is a function of the fluid potentiometric head in the case of confined flow (Kipp, 1987, p. 24) and is assumed to be constant in the case of unconfined flow. For the capacitance or storage term in the confined flow equation, the compressibility of the water and the porous medium is accounted for by the specific storage (Eagleson, 1970, p. 270; Kipp, 1987, p. 25). The specific storage includes both fluid and porous-matrix compressibility and thus is a hybrid property of both fluid and porous matrix. The fluid and porous matrix are assumed incompressible for the case of unconfined flow.

The variation in the ratio of water to solid that is caused by changes in porosity under confined flow conditions is relatively small and is neglected in the chemical calculations. Thus, for confined systems, the concentration of a solid (mol/kgw) does not vary because of changes in storage. Concentrations of solids and sorbed components only change by heterogeneous chemical reaction. (For unconfined systems, concentrations of solids also will vary as the water table rises and falls within a cell, see section D.5.7. Free-Surface Boundary.)

Hydrodynamic dispersion is characterized by the hydrodynamic dispersion-coefficient tensor (Bear, 1972; Scheidegger, 1961). For an isotropic porous medium, two parameters describe the dispersion-coefficient tensor, the longitudinal dispersivity and the transverse dispersivity. The dispersion-coefficient tensor is a function of space, through the dispersivity coefficients, and is a function of space and time through the interstitial velocity. Mathematically,

$$D_{ij} = (\alpha_L - \alpha_T)\frac{v_i v_j}{|v|} + \alpha_T |v| \delta_{ij} + D_M \delta_{ij}, \tag{D.17}$$

where D_{ij} is the dispersion-coefficient tensor element (m²/s); v_i is the interstitial velocity in the i coordinate direction (m/s); v_j is the interstitial velocity in the j coordinate direction (m/s); $|v|$ is the magnitude of the interstitial velocity vector (m/s); α_L is the longitudinal dispersivity (m); α_T is the transverse dispersivity (m); D_M is the effective molecular diffusion coefficient (m²/s); and δ_{ij} is the Kronecker delta function, $\delta_{ij} = 1$, for $i = j$ and $\delta_{ij} = 0$, for $i \neq j$. The dispersion-coefficient tensor is a nine-element tensor that is symmetric. This tensor links the solute dispersive fluxes to the concentration gradients. The off-diagonal elements give rise to cross-dispersive fluxes, which are dispersive fluxes in one coordinate direction driven by concentration gradients in another coordinate direction.

Several field studies have indicated that transverse dispersion in the vertical direction is much smaller than transverse dispersion in the horizontal direction (Robson, 1974, 1978; Garabedian and others, 1991; and Gelhar and others, 1992) for the case of a stratified porous medium with flow parallel to the stratification. Burnett and Frind (1987) formulated an *ad hoc* anisotropic dispersion-coefficient tensor for this case involving horizontal and vertical transverse dispersivities. The modified dispersion-coefficient tensor is

$$D_{xx} = \alpha_L \frac{v_x^2}{|v|} + \alpha_{T_H} \frac{v_y^2}{|v|} + \alpha_{T_V} \frac{v_z^2}{|v|} + D_M, \tag{D.18}$$

$$D_{yy} = \alpha_L \frac{v_y^2}{|v|} + \alpha_{T_H} \frac{v_x^2}{|v|} + \alpha_{T_V} \frac{v_z^2}{|v|} + D_M, \tag{D.19}$$

$$D_{zz} = \alpha_L \frac{v_z^2}{|\boldsymbol{v}|} + \alpha_{T_V} \frac{v_x^2}{|\boldsymbol{v}|} + \alpha_{T_V} \frac{v_y^2}{|\boldsymbol{v}|} + D_M, \tag{D.20}$$

$$D_{xy} = D_{yx} = (\alpha_L - \alpha_{T_H}) \frac{v_x v_y}{|\boldsymbol{v}|}, \tag{D.21}$$

$$D_{xz} = D_{zx} = (\alpha_L - \alpha_{T_V}) \frac{v_x v_z}{|\boldsymbol{v}|}, \text{ and} \tag{D.22}$$

$$D_{yz} = D_{zy} = (\alpha_L - \alpha_{T_V}) \frac{v_y v_z}{|\boldsymbol{v}|}, \tag{D.23}$$

where α_{T_H} is the horizontal transverse dispersivity (m) and α_{T_V} is the vertical transverse dispersivity (m). Here, the terms horizontal and vertical refer to the axes directions of the Cartesian coordinate system. This formulation is used in the PHAST simulator, and it reduces to the isotropic form when $\alpha_{T_H} = \alpha_{T_V}$.

Cross-derivative dispersive-flux terms may be included or excluded in a PHAST simulation. The more rigorous treatment of including the cross-derivative terms involves an explicit calculation (**SOLUTION_METHOD** data block, **-cross_dispersion true**). Explicit calculation of the cross-dispersive flux terms is lagged one iteration in the solution cycle of the solute-transport equations; this lagging preserves the seven-point stencil in the finite-difference equations, which enables the use of the same linear equation solvers for the transport equations as are used for the flow equation. When cross-derivative dispersive flux terms are included in the solution method, two iterations of transport equation solution are done for each time step to compensate for the lagged terms. The cross-derivative dispersive flux terms are recalculated for each iteration on the basis of the concentration conditions existing at the end of the previous iteration. These lagged terms are incorporated into the right-hand side of the finite-difference transport equations. This treatment requires storage of the nine finite-difference dispersion-coefficient terms for each cell.

In the solution of the transport equations, strong diagonal flows through a cell can cause the cross-dispersive solute fluxes to produce negative concentrations. Therefore, an alternative treatment of the cross-derivative dispersion terms consists of neglecting the cross-derivative dispersion fluxes entirely. No iterations of the component transport solution are required. For robustness of the PHAST simulator, the default option is to neglect the cross-derivative dispersive flux terms.

For PHAST, it is assumed that dispersion of each aqueous component can be described by the same set of dispersivity values. This yields the same set of dispersion coefficients for each chemical component. No other transport coefficients are present because, in the operator-split form of the component transport equations, no source terms from reactions appear. Each component-transport equation contains only advective, dispersive, and storage terms.

D.4. Well-Source Conditions

A point source (sink) term is used to represent injection (withdrawal) by a well in equation D.1 and D.4. Spatial discretization converts this term to a line source. Although a well is treated as a line source for the flow and transport equations, a well is a finite-radius cylinder for the local well-bore model in PHAST.

The local well-bore model calculates flow rates between the well bore and each layer of cells (Kipp, 1987). A well can be used for fluid injection or fluid withdrawal with associated solute injection or production. A well also can be used for observation of aquifer conditions. The well bore can communicate with any subset of cells along the Z-coordinate direction at a given X–Y coordinate location. That is, the well may be screened or open-hole over several intervals of its depth.

Only the two simplest options for the well-bore model from HST3D are included in PHAST. A well can be a source or sink with specified flow rate within the aquifer that is allocated by mobility or by mobility times head difference as described by Kipp (1987, p. 32–36). Mobility is the product of hydraulic conductivity and the length of open interval within the cell.

After discretization, the well flow rate to or from each penetrated cell depends on the length of open interval in that cell. The PHASTINPUT preprocessor calculates the length of the open interval above and below the node in each cell. This information on the upper and lower bounds of the open interval enables the use of the appropriate element properties in calculating the well mobility factors for each cell that communicates with a well. Variation in hydraulic conductivity among the eight elements that compose a full cell is handled by computing the arithmetic average of the conductivity in the four elements in the upper portion and the four elements in the lower portion of the cell. Then, a weighted-average hydraulic conductivity is computed by using the fractional open-interval lengths in the upper and lower portions of the cell as weighting factors.

For a well with flow-rate allocation by mobility only, every cell that communicates with the well has only injection or only production, depending on the total flow rate specified. It is not possible to have a cell with injection and a cell with production in the same well for mobility-only flow-rate allocation. For an injection well, an associated-solution composition is specified by the user by using solution indices (**WELL** data block, **-associated_solution** identifier). For a production well, the total aqueous concentrations of components produced from the well are calculated from a weighted average of concentrations from the producing cells; the concentrations are weighted in proportion to the cell flow rates to the well.

For a well with flow-rate allocation by mobility times head difference, the different cells that communicate with the well may produce or inject water. For a production well, the flow is accumulated from the bottom to the top, with the restriction that the well-bore flow at all points must be upward or the simulator terminates with an error. The total aqueous concentrations of components produced from the well are calculated by accumulating and distributing flow to each cell connected to the well, beginning from the lowermost cell. If water is produced from a cell, it changes the concentration in the well bore by mixing in proportion to the cell flow rate and the flow rate in the well bore; if water is injected into a cell, it is injected with the concentration in the well bore at that point. Similarly, for an injection well, the flow is distributed from the top to the bottom, with the restriction that the well-bore flow at all points must be downward, or the program terminates with an error. The total aqueous concentrations in the well bore are determined by accumulating and distributing flow to each cell connected to the well, beginning at the uppermost cell. If water is injected into a cell, it is injected with the concentration in the well bore at that point; if water is produced from a cell, the production changes the concentration in the well bore by mixing in proportion to the cell flow rate and the flow rate in the well bore. Thus, mixing of water within the well bore is simulated, but no chemical reactions are considered within the well bore.

The pumping or injection rate and the associated-solution composition for a well may change with time. Temporal changes in these values occur discontinuously in time; a change in value causes the beginning of a new simulation period. This discontinuity means that the effective parameter value for the first time step of a new simulation period is the new value under backward-in-time differencing and the average of the old and new values under centered-in-time differencing. For compatibility among well-source conditions and boundary conditions, see section D.5.8. Boundary-Condition Compatibility.

D.5. Boundary Conditions

To define a simulation, boundary conditions need to be specified. The types of boundary conditions available in PHAST include (1) specified potentiometric head with associated-solution composition (SPECIFIED_HEAD_BC data block, **-head** and **-associated_solution** identifiers), (2) specified potentiometric head with specified-solution composition (SPECIFIED_HEAD_BC data block, **-head** and **-fixed_solution** identifiers), (3) fluid flux and associated-solution composition (FLUX_BC data block), (4) leaky boundaries with associated-solution composition (LEAKY_BC data block), (5) rivers with associated-solution composition (RIVER data block), (6) drains (DRAIN data block), and (7) unconfined aquifer with a free surface (water table) (FREE_SURFACE_BC data block). For all boundary conditions with associated-solution compositions, inflow enters the region with the total aqueous concentrations defined by the associated-solution composition, but outflow leaves the region with the ambient total aqueous concentrations at the boundary location. A specified-head boundary condition may have a specified-solution composition instead of an associated-solution composition. For a specified-head boundary cell with a specified-solution composition, the total aqueous concentrations of all components are constant over each simulation period and any specified changes in values cause the beginning of a new simulation period.

Associated- and specified-solution compositions are designated by solution indices, which refer to solution compositions defined in the chemistry data file. Preliminary geochemical simulations defined in the chemistry data file can be used to establish the solution compositions for boundary conditions. These preliminary simulations may include mineral equilibria, ion exchange, surface complexation, solid-solution equilibria, gas-phase equilibration, and general kinetic reactions, plus mixing, irreversible reactions, and temperature variation (see section D.6. Initial Conditions).

The spatial distribution of all boundary conditions is specified on a cell basis rather than on an element basis. Specified-head boundary conditions apply to entire cells. Flux, leaky, river, and drain boundary conditions are divided into areal segments that apply to a single cell. The segment areas are less than or equal to the area of a cell face. Multiple segments of each type of boundary condition can be allocated to a given cell. The segments are modified so that there are no overlapping areas among segments of a given boundary-condition type. However, no modification is made to ensure that segments of different boundary-condition types do not overlap. The default boundary condition for the flow and transport equations is no fluid flux and no dispersive or advective solute flux through the faces of the cells at the boundary of the active grid region.

Some boundary-condition parameters may be changed with time, including head, flux, and solution composition. Temporal changes in these values occur discontinuously in time; a specified change in value causes the beginning of a new simulation period. This discontinuity means that the effective parameter value for the first time step of a new simulation period is the new value under backward-in-time differencing and the average of the old and new values under centered-in-time differencing.

D.5.1. Specified-Head Boundary with Associated-Solution Composition

A specified potentiometric-head boundary condition is available for the flow equation (SPECIFIED_HEAD_BC data block). The head value is specified relative to the Z coordinate of the grid region. Specified-head boundary conditions are incorporated into the difference equations by replacing the flow equation for those nodes by equations defining the specified heads. Therefore, it is not possible to define flux, leaky, river, or drain boundary conditions for any cell that is a specified-head cell.

The associated-solution composition option (**-associated_solution**) for a specified-head boundary defines the composition of water flowing into the boundary cells. The solution composition of a specified-head boundary cell with an associated-solution composition will vary during the simulation period depending on the flux of water through the boundary relative to the composition currently in the cell and the flux and composition of water from other faces of the cell. (An alternative concentration boundary condition for a specified-head cell is a specified solution-composition [see next section D.5.2. Specified-Head Boundary with Specified-Solution Composition], which differs from an associated-solution composition in that the solution composition is fixed for the duration of a simulation period.)

The head and associated-solution composition for a specified-head boundary condition can be a function of location and time. Zones are used to define the spatial distribution of the specified-head boundaries. The head and associated-solution composition may vary during the simulation; any specified change in value causes the beginning of a new simulation period. For compatibility among specified-head boundary conditions and other boundary conditions, see section D.5.8. Boundary-Condition Compatibility.

D.5.2. Specified-Head Boundary with Specified-Solution Composition

Specified-solution compositions are an alternative to associated-solution compositions for specified-head boundary conditions (**SPECIFIED_HEAD_BC** data block, **-fixed_solution** identifier). Unlike associated-solution composition boundary cells, the solution composition in a specified-solution-composition boundary cell does not vary during a single simulation period. The solution composition of a specified-solution-composition boundary cell is constant regardless of inflow or outflow; however, it is not physically realistic to specify concentrations at an outflow boundary. The composition of the specified solution may vary during a simulation; any specified change in composition (solution index number) causes the beginning of a new simulation period. All of the discussion in the previous section related to the flow equation for a specified-head boundary applies to boundary cells with specified-solution composition. For compatibility among specified-head boundary conditions and other boundary conditions, see section D.5.8. Boundary-Condition Compatibility.

D.5.3. Flux Boundary

Normal fluxes of fluid can be specified over parts of the boundary of the grid region as functions of location and time (**FLUX_BC** data block). Examples of physical boundary conditions that can be represented as flux boundaries include infiltration from precipitation, known flux through lateral boundaries where head gradients can be estimated, and known recharge- and discharge-boundary flow rates for simple steady-state flow problems. On fluid-inflow boundaries, the associated-solution composition of the inflowing fluid needs to be defined (**-associated_solution**), which then determines the advective flux of solute components. At fluid-outflow boundaries, the component concentrations of the outflowing fluid are equal to the ambient total aqueous concentrations. Purely diffusive solute-component flux with no fluid flow cannot be defined as a flux boundary condition.

Zones are used to define the spatial distribution of the flux boundaries. Flux boundary conditions are applied only to exterior cell faces and only to that part of an exterior cell face that is inside the zone definition. Areas of application of flux boundaries are processed to obtain segments that apply to only one cell face. Multiple flux boundary-condition segments may apply to a given cell face, but the segments are modified so that no flux segments overlap. The flux and associated-solution composition may vary during the simulation; any specified change in value causes the beginning of a new simulation period. Note that the fluid flux is specified as a volumetric flux with vector components expressed relative to the coordinate system of the grid region. Thus, fluxes are signed quantities that express direction relative to cell faces that are perpendicular to a coordinate direction. For example, precipitation recharge would have a negative sign indicating a flux in the negative Z-coordinate direction. For unconfined flow simulations, if the flux boundary applies to faces perpendicular to the X or Y axis, flux is multiplied by the fraction of the cell that is saturated. For compatibility among flux boundary conditions and other boundary conditions, see section D.5.8. Boundary-Condition Compatibility.

D.5.4. Leaky Boundary

A leaky boundary condition (or head-dependent-flow boundary condition) is represented as a fluid flux that is driven by a difference in potentiometric head across a layer of finite thickness and specified hydraulic conductivity (**LEAKY_BC** data block). Often this boundary condition is used to represent vertical or lateral flow across leaky semiconfining layers, where the hydraulic conductivity of the semiconfining layer is orders of magnitude smaller than the conductivity of the adjacent active grid region. The leaky boundary also can be used to extend the effective simulation region laterally by using hydraulic conductivity equal to that in the aquifer. A leaky boundary can be used to mimic a specified-head boundary condition by setting the hydraulic conductivity to a large number. For unconfined flow, a leaky boundary also can be used to represent surface water bodies, such as rivers, lakes, and estuaries.

Mathematical representation of leaky boundary conditions is explained in Kipp (1987). The treatment of leaky boundaries is based on the following simplifying assumptions: (1) changes in fluid storage in the semiconfining layer are neglected, (2) semiconfining-layer capacitance effects on solute transport are neglected, and (3) flow and solute component transport are affected by the leakage that enters the region, but flow and solute conditions that exist on the external side of the semiconfining layer (outside the active grid region) are not affected by fluxes that enter or leave the active grid region. Furthermore, no chemical reactions are simulated within the semiconfining layer.

For illustration, assume that part of the upper boundary surface is overlain by a semiconfining layer. Another aquifer lies above the semiconfining layer with a head distribution at its contact with the semiconfining layer that is a known function of time. The leaky boundary flux is given by

$$q_L = \frac{K_L}{b_L}(h_{eL} - h_B),$$

(D.24)

where q_L is the volumetric flux at a leaky boundary (m^3 m^{-2} s^{-1}); K_L is the hydraulic conductivity of the semiconfining layer (m/s); b_L is the thickness of the semiconfining layer (m); h_{eL} is the potentiometric head at the external side of the semiconfining layer (m); and h_B is the potentiometric head at the boundary of the active grid region (m). To calculate the leakage for a leaky boundary, the volumetric flux is multiplied by the area of application. For each leaky boundary, the external head, semiconfining layer thickness and hydraulic conductivity, and the associated-solution composition are defined by the user.

For unconfined flow simulations, leakage through upper Z faces is handled analogously to a river (D.5.5. River Boundary). For a vertical stack of cells, leakage is applied to the cell that contains the water table. The limitation for flux from a river boundary to the aquifer is applied to the leakage from the leaky boundary to the aquifer. An elevation of the top of the leaky bed is defined (**-elevation**) along with a thickness of the bed (**-thickness**), from which the bottom of the leaky bed is calculated. The maximum flux from the leaky boundary to the aquifer occurs when the head in the aquifer is at the elevation of the bottom of the leaky bed. The leakage does not increase as the head in the aquifer drops below the bottom of the leaky bed. For unconfined flow with a leaky boundary on an upper Z face, the elevation of the top of the leaky layer must be defined, in addition to the other parameters of a leaky boundary condition. In this case, it also is required that the potentiometric head assigned to the leaky boundary be greater than or equal to the elevation of the top of the leaky layer.

Zones are used to define the spatial distribution of the leaky boundaries. Leaky boundary conditions are applied only to exterior cell faces and only to that part of an exterior cell face that is contained in the zone definition. The areas of application of the leaky boundary condition are processed to produce a collection of segments that each apply only to a single cell face. Multiple leaky boundary-condition segments may apply to a given cell face, but the segments are modified so that no leaky segments overlap.

The leaky boundary conditions are transformed into source terms in the difference equations on a cell-by-cell basis. The discretized equation for one leaky segment, s, for boundary cell, m, is

$$Q_{Ls} = \frac{K_{Ls}}{b_{Ls}}(h_{eLs} - h_m^{(n)})S_{Ls} - \frac{K_{Ls}}{b_{Ls}}S_{Ls}\delta h_m, \tag{D.25}$$

where Q_{Ls} is the volumetric flow rate through a leaky boundary segment s (m^3/s); K_{Ls} is the hydraulic conductivity of the semiconfining layer for segment s (m/s); b_{Ls} is the thickness of the semiconfining layer for segment s (m); h_{eLs} is the potentiometric head at the external side of the semiconfining layer for segment s (m); h_m is the potentiometric head at the boundary of the active grid region for cell m (m); (n) is the index of the discrete time value during the simulation; S_{Ls} is the area of the leaky boundary segment s (m^2); and δh_m is the change in potentiometric head at the boundary of the active grid region over the time step for cell m (m). The volumetric flow rate given by equation D.25 has an explicit term for the right-hand side of the discretized flow equations and an implicit factor for the left-hand side. For unconfined flow simulations, if the leaky boundary applies to X or Y faces, leakage is multiplied by the fraction of the cell that is saturated.

The total leakage from leaky boundary conditions for a cell is the sum of the leakage through all of the leaky segments that apply to the cell. The associated solute component fluxes are assumed to be purely advective. They are obtained from the total aqueous concentrations of the associated-solution composition if water flux is into the active grid region, or from the ambient total aqueous concentrations at the boundary location if the water flux is out of the active grid region. The external head and associated-solution composition for a leaky boundary condition can be functions of time. New values of external head and associated-solution composition may be assigned at specified times during the simulation; any specified change in value causes the beginning of a new simulation period.

Leaky boundaries can be defined in combination with most other boundary conditions except specified-head boundary conditions. For compatibility among leaky boundary conditions and other boundary conditions, see section D.5.8. Boundary-Condition Compatibility.

D.5.5. River Boundary

A river boundary condition (**RIVER** data block) is used to simulate exchange of water and solutes between a river and the aquifer. Water may recharge the aquifer or discharge the aquifer depending on the sign of the head difference between the aquifer and the river. The rate of exchange of water between the aquifer and the river is dependent on the thickness and hydraulic conductivity defined for the riverbed. If the resistance to flow in the riverbed is small, rivers can represent piecewise linear specified-head boundaries.

The river boundary condition is similar to the leaky boundary condition but has some special characteristics: (1) this boundary condition is available only for unconfined flow; (2) the river is a land-surface feature that traverses the upper boundary of the active grid region; (3) for a vertical stack of cells, any river leakage enters or leaves the cell that contains the water table; (4) the riverbed is the semiconfining layer through which leakage occurs; and (5) the flux is limited to a maximum value when the head in the aquifer is below the bottom of the riverbed. Conceptually, all river flux occurs through the riverbed; no flux through the sides of the river channel is considered.

A simplified geometry of a cross section at a river boundary is shown in figure D.2. The flux for river leakage depends on the difference in head between the river and the aquifer, the area of the river bottom, the thickness of the riverbed, and the hydraulic conductivity of the riverbed. The maximum flux from the river to the aquifer is limited to the flux generated when the aquifer head is equal to elevation of the bottom of the riverbed. Physically, this means that if the water table declines below the bottom of the riverbed, the increased resistance to flow of the partially saturated porous medium prevents any further increase in flux from the river to the aquifer, although unsaturated flow is not simulated explicitly. The flux for a river boundary is given by

$$q_R = \frac{K_R}{b_R}(h_R - h_{wt}),$$
(D.26)

where q_R is the volumetric flux through the boundary (m^3 m^{-2} s^{-1}), K_R is the hydraulic conductivity of the riverbed (m/s), b_R is the thickness of riverbed, h_R is head in the river (m), and h_{wt} is the elevation of the water table (m). If the elevation of the water table is below the bottom of the riverbed (z_B), then the volumetric flux is equal to the flux calculated with $h_{wt} = z_B$.

In the implementation of river boundaries in PHAST, a river is a piecewise linear feature defined by its trace across the top surface of the grid region (fig. 4.9). A chain of line segments represents the river route.

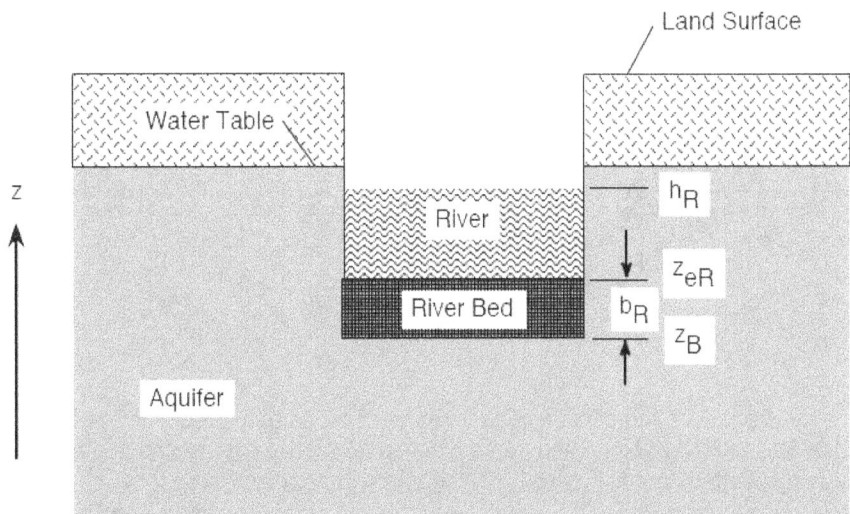

Figure D.2. Schematic section showing geometry for a river boundary, where b_R is thickness of riverbed, h_R is head of the river, z_{eR} is elevation of the top of the riverbed, and z_B is elevation of the bottom of the riverbed.

The end points of the river, and possibly other river points, have data defining the river head, width, bottom elevation, riverbed hydraulic conductivity, and associated-solution composition.

Several steps are involved in translating the data from the sequence of river points that defines a river into boundary conditions for the affected cells of the active grid region. First, each pair of river points and the associated widths are used to define trapezoidal areas as shown in figure D.3A. The collection of trapezoids has both gaps between trapezoids and overlapping areas (fig. D.3B). The gap between the end lines of adjacent trapezoids is appended to the trapezoid that was defined first in the sequence of river point definitions to produce a polygon, called a river polygon. The overlapping area of two trapezoids is assigned to the river polygon that was defined first in sequence and is removed from any other river polygon (fig. D.3C).

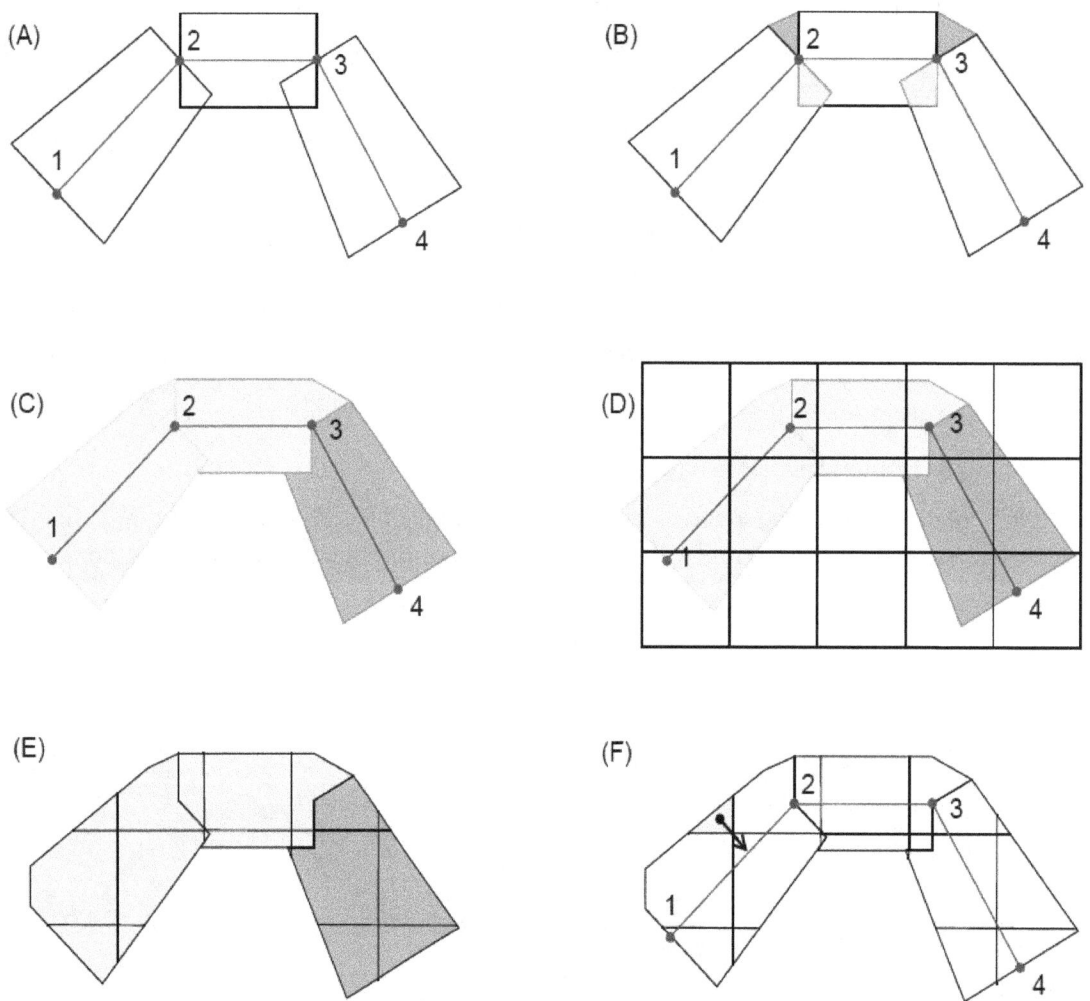

Figure D.3. Steps by which river segments and river segment properties are determined. (A) Trapezoidal areas are defined by the river points (numbered blue dots) and the river widths at each river point. (B) Gaps are filled and overlaps are eliminated. (C) River polygons are created for each line segment defined by two sequential river points. (D) River polygons are intersected with the grid to form river segments. (E) River segments are ready for final processing; color indicates association with original data. (F) Properties are assigned to river segments by finding the point on the river nearest the centroid (black dot) of the river segment and interpolating from values at river points.

This removal process is applied to all river polygons, such that overlapping areas between river polygons are assigned to the river that has the smallest integer identification number (**RIVER** *n*). No subarea of the surface of the active grid region is assigned to more than one river polygon. Each river polygon is then intersected with each cell on the surface of the active grid region (fig. D.3D). The intersection divides the area of each river polygon into subpolygons, called river segments, each of which is contained within a single cell (fig. D.3E). Finally, the centroid of each river segment is located. Properties for a river segment are determined by interpolating from the two nearest river points (one upstream and one downstream) where the property is defined, to the point on the river that is closest to the centroid of the river segment (fig. D.3F). For reactive-transport simulations, the composition of the solution associated with the river segment is determined by proportional mixing of the end-member solution compositions at the nearest river points (one upstream and one downstream) where solution compositions are defined.

The river boundary conditions are transformed into source terms in the difference equations on a cell-by-cell basis. The discretized equation for one river segment, *s*, for a water table cell, *wt*, is

$$Q_{Rs} = \frac{K_{Rs}}{b_{Rs}}(h_{Rs} - h_{wt}^{(n)})S_{Rs} - \frac{K_{Rs}}{b_{Rs}}S_{Rs}\delta h_{wt},$$ (D.27)

where Q_{Rs} is the volumetric flow rate for river segment *s* (m^3/s); K_{Rs} is the hydraulic conductivity of the riverbed for segment *s* (m/s); b_{Rs} is the thickness of the riverbed for segment *s* (m); h_{Rs} is the potentiometric head of the river for segment *s* (m); h_{wt} is the potentiometric head of the water table in the stack of cells below the river segment (m); (*n*) is the index of the discrete time value during the simulation; S_{Rs} is the area of the river segment *s* (m^2); and δh_{wt} is the change in water-table elevation over the time step (m). The river leakage rate given by equation D.27 has an explicit term for the right-hand side of the discretized flow equations and an implicit factor for the left-hand side.

Hydrostatic extrapolation of the potentiometric head is used exclusively to locate the free surface (water table) under a river. This extrapolation prevents a nonconvergent cycle of free-surface adjustment that can occur if other types of extrapolation or interpolation are used. Problems will occur if the elevation of the free surface drops to a cell that is below a cell with a partial or full saturation. This implies perched water and a zone of unsaturated flow, which PHAST cannot simulate.

Flow into or out of the aquifer is calculated for each river segment that affects a given stack of cells. The total leakage for the river boundary condition for a water-table cell is the sum of the leakage through all of the river segments that apply to that stack of cells. Any number of river segments may overlie a stack of cells, thus enabling tributaries or branches to be represented (fig. 4.9).

The associated solute component fluxes are assumed to be purely advective. They are obtained from the total aqueous concentrations of the associated-solution composition if water flux is into the active grid region, or from the ambient total aqueous concentrations at the water table if the water flux is out of the active grid region. For the case of multiple branches of rivers traversing a given stack of cells, with net recharge into the aquifer, the total aqueous concentrations of the recharge are determined as the flow-rate-weighted averages of the associated-solution concentrations of all the river segments recharging to that column of cells. It is possible that some river segments receive discharge from the aquifer; however, the recharge concentrations are calculated by using only associated-solution concentrations from river segments that recharge the aquifer.

The head and associated-solution composition for each river point may vary during the simulation; any specified change in value causes the beginning of a new simulation period. At each new simulation period, head and solution composition are reinterpolated for each river segment. A river boundary segment and a flux boundary segment can be applied to the same Z face of a cell. This feature is needed to represent river leakage and precipitation infiltration within the same cell. Numerically, the precipitation flux is distributed over the entire area of the flux segment and the river infiltration is applied over the river segment. The two areas may overlap, which is not a serious error for riverbeds covering only a small fraction of the upper surface of the grid, but could be a problem if the river area is a large fraction of the total surface area and flux is defined over the entire grid region. For compatibility among river boundary conditions and other boundary conditions, see section D.5.8. Boundary-Condition Compatibility.

D.5.6. Drain Boundaries

A drain boundary condition can be used to simulate an agricultural drain, an ephemeral stream fed by groundwater, or a spring. Water will flow to a drain whenever the head in the aquifer is greater than the elevation of the drain. However, no water flows to or from the drain if the water table is below or equal to the elevation of the drain. The conceptual model of the drain boundary condition includes the simplifying assumption that when water is flowing through the drain and the drain traverses a region of the aquifer where the water table is below the drain elevation, no recharge to the aquifer occurs.

The drain boundary condition (**DRAIN** data block) is similar to a river boundary condition. Like a river, the flux of water to the drain depends on the difference between the drain elevation and the head in the aquifer, the effective area of the drain bed, the thickness of the drain bed, and the hydraulic conductivity of the drain bed. The main difference between a drain and a river is that a constant drain elevation is used in place of the time-dependent head in the river, and there is no flow from the drain if the head in the aquifer drops below the elevation of the drain. Other than these two differences, drain leakage is calculated identically to river leakage (see equations D.26 and D.27). No solution composition is needed for a drain because no water is allowed to flow from the drain to the aquifer. Conceptually, all drain flux out of the aquifer occurs through the drain bed underlying the drain; no flux through the sides of the drain bed is considered.

In the implementation of drain boundaries in PHAST, a drain is a piecewise linear feature similar to a river. A series of points represents the drain route. At least two points must be defined. The end points of the drain must have data defining the drain elevation, bed width, bed thickness, and bed hydraulic conductivity. At intermediate drain points, the drain properties may be defined explicitly or may be interpolated from the nearest points upstream and downstream where a property has been defined.

Several steps are involved in translating the data from the sequence of drain points into boundary conditions for the affected cells of the active grid region. By assuming that drains are nearly horizontal features, the calculation of drain segments—areas of drain leakage that apply to a single cell—are identical to the calculation of river segments (see section D.5.5. River Boundary). However, the two boundary conditions differ in the application of the boundary condition to a vertical stack of cells. A river segment always is connected to the cell in a vertical column of cells that contains the water table, which may vary during a simulation. A drain segment is connected to only one cell in a vertical column as determined by the interpolated elevation for the drain segment. Water discharging to a drain segment has the ambient chemical composition of the water in the cell connected to the drain segment.

A drain boundary condition and a flux boundary condition can be applied to the same cell. This feature is needed to represent drain leakage from a cell that is receiving precipitation recharge. A leaky boundary condition should not be applied to the Z face of a cell containing a drain because the drain will tend to hold the water table at the elevation of the drain. For compatibility between drain boundaries and other boundary conditions, see section D.5.8. Boundary-Condition Compatibility.

D.5.7. Free-Surface Boundary

For an unconfined aquifer, a free-surface (water-table) boundary exists with a position in space and time that is determined as part of the simulation. The boundary condition requires that the absolute pressure on the free surface be atmospheric, which implies that the relative pressure is zero and the potentiometric head equals the free-surface elevation. Fluid and porous-matrix compressibilities are assumed to be zero for unconfined flow systems, so the specific yield is equivalent to the effective porosity.

The simulation region for saturated, unconfined, groundwater flow varies with time as the free-surface moves. The active grid region is fixed and determines the maximum extent of the simulation region. Flow, transport, and reactions are computed at any given time only for those cells of the active grid region that are fully or partly saturated. These fully and partly saturated cells form the discrete approximation to the current simulation region. Boundary potentiometric heads less than the elevation at the point of measurement (pressures less than atmospheric) imply that the free surface is below the node, whereas boundary heads greater than the elevation (pressures greater than atmospheric) imply that the free surface is above the node.

The quantities of components attached to solid phases—that is, the amounts in minerals, exchangers, surface complexes, and solid solutions—are left behind as the free surface falls below the bottom of a cell; these quantities are then outside the simulation region. Resaturation of the porous medium must occur for these quantities of solid-phase components to be reincluded in the simulation region. In other words, components attached to the porous matrix in a cell that becomes dry remain unreactive until the cell is resaturated.

For saturated media, the amounts of reactive solids are defined to be those that are contained in a volume of aquifer that contains one liter of water. Similarly, for partially saturated media, the amounts of reactive solids are defined to be those that are contained in a volume of aquifer that contains one liter of pore space. Within a cell that contains the free surface, two reservoirs of solid reactants are created: the saturated and unsaturated reservoirs. As the fraction of saturation changes as a result of the flow calculation, moles of solid reactants are transferred from one reservoir to the other. A decrease in fraction of saturation causes moles of solid reactants to be transferred from the saturated reservoir to the unsaturated reservoir. Conversely, an increase in fraction of saturation causes moles of solid reactants to be transferred from the unsaturated reservoir to the saturated reservoir. The number of moles transferred is calculated from the previous fraction of saturation and the current fraction of saturation. When the cell is fully saturated, all solid reactants are in the saturated reservoir; when the cell is dry, all solid reactants are in the unsaturated reservoir.

For chemical reactions, a volume of water (in liters) equal to the saturated fraction is allowed to react with the contents of the saturated reservoir of solid reactants. The composition of the water before reaction is equal to the water composition following the transport calculations. The amounts of solid reactants in the saturated reservoir before reaction are determined by the amounts in the saturated reservoir following the previous chemical-reaction step adjusted by transfers to or from the unsaturated reservoir to account for changes in fraction of saturation. New amounts of solid reactants are stored in the saturated reservoir after completion of the chemical-reaction calculation.

The amounts of solid reactants in the unsaturated reservoir do not participate in chemical reactions. They are only stored in this reservoir until such time as they are transferred to the saturated reservoir. Both reservoirs are assumed to be well mixed; that is, there is no accounting for individual strata of solid reactants as the water table moves up and down within the cell. A major shortcoming of the unsaturated-zone accounting is that water infiltrating to the water table does not contact the solid reactants in the unsaturated reservoirs. The unsaturated reservoir can be reintroduced to the active simulation only by a rising water table. In PHAST

Version 2, it is not possible to print or view the amounts of solid reactants and chemical components in the unsaturated reservoirs.

To allow for the definition of an active grid region with a regional topography, an inactive grid region may be defined above the active cells. That is, the active grid region of an unconfined flow simulation does not have to possess a flat upper surface.

The numerical implementation of the unconfined-aquifer, free-surface boundary condition is done by modifying the pressure-coefficient terms in the discretized equations for flow and solute-component transport and adjusting the fluid volume (saturated thickness) of the cells containing the water table (Kipp, 1997). PHAST uses a simpler algorithm than HST3D for locating the free surface. The location is established by linear extrapolation of the nodal pressure from the saturated cell below to calculate the elevation of zero (atmospheric) pressure. The algorithm is based on an assumed vertical hydrostatic pressure distribution and does not account for a flux through the bottom and sides of each cell containing the free surface. This simplification can cause the free surface to approach the incorrect elevation for a steady state, especially if strong vertical groundwater flow is present. Approaching a steady state by draining or wetting will give different results for the free-surface elevation. For example, this algorithm gives the correct elevation for a simple one-dimensional flow problem by draining to steady state, but not by filling to steady state.

The free surface can be located at any elevation within the active grid region and is allowed to rise above the upper boundary of the grid region. This rise is physically realistic only if additional porous medium exists above the upper boundary of the grid. However, large rises above the top of the grid indicate the need for the grid region to be extended vertically to enclose the entire aquifer saturated thickness. In the case of an inactive grid region above the top of the active grid region, no conversion to confined flow conditions is made if the free surface does rise above the uppermost active cells. (In the case of confined flow, no conversion to unconfined flow conditions is made if the potentiometric head drops below the elevation of the confining boundary.) The grid region can be defined only as entirely unconfined or entirely confined. Mixed confined and unconfined conditions over different subareas of the land surface cannot be represented.

The free surface may fall to any cell below the top of the active grid region. Dry cells are isolated from the groundwater flow region by assigning zero to the fluid and solute-dispersive conductances for their lateral faces. A dry cell is resaturated by flux through its bottom face. Thus, if an entire column of cells becomes totally dewatered, there is no mechanism to resaturate it. A cell that becomes dry is removed from the computational region at the end of the time step. Also, a cell that becomes resaturated is added back into the computational region at the end of the time step. The saturation of a cell that goes dry during a time step is constrained to be a small value for the remainder of that time step. Because the free-surface position is treated explicitly, it is adjusted only at the end of each time step. The conductances and source terms in the flow and transport equations have the saturated fraction parameter included, as necessary, for the cell-face area terms involving horizontal flow. The appropriate factors are given by Kipp (1997).

The case of a free-surface boundary with accretion of fluid by infiltration (precipitation recharge) is handled in an approximate fashion. The recharge fluid flux is specified at the upper boundary of the active grid region. That flux is applied to the cells that contain the free surface at any given time. The associated-solution compositions defined for the flux boundary condition combined with the fluid flux determine the amount of each solute component that enters the saturated region through the free surface. The partially saturated flow from the land surface down to the water table is not simulated in PHAST; water moves directly to the cells that contain the free surface.

Care must be taken when specifying a free-surface boundary with precipitation recharge (Z flux) for cells that overlie specified-head boundary conditions. If a specified-head boundary condition causes the free surface to drop to the specified-head cell, the flux boundary condition is no longer applied. The aggregated precipitation flux will then vary depending on the quantity of flux intercepted by specified-head cells. A more realistic specification might be to replace the specified-head boundary condition with a leaky boundary condition. A large conductance factor would cause the head in the boundary cell to approximate a specified-head boundary condition and yet also allow for precipitation recharge to the cell.

D.5.8. Boundary-Condition Compatibility

PHAST has a variety of boundary conditions that can be defined, and care needs to be taken to maintain compatibility for the boundary conditions applied to a cell. Whereas PHAST Version 1 allowed at most one leaky and one flux boundary condition on entire, separate faces of a cell, PHAST Version 2 allows multiple flux, leaky, river, and drain boundary conditions to be applied to a cell face by areal segments, which may cover all or parts of cell faces. Multiple flux and leaky-boundary-condition segments may be applied to as many as three faces of a cell (the most exterior faces that a cell may have), multiple river segments may apply to the positive Z face of a cell, and, in addition, multiple drain segments and multiple well sources or sinks may apply to a cell. Allowed combinations of boundary-condition types are summarized in table D.1.

By the use of multiple boundary segments, many combinations of boundary conditions can be applied to a single cell, but automatic checking of boundary-condition compatibility is limited to a few cases. The input processing program ensures that if a cell is a specified-head boundary-condition cell, all other flux, leak, river, drain, and well boundary conditions for that cell are ignored. River and drain boundary conditions are allowed only for unconfined flow (free-surface boundary condition).

When simulating unconfined flow, a leaky boundary on an upper Z face is handled as a special case and is equivalent to a river. An elevation of the top of the leaky bed must be defined and the flow to the aquifer is limited to the flow that occurs when the head in the aquifer is at the elevation of the bottom of the leaky layer. Leakage does not increase as the head in the aquifer drops below the elevation of the bottom of the leaky layer. For unconfined flow and leakage through an upper Z face, the leakage for a vertical stack of cells is applied to the cell that contains the water table.

Flux boundary-condition segments on a single face are not allowed to overlap; the areas of application for the segments are adjusted so that overlapping areas are removed and no two segments apply over the same area. Similarly, leaky, river, and drain segments are adjusted so that no segments of a given type overlap. A specified-head boundary condition may have either a specified-solution-composition boundary condition or an associated-solution boundary condition. The associated-solution boundary condition is the only type of solution boundary condition available for flux, leaky, river, and well boundary conditions. Any combination of boundary conditions, not explicitly excluded, is allowed (table D.1).

Even though many combinations of leaky, flux, river, drain, and well boundary conditions are allowed for a cell, some combinations may be hydrologically unreasonable. Multiple boundary conditions for a single cell may cause unanticipated local flows. For example, having both a river and a drain in the same cell may induce large intracell flows. As previously described, segments of a single boundary-condition type are adjusted to remove overlaps; however, overlaps between segments of differing boundary-condition types are allowed. Overlaps of this kind, for example, applying flux and leaky conditions to the same area of a cell face, usually do not make hydrologic sense. However, it may be convenient to allow small features to overlap large boundary-condition features. For simplicity, river and flux conditions often are defined to be overlapping because it is difficult to remove the river areas from the area of application of precipitation flux. The error caused by the overlap is expected to be small when the river is a relatively small area and precipitation flux is applied over the entire surface of the model domain.

Table D.1. Compatibility among boundary conditions and wells for a cell.

[Yes, this combination of boundary conditions can be defined for a cell; No, this combination of boundary conditions cannot be defined for a cell; Multiple, multiple boundary conditions of this type can be defined for a cell in addition to the boundary condition specified in the first column; Required, the additional boundary condition is required; —, not applicable]

Cell boundary condition	Additional boundary condition							
	Specified-solution composition	Associated-solution composition	Flux	Leaky	River[1]	Drain	Well	Free surface
Specified head	Yes	Yes	No	No	No	No	No	Yes
Flux	No	Yes	Multiple per face	Multiple per face[2]	Multiple	Multiple	Multiple	Yes
Leaky	No	Yes	Multiple per face[2]	Multiple per face	Multiple	Multiple	Multiple	Yes
River	No	Yes	Multiple per face	Multiple per face	Multiple	Multiple	Multiple	Required
Drain	No	No	Multiple per face	Multiple per face	Multiple	Multiple	Multiple	Required
Well	No	Yes	Multiple per face	Multiple per face	Multiple	Multiple	Multiple	Yes
Free surface	—	—	Multiple per face	Multiple per face	Multiple	Multiple	Multiple	Yes

[1]River leakage is always through the upper Z-face of a cell.

[2] Leaky and flux boundary conditions should not be applied over the same area of the same face.

D.6. Initial Conditions

The PHAST simulator solves only the transient forms of the groundwater flow and the component-transport equations; thus, initial conditions are necessary to begin a simulation. Initial potentiometric-head and total-aqueous-concentration distributions are required. Several options are available for specification of these initial conditions.

For the flow equation, the simplest initial condition is a hydrostatic potentiometric-head condition, with a single value specified (**HEAD_IC** data block, **-head** identifier). The most general initial conditions are zone-by-zone or node-by-node potentiometric head specification. A water-table head distribution specified over a horizontal plane of the simulation region (**HEAD_IC** data block, **-water_table** identifier) should not be used because it is grid dependent (node-by-node); use heads defined for a plane of X–Y–Z points instead.

The simulator has an option to calculate a steady-flow condition by time stepping in flow-only mode until heads are unchanging within a tolerance and flow balance is zero within a tolerance (**STEADY_FLOW** data block). If this option is used in a reactive-transport simulation, the head distribution and velocity field determined by the steady-flow calculation is used unchanged for the duration of the transport simulation. Although an initial condition for the head distribution is required for the steady-flow calculation, the final steady-flow heads do not depend on the initial condition. There is no option to specify a velocity field as an initial condition.

For the component-transport equations, the initial total aqueous concentration fields need to be specified. In PHAST, initial-solution concentrations are specified for zones by using solution index numbers (CHEMISTRY_IC data block, -solution identifier). A uniform solution composition can be defined for a zone by using a single solution index number for a zone, or a solution composition that varies linearly in a coordinate direction can be defined. For the latter, solution index numbers are specified at two locations along a coordinate axis, and solution composition is linearly interpolated from the end-member compositions to each node point between the two locations. Points outside the range of the two end-point locations are assigned the solution composition of the nearest end point. It is also possible to define solution indices or mixing fractions for two solutions at scattered X–Y–Z points, which are used for closest point interpolation. Geochemical calculations that include aqueous complexation, mineral equilibria, ion exchange, surface complexation, solid-solution equilibria, gas-phase equilibration, and general kinetic reactions, plus mixing, irreversible reactions, and temperature variation (see subsequent discussion of temperature in this section) may be defined in the chemistry data file; the results of these calculations can be used to define initial conditions for the solutions in the simulation region.

Initial conditions are specified for the type and amount of solid-phase reactants that are present in each cell of the active grid region (CHEMISTRY_IC data block). The reactants are specified for zones in the grid region by using index numbers to identify reactants that are defined in the chemistry data file. The types of reactants include ion exchange (-exchange), surface complexation (-surface), sets of phases that react to equilibrium (-equilibrium_phases), kinetic reactants (-kinetics), solid solutions (-solid_solution), and gas phases (-gas_phase). As with initial solution compositions, it is possible to use linear interpolation or closest point interpolation for an index number or a mixing fraction of two compositions at scattered X–Y–Z points to distribute reactants for cells within a zone; interpolation can be done with any type of reactant (equilibrium phases, exchange, surface, kinetics, solid solutions, or gas phase).

The units for input in the **SOLUTION** data block are concentration units; internally, all chemical calculations use molality for the concentration unit of solution. The appropriate unit for input in the data blocks **EQUILIBRIUM_PHASES, EXCHANGE, SURFACE, KINETICS, SOLID_SOLUTIONS** is moles of reactant per liter of pore space. For each cell in PHAST, the representative porous-medium volume for chemistry contains one kilogram of water, when saturated. Thus, when defining amounts of solid-phase reactants, the appropriate number of moles is numerically equal to the concentration of the reactant (moles per liter of water), assuming a saturated porous medium. In terms of moles per liter of water, the concentrations of the solid reactants vary spatially as porosity varies, which makes it difficult to define the appropriate solid reactant concentrations. To avoid this difficulty, PHAST Version 2 has options in the UNITS data block (flow and transport data file) to specify that the number of moles of solid reactants be interpreted as moles per liter of rock. As initial conditions are distributed to the finite-difference cells, the moles of solid reactants are scaled by the factor $(1-\phi)/\phi$, where ϕ is the porosity for the cell. This scaling takes into account the varying porosity and produces units of moles per liter of water. Molality (mol/kgw) is assumed to equal molarity (mol/L) for all transport calculations.

The chemical calculations of a reactive-transport simulation occur at a specified temperature for each cell. In most cases, the temperature is specified with the -temperature identifier in **SOLUTION** data blocks in the chemistry data file, but the temperature of a solution also may be defined with REACTION_TEMPERATURE data block. The solutions defined in the chemistry data file are used to define the initial conditions for reactive-transport simulations. Thus, for the chemical calculations of the reactive-transport simulation, a temperature distribution throughout the active grid region is defined by the temperatures of the solutions used for initial conditions (CHEMISTRY_IC, -solution identifier). This temperature distribution for chemical calculations remains in effect for the duration of the reactive-transport simulation. This temperature distribution has no effect on the flow or transport simulations, which are limited to a constant-density, and implicitly constant-temperature, fluid.

D.7. Method of Solution

The groundwater flow equation, the component transport equations, the equilibrium reaction equations, and the kinetic reaction equations form a coupled set of partial differential, ordinary differential, and algebraic equations. For groundwater of uniform density, the flow and transport coupling is through the interstitial velocity terms. The component transport equations are coupled to the equilibrium and kinetic reaction equations through chemical source terms, as shown in equation D.4. The chemical source terms are nonlinear functions of the chemical compositions of the aqueous and solid phases.

For groundwater of uniform density, the coupling between the flow equation and the transport equations is only in one direction, from the flow equation to the transport equations. The potentiometric head (pressure) solution to the flow equation yields the interstitial velocity field, which goes into the component transport equations through the advective transport terms and the dispersion coefficients. This one-way coupling allows the flow equation to be solved independently from all other equations.

Several methods in the literature have been used to solve the transport and reaction equations as summarized by Steefel and MacQuarrie (1996). In PHAST, the calculations for reactive transport are split into a transport calculation step and a chemical-reaction calculation step. This method is referred to as operator splitting with sequential solution (Press and others, 1989). The transport equations (and flow equation) require a sparse-matrix linear equation solver for the finite-difference equations. The chemical-reaction equations require a combination of numerical methods to solve the nonlinear algebraic equations for equilibrium and the ordinary differential equations for kinetics.

D.7.1. Operator Splitting and Sequential Solution

The three sets of equations that need to be solved simultaneously are the transport equations, the equilibrium reaction equations, and the kinetic reaction equations. The flow equation can be solved separately from the transport and reaction equations. The finite-difference equation for flow has the following form:

$$C_f \frac{h^{(n+1)} - h^{(n)}}{\Delta t} - \theta L_f(h^{(n+1)}) = (1-\theta)L_f(h^{(n)}), \tag{D.28}$$

where C_f is the flow storage factor (m^2), θ is the weighting factor for time differencing, and L_f is the spatial discretization of $\nabla \bullet K\nabla$ (m^{-1} s^{-1}).

In PHAST, operator splitting is used to separate the reactive-transport equations into transport equations and chemical-reaction equations. Each reactive transport equation is split into two equations: a solute-transport equation

$$\frac{\partial}{\partial t}(\varepsilon \rho c_j) = \nabla \bullet \varepsilon D \nabla \rho c_j - \nabla \bullet \varepsilon v \rho c_j + q\varepsilon \rho \hat{c}_j \, ; \quad j = 1, ..., N_c, \text{ and} \tag{D.29}$$

a reaction equation for heterogeneous equilibrium and kinetic equations

$$\frac{\partial}{\partial t}(\varepsilon \rho c_j) = -\sum_{e=1}^{N_E} \upsilon_{j,e}^E \frac{\partial}{\partial t}(\varepsilon \rho \bar{c}_e) + \sum_{k=1}^{N_K} \upsilon_{j,k}^K \varepsilon \rho R_k; \quad j = 1, \ldots, N_c. \tag{D.30}$$

Note that no reaction terms appear in the transport equation. Discretization of equation D.29 and conversion to units of mass fraction gives the following finite-difference equation for each component:

$$C_s \frac{w_j^{(*)} - w_j^{(n)}}{\Delta t} - \theta L_s(w_j^{(*)}) = (1 - \theta)L_s(w_j^{(n)}) + Q\hat{w}_j; \quad j = 1, \ldots, N_c, \tag{D.31}$$

where C_s is the component storage factor (m^3); w_j is the mass fraction of component j (unitless); L_s is the spatial discretization of $\rho \nabla \bullet D\nabla - \rho \nabla \bullet v$ ($kg\ m^{-3}\ s^{-1}$); $w_j^{(*)}$ is the intermediate mass fraction of component j (unitless); Q is the source flow rate for the cell (m^3/s); and $\hat{w}_j$ is the mass fraction of component j in the source water (unitless). Discretization of equation D.30 yields

$$c_j^{(n+1)} - c_j^{(*)} = -\sum_{e=1}^{N_E} \upsilon_{j,e}^E (\bar{c}_e^{(n+1)} - \bar{c}_e^{(n)}) + \sum_{k=1}^{N_K} \upsilon_{j,k}^K \int_{t^{(n)}}^{t^{(n+1)}} R_k dt; \quad j = 1, \ldots, N_c. \tag{D.32}$$

In PHAST, the set of equations D.28, D.31, and D.32 plus the algebraic equations for chemical equilibria and rate expressions are solved sequentially in three steps. First, the flow equation D.28 is solved for the potentiometric head (pressure) values. The heads are used to calculate the interstitial velocity values that are used in the component transport equations. Second, each transport equation (equation D.31) is solved individually for the total aqueous concentrations of the component. And third, the equilibrium and kinetic reaction equations (equation D.32 plus the algebraic equations for chemical equilibria and rate expressions) are solved. This completes the simulation of one time step. The process is repeated for each time step for the duration of simulation.

In PHAST, no iterations are performed between the second and third steps of the sequential solution. This combination of operator splitting without iteration is known as the sequential, noniterative approach (SNIA) as described by Yeh and Tripathi (1989). Some reactive-transport algorithms include reaction terms in the transport equations and iterate between the second and third steps in the sequential solution (Steefel and MacQuarrie, 1996), known as the sequential iterative approach (SIA).

SNIA uses much less memory than a fully coupled, simultaneous solution of all the transport and reaction equations. SNIA also is faster per time step than the fully coupled approach because the sizes of the matrices that are solved are much smaller. (It is possible that the fully coupled approach may be faster if fewer time steps are needed by that approach.) However, SNIA introduces an operator-splitting error of the order of the time-step length. This error can be reduced by iteration (SIA), Strang splitting (Steefel and MacQuarrie, 1996), or by use of smaller time steps. Only the last option is available in PHAST. In many cases, it is more critical to minimize the discretization error of the finite-difference approximations to the transport equations than to minimize the operator-splitting error.

Analysis of one-dimensional transport with constant coefficients for the special case of one linear, equilibrium sorption reaction, yields an effective dispersion coefficient from operator splitting given by

$$D_{nos} = \frac{v_x^2 \Delta t}{2R}(2\theta R - 1), \text{ with} \qquad (D.33)$$

$$R = 1 + \frac{\rho_b K_e}{\varepsilon}, \qquad (D.34)$$

where D_{nos} is the numerical dispersion coefficient from operator splitting (m²/s); R is the retardation factor for linear sorption (unitless); ρ_b is the dry bulk density of the porous medium (kg/m³); and K_e is the linear equilibrium sorption coefficient (m³/kg). The retardation factor also expresses the ratio of the effective transport velocity of the sorbed component relative to the interstitial velocity of the groundwater.

For numerical accuracy, the numerical dispersion from operator splitting should be much less than the physical dispersion. Thus, for centered-in-time differencing,

$$CrPe \ll \frac{2R}{R-1}, \text{ which reduces to} \qquad (D.35)$$

$$CrPe \ll 2, \qquad (D.36)$$

for large value of R. For backward-in-time differencing,

$$CrPe \ll \frac{2R}{2R-1}, \text{ which reduces to} \qquad (D.37)$$

$$CrPe \ll 1, \qquad (D.38)$$

for large value of R. Although these equations are from a restricted reaction case, they can give guidance for discretization of more complex reaction systems.

In SNIA, the reaction terms have been completely removed from the transport equations. Removing these terms may cause overshoot and undershoot in the intermediate concentrations in the transport solution, especially in cases of strong sorption or strong partitioning of components. The undershoot concentrations can be negative, which will result in mass-balance errors when those concentrations are set to zero in the chemical-reaction calculation. Overshoot concentrations may cause nonconvergence in the solution of the algebraic chemical equations. The remedy available in PHAST for these overshoot and undershoot problems is to use a smaller time step.

D.7.2. Linear-Equation Solvers for Flow and Transport Finite-Difference Equations

The sets of linear equations for flow or component transport include one finite-difference equation for each node in the grid region. Each set of finite-difference equations is solved in three steps: (1) the equations are row scaled using the L-infinity norm (maximum absolute value) of each row, (2) a reduced matrix is formed by renumbering the nodes and performing a partial block Gaussian elimination, and (3) the scaled, reduced matrix equations are solved either by a direct or an iterative linear-equation solver. Using a reduced matrix results in a savings in computation time and computer-storage requirements.

When using the direct solver, the reduced matrix is formed by renumbering the nodes by alternating diagonal planes (Price and Coats, 1974), followed by the partial Gaussian elimination. This reduced matrix is solved by complete Gaussian elimination. This method is referred to as the D4 solution technique described in Kipp (1987). The workload and storage requirements for the direct solver are completely defined by the number of nodes and configuration of the active grid region.

When using the iterative solver, the reduced matrix is formed by renumbering the nodes by alternating diagonal planes in a zig-zag fashion, followed by partial Gaussian elimination. This reduced matrix is solved by a generalized conjugate gradient algorithm. This method is referred to as the D4 zig-zag solution technique described in Kipp and others (1992, 1994) and Kipp (1997). The workload and storage requirements for the iterative solver depend not only on the number of nodes but also on user-specified solver parameters.

The iterative algorithm is a restarted ORTHOMIN method with incomplete lower-upper (ILU) triangular factorization preconditioning as described by Meijerink and van der Vorst (1977), Elman (1982), and Behie and Forsyth (1984). The ORTHOMIN method calculates an approximate solution vector that is based upon a set of orthogonal search directions. For the full ORTHOMIN method, the number of search directions is equal to the iteration count, which leads to workload and storage requirements that increase linearly with iteration count. To limit the workload and storage requirements, a modified algorithm, ORTHOMIN(m), is implemented by simply restarting ORTHOMIN every m iterations. To restart the ORTHOMIN(m) algorithm, the latest solution vector is used as the initial guess for the next cycle. The restart interval m is specified by the user (**SOLUTION_METHOD** data block, **-save_directions** identifier). Increasing this interval may accelerate convergence but also will increase the computer memory requirements.

The convergence criterion of the iterative solver requires that the Euclidean norm of the residual vector is small relative to the norms of the linear-equation matrix, the right-hand-side vector, and the initial solution vector (Barrett and others, 1994, p. 54). Mathematically, this is expressed as

$$\left\| \boldsymbol{r}^{(\nu)} \right\| \le \tau (\| \boldsymbol{A} \| \cdot \left\| \boldsymbol{x}^{(\nu)} \right\| + \| \boldsymbol{b} \|), \tag{D.39}$$

where $\boldsymbol{r}^{(\nu)}$ is the residual vector at iteration ν (appropriate units); τ is the convergence tolerance (unitless); $\boldsymbol{A}$ is the matrix of coefficients of the difference equations (appropriate units); $\boldsymbol{b}$ is the right hand side vector of the difference equations (appropriate units); $\boldsymbol{x}^{(\nu)}$ is the solution vector at iteration ν (appropriate units); $\| \ \|$ is the Euclidean norm for a vector or the Frobenius norm for a matrix (Stoer and Bulirsch, 1993, p. 184).

The criterion of equation D.39 yields the forward error bound of

$$\left\| \boldsymbol{e}^{(\nu)} \right\| \le \tau \left\| \boldsymbol{A}^{-1} \right\| \cdot (\| \boldsymbol{A} \| \cdot \left\| \boldsymbol{x}^{(\nu)} \right\| + \| \boldsymbol{b} \|), \tag{D.40}$$

where $\boldsymbol{e}^{(\nu)}$ is the error in the solution vector at iteration ν (appropriate units) and $\boldsymbol{A}^{-1}$ is the inverse of the coefficient matrix of the difference equations (appropriate units). Thus, the norm of the error in the solution vector is related to the norm of the residual vector under this criterion. Unfortunately, it is usually not feasible to compute the norm of the inverse of the coefficient matrix; therefore, the convergence tolerance must be determined empirically. The convergence tolerance τ is specified by the user (**SOLUTION_METHOD** data block, **-tolerance** identifier). Once the convergence tolerance τ has been specified, it is used in equation D.39 to define the criterion for convergence. Row scaling equalizes the maximum magnitude of the terms in the linear equations so that a single convergence tolerance will apply to the flow and component transport equations.

D.7.3. Solving Equilibrium and Kinetic Chemical Equations

Chemical-reaction equations are solved independently from the flow and transport equations. In addition, chemical-reaction equations are solved independently for each node in the active grid region. In the case of only equilibrium reactions, a Newton-Raphson method is used to solve the nonlinear mass-action equations and mass-balance equations that describe equilibrium. See Parkhurst and Appelo (1999) for details of the Newton-Raphson algorithm that is used in PHREEQC and PHAST.

If kinetic reactions are simulated then a set of ordinary differential equations must be integrated over the time step, in addition to solving the equilibrium equations. Two methods are available for integrating the rate equations: an explicit 5th-order Runge-Kutta algorithm (Parkhurst and Appelo, 1999) and an implicit algorithm for stiff differential equations based on Gear's method (Cohen and Hindmarsh, 1996). These methods take the time step defined for the flow and transport equations and divide it into sub time steps. The size of these sub time steps is determined automatically to satisfy a user-specified tolerance on the estimated errors of integration. At each sub time step, the rates of reaction are calculated and the kinetic mole transfers resulting from kinetic reactions for the sub time step are calculated. The kinetic mole transfers are added (or removed) from solution, after which all of the equilibrium equations (homogeneous and heterogeneous) are solved simultaneously. Additional sub time steps are simulated until the integration has completed the entire time interval specified by the time step of the flow and transport simulation.

D.8. Accuracy from Spatial and Temporal Discretization

The finite-difference equations that approximate the flow and transport equations (equation D.28 and D.31) are unconditionally stable when using either backwards-in-time and upstream-in-space or centered-in-time and centered-in-space difference approximations; numerical errors will not grow unbounded in time or space. The accuracy of the finite-difference equations depends on the choice of cell size and time-step length. Avoiding numerical oscillation is a prime consideration when using centered-in-time and centered-in-space differencing, while avoiding excess numerical dispersion is necessary for accuracy when using backward-in-time and upstream-in-space differencing. Guidelines are given in sections D.1.2. Spatial Discretization and D.1.3. Temporal Discretization for limiting cell size and time-step length to meet these objectives. The operator-splitting algorithm introduces its own temporal truncation error, but no guidelines for a general reaction system are available. Guidelines from the restricted case of linear equilibrium sorption (section D.7.1. Operator Splitting and Sequential Solution) indicate the effect of a retarded solute-transport rate on time-step selection for accuracy. Results for this case show that the effective numerical dispersion from operator splitting is always greater than or equal to the numerical dispersion from backward-in-time differencing. Note that avoiding numerical oscillations does not ensure an accurate numerical solution. The user must test for a sufficiently accurate solution by reducing the cell sizes and time step until no significant changes in the solution occur. For field-scale simulations, it may not be practical to refine the grid and time step sufficiently to avoid all numerical artifacts (oscillations or dispersion).

A free-surface boundary condition introduces nonlinearity into the finite-difference equations. Because the free-surface coefficients are calculated explicitly in time, there is a stability limit on the time-step length that can be determined only empirically. The critical regions are where the rate of movement of the free surface is large.

The river boundary condition also introduces nonlinearity into the finite-difference equations from the equation of flux as a function of head difference between the river and the aquifer. Similarly, the drain boundary condition will introduce nonlinearity into the equations for flux. Although these fluxes are computed semi-implicitly, large rates of change of head in the river or the aquifer may require small time steps for stability. Note that a stable solution is not necessarily an accurate solution.

D.9. Global Mass-Balance Calculations

The discretized groundwater flow equation and the transport equations for components represent fluid and solute mass balances over each cell. Summing over the cells and integrating over time yield global-balance equations that relate the total change of mass of groundwater and components to the net boundary flow, the net injection by wells, and the net formation by reactions. The temporal integration is done over each time step, and the cumulative balance masses are simply the sums of the corresponding masses from each time step. The fluid or component global-balance residual is defined as the imbalance between the change in the mass of fluid or solute in the simulation region and the net inflow of fluid or solute plus the net source of solute from reactions. A positive residual means that there is an excess of fluid or components present over what would be expected on the basis of transport across boundaries and sources from reactions. A fractional residual is defined as the ratio of the residual to the inflow, outflow, or magnitude of accumulation, whichever is larger.

Global balance information is calculated for each time step and for the cumulative duration of the simulation. The information produced by the balance calculation includes change in mass in the system, the mass flows of inflow and outflow, and the mass change due to reactions. Change in mass due to net inflow (inflow minus outflow) is aggregated by each type of boundary condition. Numerical integration of the boundary flow rates over a time step and over the cumulative time is done with an algorithm that depends on the weighting factor used for temporal discretization of the governing flow and component-transport equations. Thus, for centered-in-time differencing, the trapezoidal algorithm is used, and for backward-in-time differencing, the rectangular algorithm is used with integrand evaluation at the end of the time step. Having the integration algorithm depend on the temporal weighting factor makes the global mass-balance calculations consistent with the formulation of the finite-difference equations and thus gives an exact mass-balance for each cell over a time step.

The fluid-flow at specified-head boundary cells is obtained by evaluating the residual of the finite-difference flow equation. The component mass flows at specified-concentration boundary cells are obtained by evaluating the residual of each transport equation. The residual is defined as the rate of change in groundwater storage minus the net inflow rate through the interior cell faces minus any source flow rate terms. Any residual flow rate is deemed to occur through the exterior cell faces—that is, those cell faces that form the boundary of the simulation region. These residuals are the flow rates of fluid and components that are necessary to satisfy the fluid-balance and solute-balance equations. Thus, the fluid mass-balance equation is satisfied exactly for each of the specified-head boundary cells, and the component mass-balance equation is satisfied exactly for each of the specified-concentration boundary cells. To make the global-balance calculations for the components compatible with the operator-splitting algorithm, the exterior boundary mass flows are based on the intermediate concentrations resulting from equation D.31. The change in component mass from reactions is taken to be the difference between the mass of the component before and after the reaction step.

The information produced by the global-balance calculation is printed to the file *prefix*.**bal.txt**. The primary use of the global-balance calculations is to identify changes in mass in the system relative to reactions and mass flows (kg) through the various types of boundary conditions. Mass-accumulation and mass-flow rates (kg per time unit) also are tabulated. The utility of the fractional residuals is not great. It is more informative to compare the residuals with each of the various flows, sources, and accumulations in the region.

Flow and component mass balances with small residuals are a necessary, but not a sufficient, condition for an accurate numerical simulation. Because the system equations are a balance for each cell, and the methods used for integrating boundary flows and reaction sources over a time step are compatible with the finite-difference equations and the operator-splitting method, and because the fluxes between the cells are conservative, errors in the global-balance equations may result from the following remaining causes: (1) explicit-in-time treatment of the water-table elevation for unconfined flow, (2) explicit-in-time treatment of well flow rates, (3) explicit-in-time treatment of the nonlinear river and drain leakages, (4) explicit-in-time treatment or neglect of the cross-dispersive flux terms, (5) use of the iterative-matrix equation solver to obtain an approximate numerical solution, and (6) roundoff error in special cases, such as wide variation in parameter magnitudes. Explicit-in-time treatment means that the terms are evaluated at the beginning of a time step using only known values of the dependent variables and parameters. The first three items are usually the major cause of mass-balance errors. Errors caused by discretization in time or space will not be revealed by these global-balance calculations. However, the inaccuracies resulting from a time step that is too long under conditions of significant nonlinearity will be evident. Significant nonlinearity can be caused by large rates of change in the water-table location or by nonlinear reaction mechanisms.

When using weighting other than upstream-in-space and backward-in-time, numerical oscillations may occur, which may lead to erroneous balances, even though the simulation results appear to be in balance. Oscillation may result in negative concentrations, which are balanced by excess positive concentrations. However, the negative concentrations are set to zero when data are sent to the chemistry calculation, while the excess positive concentrations are preserved. It then appears that chemical reactions have increased concentrations. For a conservative component, the increase from reactions should always be zero, and any nonzero accumulations from reactions must be attributed to numerical errors. For reactive components, it may be hard to separate the artificial increase due to oscillations from a true chemical reaction increase, but the presence of errors in a conservative component implies that there also may be errors in reactive components.

D.10. Zone Flow-Rate Calculation

Provision has been made to define subregions of the active grid for the purpose of calculating the incoming and outgoing flow rates of water and chemical components. These subregions are denoted as zones and are defined according to the options described in section 4.5. Description of Input for Zones. For each zone, the inflow rates and outflow rates are calculated for boundary cells of the zone that are adjacent to active cells in the grid region; inflow rates and outflow rates for each type of boundary condition found within the zone also are calculated. For unconfined flow systems, the precipitation flux and the river leakage boundary flow rates for a stack of cells are included in the zone flow-rate calculation if the water table for the stack is within the defined zone.

The primary use of the zone flow-rate calculation is to aid in the interpretation of the magnitudes of transport that are occurring with respect to particular subregions (zones) of the grid region. Also of interest is the distribution of transport rates among the various types of boundary conditions and sources associated with a given zone.

In the zone flow-rate calculations, no accounting is made for the appearance or disappearance of a component due to chemical reactions. Also, the zone flow-rate calculations do not consider the accumulation of water or component within the zone. Thus, the flow-rate tables (files *prefix*.**zf.txt** and *prefix*.**zf.tsv**) for water and reactive components are useful only for observing total inflow and outflow rates through the various boundaries.

D.11. Nodal Velocity Calculation

A secondary dependent variable is the interstitial velocity field, which is obtained from the gradient of the potentiometric head field. The numerical implementation of PHAST uses finite differences of the head values, which give velocity values at the cell boundaries for all cell faces internal to the boundary of the active grid region. For all internal cells, interstitial velocities are calculated at the six cell faces. Vector components of interstitial velocity at internal nodes of the active grid region are interpolated from velocity values at the six cell faces. The velocity components can be written to the files *prefix*.**vel.txt**, *prefix*.**vel.xyz.tsv**, and *prefix*.**h5**.

For cells with one face at the boundary of the active grid region, the velocity-vector component normal to that face is calculated from the boundary flow rate. These flow rates have been calculated previously for each type of boundary condition. Because all boundaries of the active grid region are at planes of nodes, the velocity component normal to the face needs no interpolation.

Only the net boundary flow rate for all boundary faces of a cell are known for a cell with a specified-head boundary condition. For specified-head cells with two or three boundary faces, the net boundary flow rate is apportioned among the faces in proportion to the magnitude of the flow rate through the cell face opposite to the given boundary face.

For nodes along inside edges of the boundary of the active grid region with a no-flux boundary condition, the nodal velocity is taken to be zero. In actuality, the velocity components have discontinuous jumps from zero to finite values at the inside edges of the boundary.

For unconfined flow systems, the cells containing the free-surface boundary are partially filled. No velocity-vector components are interpolated to nodes that are located above the free surface because these nodes are outside the boundary of the saturated computation region. If the node is at or below the free surface, the Z-component of nodal velocity is interpolated from the velocity at the bottom cell face and the vertical velocity of the free surface. The free-surface vertical velocity is calculated from a finite-difference approximation to the rate of movement of the free surface over the time step. The free-surface velocity for a cell is not calculated for the time step during which the node becomes covered by the free-surface boundary.

Finally, interpolation to the node is done for velocity components parallel to the boundary faces of the active grid. This computation of nodal velocity-vector components enables visualization of the interstitial velocity field over the entire simulation region. These nodal velocity calculations are not conservative, but they are used only for visualization.